SUNSPOTS AND STARSPOTS

The past two decades have seen remarkable advances in observations of sunspots and their magnetic fields, in imaging of spots and fields on distant stars, and in associated theoretical models and numerical simulations.

This volume provides the first comprehensive combined account of the properties of sunspots and starspots. It covers both observations and theory, and describes the intricate fine structure of a sunspot's magnetic field and the prevalence of polar spots on stars. The book includes a substantial historical introduction and treats solar and stellar magnetic activity, dynamo models of magnetic cycles, and the influence of solar variability on the Earth's magnetosphere and climate.

This book conveys the excitement of its subject to graduate students and specialists in solar and stellar physics, and more broadly to astronomers, geophysicists, space physicists and experts in fluid dynamics and plasma physics.

JOHN H. THOMAS is Professor of Mechanical and Aerospace Sciences, and Astronomy, at the University of Rochester. He has been the Chair of the Solar Physics Division of the American Astronomical Society, and was a Scientific Editor of the *Astrophysical Journal* for ten years.

NIGEL O. WEISS is Emeritus Professor of Mathematical Astrophysics at the University of Cambridge. He is a former President of the Royal Astronomical Society, which awarded him a Gold Medal in 2007.

Cambridge Astrophysics Series

Series editors
Andrew King, Douglas Lin, Stephen Maran, Jim Pringle and Martin Ward

Titles available in this series

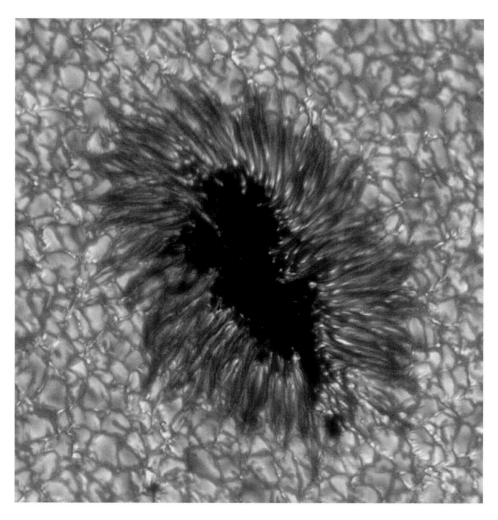

High-resolution image of a sunspot, showing the dark umbra enclosed by a filamentary penumbra, as well as tiny bright points in the surrounding granulation. This image was obtained in the continuum near 436 nm with the Swedish Solar Telescope on La Palma. (Courtesy of L. H. M. Rouppe van der Voort and the Royal Swedish Academy of Sciences.)

SUNSPOTS AND STARSPOTS

JOHN H. THOMAS

*Department of Mechanical Engineering and Department of Physics
and Astronomy, University of Rochester, USA*

NIGEL O. WEISS

*Department of Applied Mathematics and Theoretical Physics,
University of Cambridge, UK*

CAMBRIDGE
UNIVERSITY PRESS

CAMBRIDGE UNIVERSITY PRESS
Cambridge, New York, Melbourne, Madrid, Cape Town, Singapore, São Paulo, Delhi

Cambridge University Press
The Edinburgh Building, Cambridge CB2 8RU, UK

Published in the United States of America by Cambridge University Press, New York

www.cambridge.org
Information on this title: www.cambridge.org/9780521860031

First published 2008

Printed in the United Kingdom at the University Press, Cambridge

A catalogue record for this publication is available from the British Library

Library of Congress Cataloguing in Publication data
Thomas, John H., 1941–
Sunspots and starspots / John H. Thomas, Nigel O. Weiss.
   p.   cm. – (Cambridge astrophysics series)
ISBN 978-0-521-86003-1
1. Sunspots.   2. Starspots.   I. Weiss, N. O. (Nigel Oscar)   II. Title.
QB525.T56   2009
523.7′4–dc22
                                                            2008037116

ISBN 978-0-521-86003-1 hardback

To Lois and Judy

Contents

Preface

In 1858, Richard Carrington wrote, "Our knowledge of the Sun's action is but fragmentary, and the publication of speculations on the nature of his spots would be a very precarious venture." Fifty years later, George Ellery Hale's discovery of the magnetic field in a sunspot ushered in the modern era of research into solar, stellar and cosmical magnetic fields. This book, coincidentally, marks the hundredth anniversary of his discovery. The past century has seen enormous and rapidly accelerating progress in our understanding not only of sunspots but also of starspots and the whole solar–stellar connection. Our purpose here is to bring these advances together and to offer a unified account of sunspots and starspots in the context of solar and stellar magnetic activity.

Our own collaboration goes back more than 40 years, to the academic year 1966–67 when JHT was a NATO postdoctoral fellow in the Department of Applied Mathematics and Theoretical Physics at Cambridge, where NOW was a recently appointed lecturer. In 1991 we organized a NATO Advanced Research Workshop on *Sunspots: Theory and Observations*, which produced an edited volume (Thomas and Weiss 1992a) designed to serve as a monograph on the subject. Progress on sunspots has been very rapid since then, especially in high-resolution observations (both ground-based and from space) and in numerical modelling; meanwhile, with new techniques such as Doppler and Zeeman–Doppler imaging, the study of starspots (treated briefly in the 1992 volume) has emerged as a fully fledged subject of its own. Hence it seems to us that the time has come for a new, comprehensive book on sunspots and starspots that emphasizes recent developments.

Our aim in this book is to convey the excitement of this subject to a readership consisting not only of specialists in solar and stellar physics, but also more generally of astronomers, astrophysicists, geophysicists and space physicists, as well as experts in plasma physics and fluid dynamics with interest in astrophysical applications. We have attempted to give a balanced and self-contained account of observations and of theory. In composing this book, we have had in mind the needs both of established experts and, particularly, of graduate students entering the field.

We wish to thank all those collaborators and close colleagues who, over the span of many years, have enriched our knowledge of sunspots and solar and stellar magnetic activity and have added immeasurably to our pleasure in doing research. They include Tom Bogdan, Tim Brown, Nic Brummell, Paul Cally, Fausto Cattaneo, Al Clark, Lawrence Cram, Leon Golub, Bruce Lites, Leon Mestel, Friedrich Meyer, Benjamin Montesinos, Steve Musman, Bob Noyes, Mike Proctor, Bob Rosner, Hermann Schmidt, George Simon, Alan Title,

Steve Tobias, the late Peter Wilson, and the late Yutaka Uchida. Special thanks go also to our former graduate students and postdoctoral fellows: (JHT) Toufik Abdelatif, Andrew Markiel, Alan Nye, Colin Roald, Mark Scheuer, and Don Stanchfield; (NOW) Paul Bushby, Dave Galloway, David Hughes, Neal Hurlburt, Paul Matthews, Dan Moore, Neil Roxburgh, Alastair Rucklidge, and Mike Tildesley.

Tom Bogdan, Andrew Collier Cameron and Steve Tobias each read portions of this book in draft form and made many valuable suggestions. The convenient electronic access to scientific papers provided by the NASA Astrophysics Data System has been a very significant aid to us in writing this book. We also wish to thank the numerous authors whose figures are reproduced here, and Amanda Smith for assistance in preparing some of the illustrations.

Our work on this book began during the programme on 'Magnetohydrodynamics of Stellar Interiors' at the Isaac Newton Institute for Mathematical Sciences in Cambridge in the autumn of 2004. In the course of writing the book, we have been supported in many ways by our own institutions: the Department of Mechanical Engineering and the Department of Physics and Astronomy at Rochester, and the Department of Applied Mathematics and Theoretical Physics, Clare College, and Clare Hall at Cambridge. Our own research in the areas covered in this book has been supported over many years by grants from the National Aeronautics and Space Administration and the National Science Foundation in the USA, and the Particle Physics and Astronomy Research Council (now STFC) in the UK.

1

The Sun among the stars

Our Sun is a typical, middle-aged star, but it occupies a special place in astronomy as the only star that we can observe in great detail. Conversely, it is only by studying other stars with different properties, whether of age, mass or angular momentum, that we can fully explain the behaviour of the Sun. This book is concerned with dark spots on the surfaces of the Sun and other stars, which result from the interplay between magnetic fields and convection. In this opening chapter we provide a brief introduction to the properties of these spots, a summary of the important overall properties of the Sun and other stars, and an overview of the topics that will be covered in the remainder of the book.

1.1   Sunspots and solar magnetic activity

In this section we introduce a variety of features and phenomena associated with sunspots and solar activity, all of which will be discussed in greater detail in later chapters.

In images of the full solar disc, such as that shown in Figure 1.1, sunspots appear as dark patches at low latitudes. The fact that sunspots are associated with strong magnetic fields emerging through the solar surface is readily apparent in the accompanying *magnetogram* in Figure 1.1, which shows the strength and polarity of the longitudinal (line-of-sight) magnetic field.

In a close-up image, such as the one in Figure 1.2, a typical sunspot is seen to consist of a dark central region called the *umbra* surrounded by a less dark, annular region called the *penumbra*. Some sunspots are remarkably circular and axisymmetric (favourites of theoreticians), while others have very irregular shapes with perhaps only partial penumbrae. There are also smaller dark features known as *pores* that are essentially naked umbrae, or spots without penumbrae. Examples of both sunspots and pores can be seen in Figure 1.2. Also evident in this image is the pattern of *granulation* in regions outside of sunspots, caused by thermal convection just below the solar surface. This pattern consists of bright *granules* corresponding to hot, rising plumes of gas, surrounded by dark lanes corresponding to cooler downflows. A typical bright granule is about 700 km, or $1''$, across.[1]

The most conspicuous feature of a sunspot is, of course, its darkness relative to the surrounding photosphere. In an absolute sense, a sunspot is not so dark; indeed, if it were placed alone in space at the same distance from us as the Sun, it would shine about as brightly as the full Moon, a fact already understood by Galileo. A sunspot appears dark on the solar

---

[1] As viewed from the Earth at the mean Earth–Sun distance, an angle of one arcsecond ($1''$) is subtended by a distance of 726 km on the surface of the Sun at the centre of the solar disc.

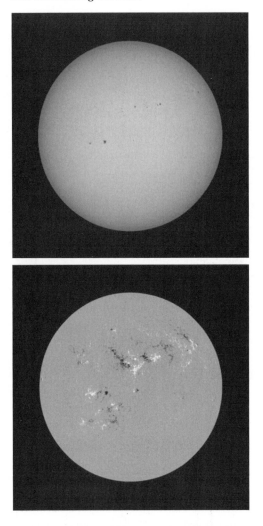

Fig. 1.1. SOHO images of the full solar disc in white light (above) and as a magnetogram (below) on 8 February 2001. In each image, North is upward and East and West are to the left and right, respectively, according to the usual convention. (Courtesy of Lockheed-Martin Solar and Astrophysics Laboratory.)

disc because it is relatively cooler than its surroundings, and we understand that the reduced temperature is due to the inhibiting effect of the spot's strong magnetic field on the vertical convective transport of heat just below the solar surface.

Sunspots come in a wide range of sizes. The largest have diameters of 60 000 km or more and are visible to the naked eye. At the other end of the scale are the pores, which have diameters typically in the range 1500–3500 km but can be as small as a single granule (about 700 km) or as large as a small sunspot (7000 km). Indeed, the largest pores are bigger than the smallest sunspots, a fact of some importance when we come to consider the formation and maintenance of the penumbra.

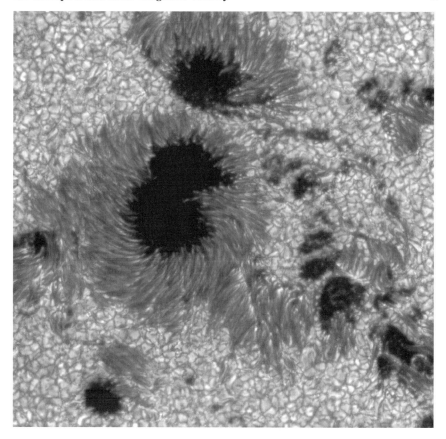

Fig. 1.2. High-resolution G-band image, obtained with the Swedish Solar Telescope, showing an active region with sunspots and pores. Penumbral filaments are clearly visible, as is the surrounding pattern of granular-scale convection. (Courtesy of the Royal Swedish Academy of Sciences.)

Within a sunspot there is a good deal of fine structure that becomes evident with higher resolution and suitable exposure times. The penumbra displays a characteristic pattern of elongated, bright and dark radial *penumbral filaments*, while the dark umbra contains a number of small, bright features known as *umbral dots* (which are not very evident in Fig. 1.2 because the umbra is under-exposed). This fine structure in the intensity of light emerging from a spot is a consequence of the pattern of thermal convection as influenced by the spot's magnetic field (*magnetoconvection*).

A sunspot marks a patch of the solar surface through which a close-packed bundle of nearly vertical magnetic flux (a magnetic flux tube) emerges from the solar interior. The magnetic field strength in the centre of a sunspot is typically about 2800 G (or 0.28 T) and can be as high as 3500 G or more. The magnetic field exerts a force on the solar plasma, consisting in general of a tension force along the field lines and an isotropic pressure. The total pressure, gas plus magnetic, within the spot must be in balance with the gas pressure in the field-free surroundings. As a consequence, the spot's magnetic flux tube must expand (in cross-section) rapidly with height above the solar surface, thus reducing its magnetic pressure, in order to be in pressure balance with the external atmosphere, in which the gas pressure is decreasing rapidly (nearly exponentially) with height.

In addition to dark sunspots and pores, there are also localized patches of excess brightness on the solar surface, known as *faculae*, which are also sites of strong emerging magnetic field. (Faculae are visible in the full-disc image in Fig. 1.1.) There are several other visible manifestations of the Sun's magnetic field, including enhanced emission from the upper layers of the solar atmosphere and transient events such as flares, surges and radio bursts. These phenomena, known collectively as *solar magnetic activity* (or simply solar activity), are not distributed uniformly across the solar surface, but instead are concentrated into *active regions* containing one or more sunspots, pores and surrounding faculae. Figure 1.3 shows

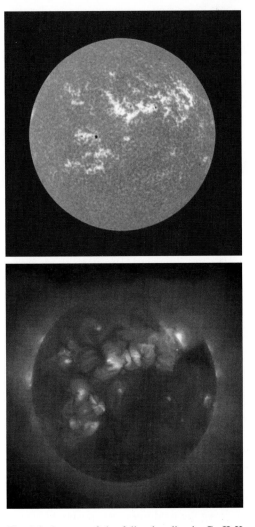

Fig. 1.3. Images of the full solar disc in Ca II K emission (above, from Big Bear Solar Observatory) and in coronal X-ray emission (below, from Yohkoh), again on 8 February 2001. (Courtesy of Lockheed-Martin Solar and Astrophysics Laboratory.)

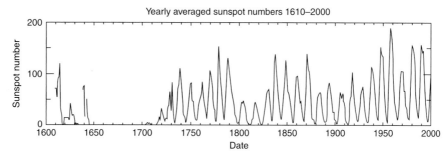

Fig. 1.4. Cyclic solar activity from AD 1610 to 2000, as shown by annual values of the group sunspot number. Note the interval of inactivity in the seventeenth century (the Maunder Minimum). (Courtesy of D. H. Hathaway.)

full disc images of Ca II emission (from the chromosphere) and X-ray emission (from the corona) in which active regions are clearly identifiable.

Solar activity is also not uniformly distributed in time: it varies in a nearly cyclic fashion with a period of about 11 years. This behaviour is readily apparent in the record of the number of sunspots appearing on the solar disc, as shown in Figure 1.4. Here one can see a somewhat irregular cyclic variation in the number of sunspots, with an average period between maxima of about 11 years, and a longer-term modulation of this cyclic variation. Of particular interest is the period from about 1645 to 1715 during which there were very few sunspots (the so-called *Maunder Minimum*).

The sunspot cycle has a period of about 11 years, but the magnetic polarity arrangement of the spots reverses in each successive cycle, indicating a signed magnetic cycle with a period of about 22 years. It is generally understood that the Sun's magnetic field and its cyclic behaviour are generated by a fluid *dynamo* acting in the solar interior through the interaction between the Sun's internal differential rotation and turbulent convection.

1.2 The Sun as a star

In this section we first describe the overall properties of the Sun and the structure of its interior and atmosphere, and then go on to summarize the overall properties of other stars to set the stage for our discussions of starspots and stellar activity.

1.2.1 Solar structure

Table 1.1 provides a list of important properties of the Sun that will prove useful in our discussions of sunspots and solar magnetism. For a clear discussion of how the values of these various quantities are determined, see the book by Stix (2002).

The radial structure of the Sun is depicted schematically in Figure 1.5. The energy generated by nuclear reactions in the *core* (where the temperature is of order 10^7 K) is carried radially outward by radiation across the *radiative zone* extending out to roughly $0.7\,R_\odot$, where the temperature of the solar plasma has decreased to the point where the increased opacity no longer permits the energy flux to be carried by radiation alone and thermal convection sets in. The energy is then carried outward almost exclusively by convection across the *convection zone* extending up to the solar surface. The relatively sharp visible surface of

Table 1.1 *Properties of the Sun*

Age	4.5×10^9 yr
Mass	$M_\odot = 1.99 \times 10^{30}$ kg
Radius	$R_\odot = 6.96 \times 10^8$ m = 696 000 km
Luminosity	$L_\odot = 3.84 \times 10^{26}$ W = 3.84×10^{33} erg s^{-1}
Effective temperature	$T_{\text{eff}} = 5785$ K
Spectral type	G2 V
Mean density	1.4×10^3 kg m^{-3}
Surface gravity	$g_\odot = 274$ m s^{-2}
Rotation period (equatorial)	26 days
Distance from Earth	1 AU = 1.50×10^{11} m = $215\,R_\odot$

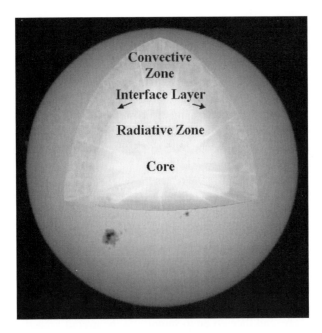

Fig. 1.5. Cutaway image of the Sun's internal structure, showing the photosphere (with sunspots), the convective zone beneath it, the inner radiative zone and the central core, where energy is generated by thermonuclear reactions. The tachocline is located at the interface between the radiative and convective zones. (Courtesy of D. H. Hathaway.)

the Sun occurs where the mean free path of photons increases abruptly and radiation is permitted to escape unimpeded into space; this transition takes place across a thin layer known as the *photosphere*, which is only a few hundred kilometres thick. In this layer, and higher layers of the solar atmosphere, a geometric height is determined from a measured *optical depth* τ_λ: a layer of optical depth $\tau_\lambda = 1$ reduces the intensity of radiation at wavelength λ by a factor of e^{-1}, and each further unit of τ_λ reduces the intensity by a further factor of e^{-1}. The most commonly used optical depth is τ_{500}, for radiation in a continuum window at wavelength 500 nm near the centre of the visible range.

At the base of the photosphere the temperature is about 6000 K, close to the Sun's *effective temperature* $T_{\rm eff} = 5785$ K based on its luminosity. The temperature reaches a minimum value of about 4200 K in the upper photosphere and then begins climbing again, reaching some 10 000 K or so in the *chromosphere*, named for its coloured appearance during a solar eclipse. Above the chromosphere, the temperature climbs steeply across a relatively thin *transition region* and reaches values of 2×10^6 K or more in the corona. The temperature then decreases in the outer corona, which expands and flows outward into space as the *solar wind*.

Direct observations are limited to the surface and atmosphere of the Sun (containing only $10^{-10} M_\odot$), while its magnetic fields are generated in the solar interior. Fortunately, this interior is accessible to *helioseismology*: the frequencies and horizontal wavenumbers of acoustic *p*-modes (with typical periods around 5 minutes) can be measured with great accuracy and used to establish the internal structure of the Sun with previously unattainable precision. The most remarkable achievement has been the determination of the Sun's internal rotation profile. It has long been known that equatorial regions rotate faster than polar regions at the surface, but helioseismology has shown that this differential rotation persists throughout the convection zone. There is then an exceedingly abrupt transition, across a thin layer – the *tachocline* (less than 0.04 $R_\odot$ thick) – to an almost uniformly rotating radiative interior (Thompson *et al.* 2003).

1.2.2 *Properties and classification of stars*

Although the most fundamental property of a star is its total mass, stars are usually classified according to their directly observable properties of luminosity and colour, or luminosity and surface temperature (which can be inferred from the colour). This classification is most often displayed in a *Hertzsprung–Russell (H–R) diagram*, in which the absolute magnitude (or the logarithm of the total luminosity) is plotted against the logarithm of the surface temperature (by tradition, decreasing to the right). When a large sample of stars is plotted in such a way, as in Figure 1.6, it is evident that most of the stars lie along a relatively narrow

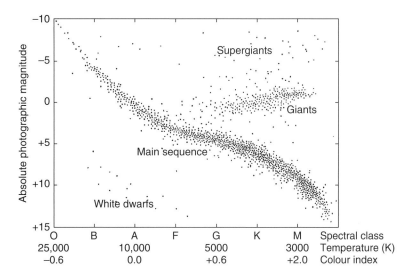

Fig. 1.6. The Hertzsprung–Russell diagram for a sample of stars of known distance. (From Abell 1964.)

band known as the *main sequence*. From the point of view of stellar evolution, the main sequence represents a long, intermediate stage in which stars spend most of their lifetime.

The Harvard classification scheme of stars into *spectral types* O, B, A, F, G, K, M, R and S represents a sequence of decreasing mass and decreasing surface temperature, and generally increasing complexity in their spectra. Each letter class has ten subdivisions (e.g. G0 through G9) in the order of decreasing surface temperature. Stars are also divided into *luminosity classes* according to their size, or phase of evolution, as follows: I for supergiants, II for bright giants, III for giants, IV for subgiants, V for main-sequence (or dwarf) stars, and VI for subdwarfs. The Sun, for example, is a G2 V star, and the star SU Aurigae, of class G2 III, is a giant star with a solar-like spectrum.

The *bolometric* (or total) *luminosity L* of a star is its total rate of energy output, integrated over all wavelengths; it is often expressed in units of the solar luminosity $L_\odot = 3.84 \times 10^{33}$ erg s^{-1}. Because the Earth's atmosphere is opaque at many wavelengths, L is in general difficult to determine, and instead the luminosity is often measured within certain wavelength bands, such as the V (visual) band centred on wavelength $\lambda = 555$ nm, the B (blue) band centred on $\lambda = 435$ nm, and the U (ultraviolet) band centred on $\lambda = 350$ nm, each band having relative width $\Delta\lambda/\lambda = 0.2$. The luminosity L in each band is often expressed on a logarithmic scale in terms of the *absolute magnitude M* in that band, defined by

$$M \equiv -2.5 \log_{10} L + C, \tag{1.1}$$

where C is a constant that is different for each wavelength band. For the B and V bands the constants are chosen so that the Sun has magnitudes $M_{B\odot} = 5.48$ and $M_{V\odot} = 4.83$. The magnitudes M_B and M_V are usually denoted simply as B and V.

The *colour* of a star is measured by the ratio of its luminosity in two wavelength bands, most often by L_V/L_B or equivalently by $B - V$. To the extent that a star's spectrum matches that of a black body, the colour $B - V$ is a measure of the surface temperature of the star. A more precise measure of a star's surface temperature is given by its *effective temperature* T_{eff}, defined as the temperature of a spherical black body having the same radius and bolometric luminosity as the star. Thus, according to the Stefan–Boltzmann law, T_{eff} is given in terms of a star's bolometric luminosity L by the relation

$$L = 4\pi R^2 \sigma T_{\text{eff}}^4, \tag{1.2}$$

where R is the star's radius and $\sigma = 5.67 \times 10^{-5}$ erg s^{-1} cm^{-2} K^{-4}.

From a theoretical viewpoint, the fundamental properties of a star that determine its position on the H–R diagram are its mass, its chemical composition, and its age. The theory of stellar evolution seeks to predict the track the star will follow in the H–R diagram as it evolves, given its initial mass and chemical composition (see Kippenhahn and Weigert 1990 or Hansen and Kawaler 1994). Another property that can affect a star's evolutionary track is its rotation rate, especially when the rotation is rapid.

Stars form through the condensation of interstellar gas by gravitational collapse into 'protostars', somehow shedding angular momentum in the process (for otherwise they would break apart due to centrifugal forces). Once a typical protostar reaches a state of hydrostatic equilibrium, its luminosity is maintained by the liberation of gravitational energy through slow contraction, and energy transport within the star is fully convective. As slow contraction proceeds, the luminosity decreases while the central temperature increases and the star

Table 1.2 *Typical properties of stars on the main sequence*

Spectral class	Mass $M/M_\odot$	Radius $R/R_\odot$	Luminosity $L/L_\odot$	$B - V$	T_{eff} (K)
O5	58	14	800 000	−0.32	46 000
B0	16	5.7	16 000	−0.30	29 000
B5	5.4	3.7	750	−0.16	15 200
A0	2.6	2.3	63	0.00	9 600
A5	1.9	1.8	24	0.15	8 700
F0	1.6	1.5	9.0	0.33	7 200
F5	1.35	1.2	4.0	0.45	6 400
G0	1.08	1.05	1.45	0.60	6 000
G2	1.0	1.0	1.0	0.64	5 780
G5	0.95	0.98	0.70	0.68	5 500
K0	0.83	0.89	0.36	0.81	5 150
K5	0.62	0.75	0.18	1.15	4 450
M0	0.47	0.64	0.075	1.41	3 850
M5	0.25	0.36	0.013	1.61	3 200

develops a radiative zone that grows outward from its centre. Eventually, provided the mass of the star is greater than about 0.1 $M_\odot$, the central temperature and density reach the point where the thermonuclear fusion into helium begins and soon takes over from gravitational contraction as the primary energy source. At this point, the star begins its long life on the main sequence, and all the stars of different masses but at this same stage of evolution are said to define the *zero-age main sequence* (ZAMS). Many observed stars are known to lie along pre-main-sequence evolutionary tracks, especially the so-called T Tauri stars, which may be associated with proto-planetary accretion discs.

Stars spend a large fraction of their lifetime on the main sequence, during which they are sustained by the fusion of hydrogen in their cores and move only slightly away from their ZAMS position on the H–R diagram. Table 1.2 lists typical properties of stars of different spectral types on the main sequence (luminosity class V). Note that the radius, luminosity, effective temperature and spectral class of a star of a given mass all vary over the star's lifetime on the main sequence: these variations are small for the lower-mass stars but are significant for the higher-mass stars (e.g. radius and luminosity can vary by as much as 20–30%).

Along the main sequence, stars of greater mass have higher central temperature but lower central pressure and density, and as a consequence, high-mass and low-mass stars have quite different structures. For this reason it is convenient to divide the main sequence into an *upper* main sequence (roughly $M > 2\,M_\odot$) and a *lower* main sequence ($M < 2\,M_\odot$). The high-mass stars evolve faster and thus spend a shorter time on the main sequence, while low-mass stars evolve more slowly, and indeed stars of mass less than about 0.8 $M_\odot$ have main-sequence lifetimes greater than the age of the Milky Way and hence have not yet left the main sequence. When a star exhausts the hydrogen fuel in its core, rapid changes ensue and the star moves off the main sequence onto its post-main-sequence evolutionary track, ending up eventually as some sort of degenerate star. Post-main-sequence tracks differ greatly for stars of different masses, involving different fusion reactions in different layers of the star. For example, a star of mass 1 $M_\odot$ depletes the hydrogen in its core in about 7×10^9 years, after which its helium core contracts and becomes degenerate while hydrogen burning takes place in a shell outside the core. When the hydrogen in this shell is mostly depleted, the star evolves up the red giant branch, with a degenerate core and greatly extended convective envelope. Its evolution up the red giant branch ends when helium suddenly begins to burn in the core (the helium flash). Eventually all of its nuclear fuel is exhausted and the envelope is blown off, leaving the core to continue life as a white dwarf.

1.3   Starspots and stellar magnetic activity

1.3.1   *The solar–stellar connection*

Since the Sun is not unique among stars, we must expect to find signs of magnetic activity in other stars that are similar to the Sun. Although individual starspots cannot be directly resolved, effects that are associated with magnetic activity on the Sun – X-ray and radio emission, Ca II emission, and signs of flaring – can certainly be detected in such stars. These indications are most obvious in cool, late-type stars (of spectral types F, G, K and M), which possess deep outer convection zones. (There is also a group of B and A type stars that can possess very strong magnetic fields, but their behaviour is very different from that of the Sun.) Measurements of Ca II H and K emission from these cool stars show that their magnetic activity decreases with age, and is closely correlated with their rotation rates. Young, rapidly spinning stars are far more active than the Sun.

Ca II emission from a group of these stars has been monitored for almost 40 years at Mount Wilson Observatory (Baliunas *et al.* 1995, 1998). The relative Ca II emission flux S provides a robust measure of activity, and Figure 1.7 shows three very different records of

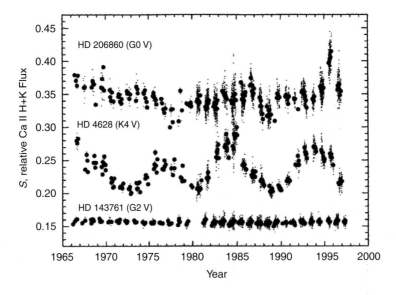

Fig. 1.7. Records of the relative Ca II emission flux S for the very active star HD 206860 (spectral type G0 V), the cyclically active star HD 4628 (type K4 V) and the weakly active star HD 143761 (ρ CrB, type G2 V). (From Baliunas *et al.* 1998.)

the temporal variation of S over a 30-year interval. The K4 V star HD 4628, with a rotation period of 39 days, exhibits quasiperiodic cyclic activity like that of the Sun, with a period of 8.4 years; it belongs to a family of about a dozen stars that have been found to show such behaviour. By contrast, the rapidly rotating G0 V star HD 206860, with a rotation period of only 4.7 days, is much more active but displays only aperiodic modulation of its activity. The third record, for the G2 V star HD 143761, with a rotation period of 17 days, shows only a very low, unvarying level of activity, suggesting that this star may have entered a Maunder-like grand minimum.

Such observations make it possible to reconstruct the magnetic evolution of a single star like the Sun. Prior to arriving on the main sequence it appears as a strongly magnetized, rapidly rotating T Tauri star, typically surrounded by a magnetized accretion disc. As it collapses onto the main sequence, conserving its angular momentum, the star spins up and may arrive rotating 100 times faster than the present Sun. As a result it displays extremely strong activity, associated with a vigorous stellar wind. The magnetic field interacts with the wind to exert a moment that brakes the rotation of the star and causes it to spin down. Eventually, when (like the Sun) it has a rotation period of weeks to a month, the star exhibits cyclic activity, interrupted perhaps by grand minima. Thereafter its rotation rate, and hence its magnetic field, gradually decay until the star ultimately evolves off the main sequence to become a red giant.

Apparent variations in Ca II emission not only facilitate the determination of a star's rotation rate but also provide evidence of differential rotation at its surface. The combination of differential rotation and turbulent convection in these magnetic stars provides a firm observational basis for assuming that there is a common dynamo mechanism responsible for generating magnetic fields not only in the Sun but also in other late-type stars. The variation of cycle periods and amplitudes with different stellar parameters then offers a means of testing theoretical models of the dynamo.

1.3.2 Spots on stars

The most vigorous magnetic activity appears in stars that rotate with periods of only a few days, either because they have only recently arrived on the main sequence or because they are in close binaries, with tidal synchronization of spin and orbital rotation. The most prominent examples of the latter group are the RS Canum Venaticorum variables, which have already evolved off the main sequence. It is in such rapid rotators that starspots are detected. Whereas sunspots never fill more than 0.5% of the solar surface, dark spots may occupy more than 20% of the surface of a star. As one of these active stars rotates, carrying the spots around with it, its apparent magnitude waxes and wanes (by up to 0.6 mag). Such variability can readily be detected and photometric measurements offer the most straightforward evidence for the presence of starspots.

More recently, the technique of *Doppler imaging* – which takes advantage of latitude-dependent blueshifts and redshifts as the star rotates – has made it possible to obtain images of the spots themselves. Unlike sunspots, which only appear at low latitudes, starspots are typically sited at or near the poles. Figure 1.8 shows a striking example of a huge polar spot on the RS CVn star HD 12545, a K0 III giant (far larger than the Sun); the temperature variation of 1300 K across its surface is comparable to the temperature deficit in the umbra of a sunspot. Although spots that have been observed on other stars are vastly larger than those on the Sun, it is only by reference to sunspots that the behaviour of starspots

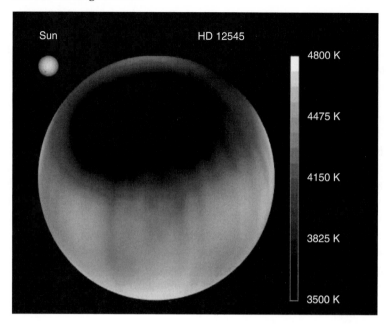

Fig. 1.8. Doppler image of the giant star XX Trianguli (HD 12545), showing a huge starspot near the pole. The radius of this RS CVn variable is 11.5 $R_{\odot}$. (Courtesy of K. G. Strassmeier.)

Fig. 1.9. Zeeman–Doppler image of the surface magnetic field on the classical T Tauri star V2129 Oph, together with a potential extrapolation of its external field, with both closed and 'open' field lines. There is a large starspot near the pole. (From Donati *et al.* 2007.)

can be explained. Thus we expect them to be associated with strong magnetic fields and, indeed, the development of *Zeeman–Doppler imaging* actually allows these magnetic fields to be measured. Figure 1.9 shows the radial magnetic field at the surface of the classical T Tauri star V2129 Ophiuchi, together with an extrapolated reconstruction of field lines that either close near the surface of the star or extend outwards into its accompanying accretion disc.

1.4   Plan of the book

We begin, in the next chapter, with a historical account, explaining how the complex properties of sunspots and starspots have gradually been revealed. Then we focus on sunspots and solar magnetic activity before broadening our attention to cover starspots and stellar activity.

In the next three chapters we examine the structure and dynamics of an individual sunspot, beginning with the overall structure (Chapter 3) and then going into the details of the fine structure of the umbra (Chapter 4) and the penumbra (Chapter 5) and the various oscillation modes present in a sunspot (Chapter 6). Then we discuss sunspots and groups of sunspots in the context of an active region (Chapter 7). Next we turn to the subject of magnetic activity in other stars (Chapter 8) and the basic properties of starspots (Chapter 9). After reviewing the main features of both solar and stellar magnetic activity cycles (Chapter 10), we discuss our theoretical understanding of the dynamos that produce this activity (Chapter 11). Finally, we discuss the effects of solar activity and variability on the near-Earth and interplanetary space environment and on the Earth's climate (Chapter 12), and then close with a brief look ahead to the observational and theoretical advances we might expect in the near future (Chapter 13).

In many ways, this book reflects our own view of sunspots as an ideal laboratory for magnetohydrodynamics (MHD) under astrophysical conditions. (We summarize some important aspects of MHD theory in Appendix 2.) MHD phenomena are ubiquitous in the Universe, but nowhere else in astrophysics is MHD theory confronted with such a wealth of observational data as it is for sunspots. It has turned out that the gross structure of a sunspot depends in many ways on the finer structure within it, and theoretical models must account for this fine structure. Almost any aspect of MHD that one can think of, including magnetohydrostatics and stability, MHD waves, MHD turbulence, magnetoconvection, and dynamo theory, is put to the test in our quest to understand sunspots, and therefore starspots too.

1.5   References for background reading

An excellent general introduction to the Sun is the book by Stix (2002). The book by Foukal (2004) provides a broad introduction to solar astrophysics, while the introductory text by Tayler (1997) treats the Sun from the perspective of stellar astrophysics. A comprehensive account of stellar magnetism is given in the book by Mestel (1999), and Priest (1982) provides an appropriate introduction to MHD.

On the subject of sunspots themselves, the classic monograph by Bray and Loughhead (1964) and the volume edited by Thomas and Weiss (1992a) are still essential reading. The conference proceedings edited by Cram and Thomas (1981) and Schmieder, del Toro Iniesta and Vázquez (1997) contain many papers that are still relevant. Recent reviews of sunspots include the comprehensive treatment by Solanki (2003) and the review devoted to fine structure by Thomas and Weiss (2004).

There have been no books devoted to starspots before the present volume, but much useful information may be found in the conference proceedings edited by Strassmeier, Washuettl and Schwope (2002) and the review by Berdyugina (2005). The volumes in this series by Wilson (1994) and Schrijver and Zwaan (2000) are also relevant.

2

Sunspots and starspots: a historical introduction

In that part of the sky which deserves to be considered the most pure and serene of all – I mean in the very face of the Sun – these innumerable multitudes of dense, obscure, and foggy materials are discovered to be produced and dissolved continually in brief periods. Here is a parade of productions and destructions that does not end in a moment, but will endure through all future ages, allowing the human mind time to observe at pleasure and to learn those doctrines which will finally prove the true location of the spots.

Galileo Galilei[1]

In this chapter we offer a brief historical introduction to sunspots, from the earliest naked-eye observations up to the remarkable advances in high-resolution observations and numerical simulations of recent years. We also discuss early speculations about starspots and their first observational detections. While we aim to give a balanced account, covering both observations and theory, our presentation does not follow a strict chronological path, but is instead arranged by topic. We do, however, provide a list of major advances in strict chronological order at the end of this chapter.

2.1   Early observations of sunspots

Seasonal changes were all-important to early agrarian societies and so they naturally worshipped the Sun. Indeed, the sun-god headed the pantheon in many cultures, ranging from Egypt to Peru. Since astronomy was also practised in these cultures, it seems likely that sunspots must occasionally have been detected with the naked eye, which is possible when the Sun is low on the horizon and partially obscured by dust storms, volcanic dust or smoke. Thus Needham (1959) conjectured that the traditional Chinese image of a red sun with a black crow superimposed upon it was derived from early sunspot observations. The first recorded mention of sunspots comes, however, from Greece: around 325 BC, Theophrastus of Athens, who was a student of Aristotle and succeeded him as leader of the Lyceum, referred in a meteorological treatise to black spots on the Sun as indicators of rain.[2] His casual references to spots suggest that they were already well known. Written records of non-telescopic observations of sunspots made in China date back to 165 BC (Wittmann and

---

[1] From *Letters on Sunspots*, translated by Stillman Drake (1957).

[2] The brief comments in Theophrastus's *De Signis* were amplified in the following century by Aratus in his *Phaenomena*, a didactic poem that was later translated into Latin by Cicero and Germanicus (Sider and Brunschön 2007). Virgil and Pliny the Elder both mention dark spots on the Sun as portents of rain, as does Bede much later.

Xu 1987; Yau and Stephenson 1988). These accounts range from straightforward ("Within the Sun there was a black spot") to fanciful ("The Sun was orange in colour. Within it there was a black vapour like a flying magpie. After several months it dispersed."). The series of naked-eye observations continues until the early twentieth century, overlapping the telescopic records that begin in 1611. However, the total number of recorded sunspot sightings before 1611 is less than 200, fewer than one sunspot per decade. This is only a small fraction (less than 1%) of the total number of sunspots that should have been visible to the naked eye during this period (Eddy, Stephenson and Yau 1989; Mossman 1989), and the observations were clearly sporadic, depending on phases of the Moon and seasonal dust storms (Yau 1988), as well as on political tact and revolutionary upheavals (Stephenson 1990).

Aristotle had taught that the Sun was perfect and immaculate, and the few sunspots observed by Arab astronomers were interpreted as transits of Mercury or Venus, as was the earliest European observation in AD 807 (Wittmann and Xu 1987), which Einhard regarded as a portent in his *Life of Charlemagne*. Indeed, Kepler himself thought that he had observed a transit of Mercury when he detected a sunspot in 1607. The oldest known drawing of sunspots appears in the manuscript chronicle of John of Worcester, depicting two spots observed in December 1128, when the Sun was low in the English winter sky (Stephenson and Willis 1999). Two centuries later, when the sky was obscured by smoke from forest fires in Russia, chroniclers there reported, "there were dark spots on the Sun as if nails were driven into it" (Wittmann and Xu 1987).

Progress since then has relied on technological advances. The invention of the telescope in the Netherlands in 1608 opened the possibility of detailed astronomical observations. News of the invention spread rapidly and reached Galileo Galilei (1564–1642) at Padua in June 1609. He used his improved instrument to observe the Moon and the Milky Way and to discover the satellites of Jupiter. In May 1612 he stated that he had been observing sunspots for 18 months (i.e. since November 1610) and there is no reason to doubt either this or his later claim that he had noticed them earlier that year, before he moved from Padua to Florence. The Sun was active at that time: Galileo may have been the first to see spots through a telescope but he was rapidly followed by others. Thomas Harriot (1560–1621), in England, was the first to record his observations though his manuscripts, with drawings made in December 1610, lay undiscovered at Alnwick Castle until 1786. The credit for publishing the first account goes to Johann Fabricius (latinized from Goldsmid, 1587–1616) who came from East Friesland. In his book *An Account of Spots Observed on the Sun and their Apparent Rotation with the Sun*, published at Wittenberg in June 1611, he describes how he and his father saw several spots in March 1611, first through a telescope and then using a camera obscura. They followed the spots as they moved across the solar disc and recognized one when it reappeared again; noticing that the spots were foreshortened at the limb, Fabricius concluded that they lay on the surface of a rotating Sun (Casanovas 1997; Hoyt and Schatten 1997).

A few days before Fabricius's first observation, Christoph Scheiner (1575–1659), a Jesuit professor at Ingolstadt in Bavaria, had also noticed some sunspots, through a smoke-filled sky; he made more systematic observations during the last few months of 1611 but was persuaded to publish them under a pseudonym, in the form of three letters addressed to Mark Welser, a wealthy patrician in Augsburg. In these letters, Scheiner asserted that the dark spots were caused by small bodies orbiting around the Sun and blocking its light; thus he was able to avoid any contradiction with the Aristotelian notion of a perfect Sun.

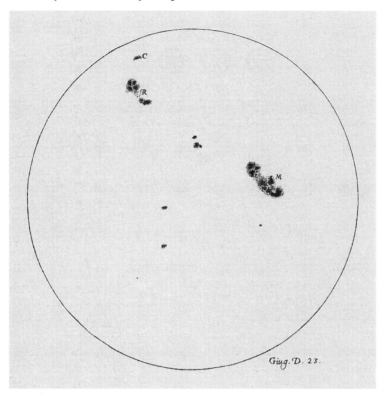

Fig. 2.1. Sunspot drawing by Galileo.

Welser forwarded these letters, published early in 1612, to Galileo in Florence and sought his comments. This set Galileo making his own systematic observations; he too recognized that the spots were foreshortened as they approached the limb and rapidly concluded that they were on the surface of the Sun and probably produced by clouds. Since he had just written a treatise on hydrostatics that contradicted Aristotelian notions, he eagerly inserted a paragraph on sunspots into the second edition. Then he went on to make a prolonged series of observations in the summer of 1612, using a projection technique developed by his colleague and former student, Benedetto Castelli. Galileo described his findings in three letters, addressed to Welser but written eloquently in Italian; they were published in 1613, as *Istoria e Dimostrazioni intorno alle Macchie Solari*[3] by the Accademia dei Lincei in Rome. His new observations (see Fig. 2.1) confirmed that the spots rotated with the Sun. He noticed that sunspots always lie near the solar equator, "in a narrow zone of the solar globe corresponding to the space in the celestial sphere that lies within the tropics", and he also realized that sunspots are dark only in a relative sense and by themselves are "at least as bright as the brightest parts of the Moon". He also noted the existence of bright patches near sunspots (later to be named faculae). After mentioning his discovery of the satellites of Jupiter and the phases of Venus, he concluded the book with a firm statement of support for Copernicus's heliocentric system.

---

[3] For an English translation, see Drake (1957), from which the quotations in this paragraph are taken.

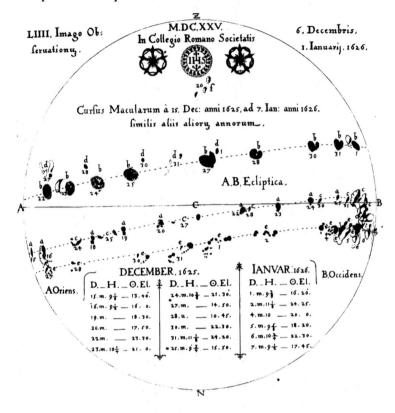

Fig. 2.2. Sunspot drawing from Scheiner's *Rosa Ursina*.

Later, there was a dispute between Galileo and Scheiner (who moved to the Jesuit College in Rome in 1624) over priority in the discovery of sunspots. Scheiner observed sunspots meticulously from 1625 to 1627, again projecting the telescopic image on to a screen or sheet of paper. By this time he had discarded some of Aristotle's teaching and had come to accept that the spots were on the solar surface and rotated with the Sun. From their apparent motion he deduced that the Sun's axis of rotation is not quite perpendicular to the plane of the ecliptic, but is inclined at about $7\frac{1}{2}°$ to the normal. In 1630 he published his results in a sumptuous volume, entitled *Rosa Ursina sive Sol* and dedicated to the Orsini family whose emblem was a rose, that remained the standard text on sunspots for a century and more. The first drawings by Scheiner and Galileo had already shown some sunspots with dark cores but Scheiner was now able to emphasize the distinction between the dark nucleus of a sunspot and the shadowy ring that surrounds it (which he confusingly referred to as the "umbra"), as shown in Figure 2.2. Scheiner continued to support a geocentric cosmology, though in Tycho's variant form (which allowed the other planets to rotate about the Sun) rather than Ptolemy's original version, and he remained a firm opponent of the Copernican system until the end of his life.

Galileo, in his *Dialogo sopra i Due Massimi Sistemi del Mondo*, ignored Kepler's discoveries but brought in the motion of sunspots as an argument in favour of a heliocentric system. He realized that there would be two occasions in the year, separated by six months, when the line of sight to the Sun was perpendicular to the plane through the solar centre containing the Sun's rotation axis and the normal to the ecliptic plane. At those times sunspots would appear to move in a straight line as the Sun rotated; in between, their paths would appear convex, pointing alternately up and down. This was confirmed by observations. In a geocentric system, on the other hand, Scheiner had to invoke an additional precession of the Sun's axis of rotation, with a period that just happened to be one year, which seemed less plausible. Galileo's disputes with Scheiner and Grassi, both professors at the Jesuit Collegio Romano, contributed to the events that led to his trial and sentence by the Inquisition.

Sunspots continued to be observed throughout the rest of the seventeenth century (Hoyt and Schatten 1997), though the non-achromatic refractors that were used had spatial resolution no better than 10 arcseconds. Among the most active observers was Johannes Hevelius (1611–1687) in Gdansk: in his *Selenographia* (1647) he included observations made in 1642–5, when sunspots were plentiful, but 20 years later, in his *Cometographia* (1667), he complained, "For a good many years recently, ten and more, I am certain that absolutely nothing of great significance (apart from some rather unimportant and small spots) has been observed either by us or by others. On the other hand, in former times (as *Rosa Ursina* and *Selenographia* confirm) a great many spots, remarkable for their size and density of distribution, appeared within a single year" (Weiss and Weiss 1979). He was not the only one to notice this dearth of sunspots, now referred to as the Maunder Minimum (Eddy 1976), which lasted from 1645 to 1715 and coincided with the reign of Louis XIV, the *Roi Soleil*.[4] Boyle (in 1660) and Fogelius (in 1661), as well as Picard and Cassini in Paris (in 1671) all reported excitedly when new spots occasionally appeared. That the seventeenth-century observers were both assiduous and competent is amply demonstrated by the records of the Paris Observatory, where systematic observations were carried out from 1667 on a daily basis, whenever the skies were clear. Ribes and Nesme-Ribes (1993) found that there were generally at least 15 days of observation each month. Between 1670 and 1700 there were never more than eight spots visible and long intervals when none were seen at all; indeed, only a single, short-lived spot was seen during the last decade of the century. Moreover, almost all the spots that did appear after 1660 were in the southern hemisphere only, and it was not until 1715 that spots were once more detected in both hemispheres, as they had been at the time of Galileo.

The next step forward came in 1769, when Alexander Wilson (1714–1786), professor of astronomy at Glasgow, discovered that, as a sunspot approaches the solar limb, the width of the penumbra on the side farthest from the limb decreases faster than the width of the penumbra on the side nearest the limb. From this result, now called the *Wilson effect*, he deduced that a sunspot corresponds to a saucer-shaped depression of the visible surface of the Sun or, as he put it, "a vast excavation in the luminous matter of the sun". This so-called *Wilson depression* is now understood to be a consequence of the fact that in the cooler

---

[4] By 1667 the change was sufficiently well known to be referred to by the poet Andrew Marvell in his satire *Last Instructions to a Painter* (Weiss and Weiss 1979):

> So his bold Tube, Man to the Sun apply'd,
> And Spots unknown to the bright Star descry'd;
> Show'd they obscure him, while too near they please,
> And seem his courtiers, are but his disease.
> Through Optick trunk the Planet seem'd to hear,
> And hurls them off, e'er since, in his Career.

and less dense (and hence more transparent) atmosphere in the penumbra, and especially in the umbra, the emergent radiation comes from a deeper geometric level.[5] Wilson also conjectured that the "excavations" might actually be revealing the dark interior of the Sun. This notion was taken up by William Herschel (1738–1822), who suggested that the Sun is a cool body covered in bright clouds, and that sunspots are holes in these clouds ("Places where the luminous Clouds of the Sun are Removed") revealing the cooler surface beneath, which he further suggested could be "richly stored with inhabitants" (Herschel 1795, 1801)! (This picture of the Sun as a cool body enveloped in layers of hot clouds persisted for a surprisingly long time, still appearing in the 1860s in John Herschel's standard textbook, but gradually gave way in the face of new results from spectroscopy.) With his 10-foot (focal length) reflecting telescope, William Herschel found that the sunspots, which he regarded as "openings" were surrounded by "shallows" which were "tufted": this is the first indication of fine structure in penumbrae. He also noted that sunspots had been scarce between 1795 and 1800 or, as he put it, "that our sun has for some time past been labouring under an indisposition, from which it is now in a fair way of recovering." He then went on to argue that there was a link between climate and the incidence of sunspots, supporting his case by citing variations in the price of wheat – conveniently gleaned from Adam Smith's *Wealth of Nations* – between 1650 and 1717; thus there were material advantages to be gained from solar observations.[6]

2.2   The sunspot cycle

In 1826 the German amateur astronomer Heinrich Schwabe (1789–1875) began his systematic observations of the Sun in search of the transit of a possible planet inside the orbit of Mercury. As part of this search, which continued for 43 years, Schwabe carefully recorded the occurrence of sunspots and in the process discovered the sunspot cycle. His first announcement (Schwabe 1843) of a possible 10-year periodicity in the number of sunspots attracted little notice, but in 1851 his table of observations over 25 years, clearly showing the periodicity, was published by Humboldt in the third volume of his *Kosmos* and attracted widespread attention.[7] The following year, in 1852, Edward Sabine (1788–1883) in England, and Rudolf Wolf (1816–1893) and Alfred Gautier (1793–1881) in Switzerland, independently found relations between the sunspot cycle and various magnetic disturbances on the Earth.

Inspired by Schwabe's discovery, the wealthy English amateur astronomer Richard Carrington (1826–1875) made regular sunspot observations during the years 1853–1861. By 1858, Carrington had made two important discoveries. First, he found that the latitudes (N and S) at which sunspots appear drift equatorward during the course of the sunspot cycle, from about 40° at the beginning of the cycle (at solar minimum) to about 5° at the end. Second, he found that spots at lower latitudes are carried around the Sun more quickly than

---

[5] Since the most reasonable definition of the solar surface is the surface at optical depth unity, there really should be no controversy as to whether the Wilson effect is due to a true depression of the surface.

[6] This notion was later taken up by W. S. Jevons, who attempted to relate economic cycles to solar activity. Jevons credited Arthur Schuster with having recognized a connection between the sunspot cycle and good vintages of wine.

[7] It is interesting to note that the sunspot cycle, which with hindsight is readily apparent in data available well before Schwabe's discovery (e.g. the records of the Observatoire de Paris or in W. Herschel's observations), was only discovered serendipitously as a result of a search for another phenomenon.

spots at higher latitudes, indicating that the Sun does not rotate as a rigid body.[8] This differential rotation provided direct evidence of the fluid nature of the Sun, at least in its outer layers. On 1 September 1859, Carrington (and independently Richard Hodgson) recorded the first observation of a solar flare. His important monograph *Observations of the Spots on the Sun*, containing all these results, was published by the Royal Society (Carrington 1863). Carrington also used his sunspot observations to determine the inclination i of the Sun's rotation axis to the normal to the ecliptic plane with unprecedented accuracy, obtaining the value $i = 7.25°$, which agrees remarkably well with a recent determination $i = 7.137° \pm 0.017°$ (Balthasar *et al.* 1986). At about the same time, Gustav Spörer (1822–1896) began his systematic observations of sunspots in Germany in 1860 and independently discovered the equatorward drift of the sunspot zones, a result that is now often called *Spörer's law*; he also independently noted the differential rotation revealed by east–west sunspot motions.

Schwabe's discovery of the 11-year sunspot cycle naturally raised the question of whether the cycle could be traced backwards in time using existing sunspot records. This question was pursued most actively by Rudolf Wolf in Switzerland. As a means of normalizing the sunspot observations carried out by different observers using different instruments at different locations, Wolf in 1848 introduced his *relative sunspot number* $\mathcal{R}$, defined by $\mathcal{R} = k(10g + f)$, where g is the number of sunspot groups visible on the disc, f is the number of individual spots on the disc (including those in the groups), and k is a correction factor with a different value for each set of observations (with $k = 1$ for Wolf's own original observations, so that all sunspot numbers are reduced to the values that he would have obtained had he made similar observations with his original telescope in Zurich in 1848). By computing values of $\mathcal{R}$ from various sunspot records, Wolf was able to trace the sunspot cycle back to the cycle occurring in 1755–66. (By convention, this cycle is denoted as 'Cycle 1' and subsequent sunspot cycles are numbered consecutively; at the time of this writing we are coming to the end of Cycle 23.) In 1855, when he became director of the new Zurich Observatory, Wolf began a programme of daily determinations of the relative sunspot number $\mathcal{R}$ (now also known as the Wolf, or Zurich, sunspot number). In order to avoid gaps in the record due to cloudy weather or instrument failures, Wolf enlisted the help of other observatories around the world. This programme has continued until today and now involves more than 30 observatories, with the data tabulated by the Solar Influences Data Analysis Center (SIDC) in Belgium. Other tabulations of sunspot numbers are carried out by the US National Oceanic and Atmospheric Administration (NOAA) and by the American Association of Variable Star Observers (an organization of amateur astronomers).

The long record of the Wolf sunspot numbers, already shown in Figure 1.4, provides a useful measure of the level of solar activity over time, which is similar to that obtained using the group sunspot number defined by Hoyt and Schatten (1998). Despite its arbitrary definition, the sunspot number $\mathcal{R}$ correlates well with other, physically more plausible measures of solar activity. In particular, it is nearly proportional to the total surface area of the Sun covered by spots, and since the magnetic field strength does not differ greatly among sunspots, $\mathcal{R}$ also provides a rough measure of the total magnetic flux emerging in sunspots on the visible hemisphere of the Sun. The most extensive compilation of sunspot numbers is that of Waldmeier (1961).

---

[8] Cassini had already been aware, a century earlier, that the Sun's rotation varied systematically with latitude (Ribes and Nesme-Ribes 1993).

2.2.1   The Maunder Minimum

Spörer did historical research on earlier sunspot records and drew attention to the absence of sunspots during the late seventeenth century (Spörer 1889). However, this discovery attracted little attention until it was taken up again by Ernest Maunder somewhat later (Maunder 1890, 1922b). Then it was ignored again until interest was revived by John Eddy (1976), who dubbed this episode of reduced solar activity the *Maunder Minimum* and demonstrated that the "prolonged sunspot minimum" was also associated with reductions in various proxy measures of solar activity. Eddy also revived the issue of a possible connection between solar activity and climate (first mooted by Herschel) and, in particular, the relation between the Maunder Minimum and the Little Ice Age (Grove 1988) that extended from the seventeenth century to about 1850. Conversely, Eddy suggested that the Medieval Warm Period, when Greenland was settled and good wine was produced in England, was associated with an episode of enhanced solar activity, followed by a reduction in the sixteenth century (the Spörer Minimum) that preceded the Maunder Minimum.

2.2.2   Photographic studies of sunspots

The development of photography in the 1840s allowed the Sun's surface to be examined more objectively and in more detail. The first photograph of the Sun was a daguerrotype taken in 1845 by Hippolyte Fizeau and Léon Foucault at the Paris Observatory; it showed several sunspots. One of the early photographic results was to confirm the darkening of the solar disc toward the limb noted by earlier visual observations; this *limb darkening* was later explained by Karl Schwarzschild (Schwarzschild and Villiger 1906) as being an effect of radiative transfer due to the decrease of temperature with height in the photosphere. Photographs of the 1860 solar eclipse in Spain firmly established the reality of the chromosphere and its prominences.

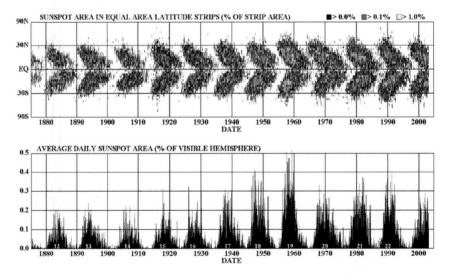

Fig. 2.3. Sunspot cycles from 1878 to 2005: daily observations, averaged over individual solar rotations. The upper panel shows the incidence of sunspots as a function of latitude and time, with new cycles beginning around latitudes of $\pm 30°$ as the previous cycle decays at the equator. The lower panel shows the proportion of the area of the visible hemisphere occupied by sunspots as a function of time. (Courtesy of D. H. Hathaway.)

In England, Warren de la Rue (1815–1889) began by obtaining photographs of the Moon (Bray and Loughhead 1964) and then – encouraged by John Herschel, who had himself coined the word *photography* and invented the process of fixing with hypo – constructed a photoheliograph (a telescope that projected its image onto a photographic plate) which was installed at Kew Observatory but later moved to Greenwich. The Greenwich photo-heliographic record of the Sun provided daily images (supplemented, if necessary, from observations made in India or at the Cape of Good Hope) from 1874 to 1976 and the record has been continued at Debrecen in Hungary since then.

From this series of photographic images it is possible to calculate the fraction of the visible solar disc that is covered by spots on any given day and so to obtain a detailed record of solar activity, as shown in Figure 2.3. It is apparent that the variation is not exactly periodic, though there is a well-defined mean period of just over 11 years. Maunder (1904, 1922a) also devised his famous *butterfly diagram*, which displays the incidence of sunspots as a function of solar latitude and time. Figure 2.3 clearly shows successive cycles starting at high latitudes and spreading towards the equator, where they eventually decay away, just as the next cycle starts. Moreover, this pattern is very nearly symmetrical about the equator.

2.3 Fine structure in sunspots

Over the past two centuries the study of sunspots has emphasized more and more the determination of the detailed structure of an individual spot, based on observations with increasing resolution. The invention of achromatic lenses made it possible to construct refracting telescopes with much higher resolution. In 1826 several large spots were observed by Ernesto Capocci at Naples, with a 9-foot Fraunhofer refractor; he noted for the first time the characteristic filamentary structure of the penumbra (Capocci 1827), which appears both in his drawings and in those of Johann Wilhelm Pastorff (1828), as illustrated in Figure 2.4. When John Herschel succeeded in making a similar observation, while at the Cape of

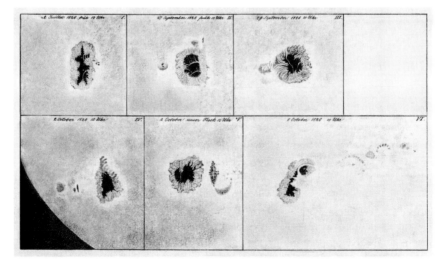

Fig. 2.4. Sunspots observed by E. Capocci in July–October 1826. These are the first observations that revealed the filamentary structure of sunspot penumbrae. Note also the light bridges. (From Capocci 1827.)

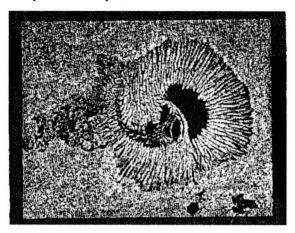

Fig. 2.5. Sunspot drawing by A. Secchi (1870).

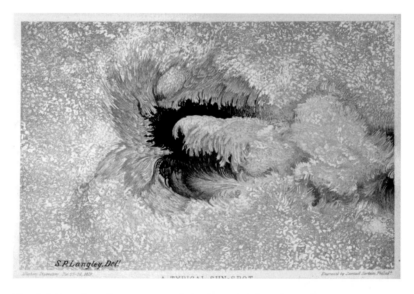

Fig. 2.6. Drawing of a sunspot made by Samuel P. Langley at the Allegheny Observatory on 23–24 December 1873. This drawing appears as the frontispiece in books on the Sun by C. A. Young (1881) and C. G. Abbot (1929).

Good Hope in 1837, he commented on the "remarkable radiated or striated apparent structure of the penumbra, ... which is obviously connected very intimately with the physical cause of the spots" (Herschel 1847; Tobias and Weiss 2004). By the mid nineteenth century this structure was revealed in considerable detail, as shown by the image from Angelo Secchi's masterly monograph *Le Soleil* (1870) in Figure 2.5, while the remarkable drawing by Langley shown in Figure 2.6 shows many fine details of the filamentary penumbra. Such patterns led Herschel to suggest that sunspots were sites of cyclonic motion and this notion of tornadoes (associated with upward or downward motion) dominated nineteenth-century discussions.

By the 1870s, photography was available to record the structures of individual pores and sunspots. A systematic series of photographic observations was initiated by Jules Janssen (1824–1907) at Meudon, near Paris, and similar images were obtained by Hansky at Pulkovo Observatory, near St Petersburg, and by Chevalier at Zô-Sè Observatory, near Shanghai. Similar programmes were gradually established at Mount Wilson and elsewhere: Bray and Loughhead (1964) reproduce many images obtained at Sydney. The greatest obstacles to resolving the smallest features on the Sun are local convection in or near the telescope and atmospheric 'seeing', the distortion of an image due to inhomogeneities in the refractive index of air in the Earth's atmosphere. These effects can be reduced by establishing an observatory at high altitude, as at the Pic-du-Midi in the Pyrenees, where the atmosphere can sometimes be exceptionally clear and stable, and by selecting the best images from series of rapid exposures (Rösch 1959).

Before the era of space missions, there were attempts to reduce the effect of seeing by flying balloon-borne telescopes to high altitudes. Blackwell, Dewhirst and Dollfus (1957, 1959) themselves undertook a balloon flight to an altitude of 20 000 feet (6000 m) with a 29-cm aperture telescope to photograph the solar granulation, but their results were still not as good as the best ground-based observations. Shortly thereafter, in 1957 and 1959, a team under the direction of Martin Schwarzschild (1912–1997) flew an unmanned balloon with a 12-inch telescope (Project Stratoscope) to a height of 80 000 feet (24 000 m), producing photographs of sunspots with unprecedented spatial resolution, as shown in Figure 2.7. These images revealed not only the detailed pattern of penumbral filaments but also the presence of bright pointlike features (umbral dots) within the umbra (Danielson 1961a, 1964).

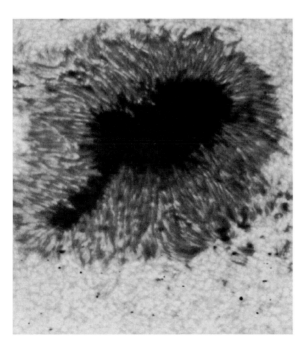

Fig. 2.7. Sunspot image obtained by Project Stratoscope. (From Danielson 1961a.)

2.4   Spectroscopy and the sunspot magnetic field

Absorption lines in the photospheric spectrum, first noticed by Wollaston in 1804, were classified by Fraunhofer, Kirchhoff and Ångström and identified by comparison with laboratory spectra. Kirchhoff's interpretation of the dark Fraunhofer lines in the solar spectrum as absorption lines required that the interior of the Sun be hotter than the overlying atmosphere where the absorption lines are formed, thus refuting the long-standing view originated by Herschel that the Sun was a relatively cool body enveloped by hot, luminous clouds. Emission lines from the chromosphere and corona were first observed by Jules Janssen and Norman Lockyer (1836–1920) in 1868 (see the accounts by Bray and Loughhead 1974 and Golub and Pasachoff 1997) and led to the identification of helium. In the early 1890s the spectroheliograph was invented independently by George Ellery Hale (1868–1938) at his Kenwood Observatory in Chicago and by Henri Deslandres (1853–1948) in Paris. By the end of the nineteenth century the development of spectroscopy had given rise to the new discipline of astrophysics. The modern era of solar physics began shortly after 1900 with rapid advances both in atomic physics and in the application of spectroscopy to the study of the solar atmosphere.

2.4.1   *Hale's discovery of the sunspot magnetic field*

The most important breakthrough in our understanding of sunspots came in 1908 with the discovery by Hale of the strong magnetic field in sunspots. Hale (by then at Mount Wilson in California) measured the Zeeman splitting in magnetically sensitive spectral lines formed in sunspots, obtaining a field strength of some 3000 G, and he also detected the polarization of the split components of these lines (Hale 1908b). This was the first detection of an extraterrestrial magnetic field,[9] which opened the way for the discovery of magnetic fields in other stars and other astronomical objects.

Hale's discovery of the sunspot magnetic field came soon after Zeeman's discovery (in 1896) of magnetic splitting of spectral lines in the laboratory. Hale was led to his search for Zeeman splitting in sunspots because he had previously observed vortical structure in sunspot chromospheres (in Hα; Hale 1908a), suggesting vortical motions of the ionized gas in the sunspot atmosphere, and, by analogy with the magnetic field produced by a rotating charged body in the laboratory, he inferred that there should be a magnetic field in sunspots. We now know that his reasoning was faulty on two counts: most sunspots show little or no vortical motion, and, more importantly, the solar plasma maintains charge neutrality and any magnetic field must be generated by an electrical current flowing in the plasma, not by a net charge carried by the fluid motion. This was certainly not the first (or the last) time an important discovery has been made for the wrong reasons.

In the decade following Hale's discovery of the magnetic field in sunspots, he and his collaborators at Mount Wilson discovered the systematic polarity pattern of the magnetic field (Hale and Nicholson 1925, 1938; see Fig. 2.8). Large sunspots generally occur in pairs of opposite magnetic polarity. The polarity of the leading spot (in the direction of the Sun's rotation) is the same for all pairs in the same hemisphere, but opposite in the other hemisphere. This arrangement reverses sign in the next solar cycle, and hence the Sun's *magnetic*

---

[9] In 1866, Lockyer first observed the doubling of spectral lines in a sunspot, but he interpreted it as a reversal. According to Kiepenheuer (1953), "This observation appears little known and may have been overlooked soon after it was made."

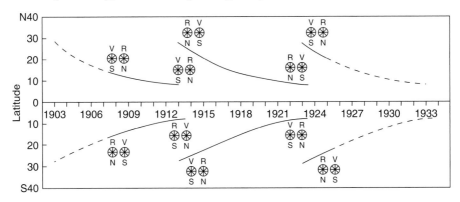

Fig. 2.8. Diagram from Hale and Nicholson (1938) showing the systematic polarity arrangement of sunspot pairs in consecutive solar cycles, known as *Hale's polarity laws.*

cycle has a period twice that of the sunspot cycle. These rules have stood the test of time and have come to be known as *Hale's polarity laws.* This regularity in the polarity arrangement of sunspots reveals the existence of an organized, large-scale magnetic field in the solar interior that reverses its polarity at the end of each sunspot cycle.

Sunspot fields are strong enough to produce directly measurable splitting of a Zeeman triplet, but weaker fields can only be detected by comparing overlapping components with opposite circular polarizations. Using this technique, Babcock and Babcock (1955) constructed a magnetograph at Mount Wilson that could measure fields of up to 10 G. They found that sunspots are associated with bipolar magnetic regions that obey the same polarity laws, and they also detected a weak dipolar field at high latitudes, which was later found to reverse at sunspot maximum (Babcock 1959).

2.4.2 The Evershed effect

Sunspots are not static; they exhibit a variety of internal flows and oscillations. The first of these motions to be discovered, by John Evershed (1864–1956) at Kodaikanal Observatory in India in 1909, is the flow associated with the *Evershed effect*, a characteristic wavelength shift and asymmetry of a photospheric absorption-line profile in the penumbra (see Fig. 2.9).[10] Although other mechanisms were later proposed, Evershed's interpretation of this effect as a Doppler shift caused by a radial outflow of gas across the penumbra has proved to be correct (Thomas 1994).

In his first report of the effect that now bears his name, Evershed (1909a) presented results for eleven sunspots which he followed across the solar disc. He established that the line displacements are invariably toward the red in the limb-side penumbra and toward the blue in the centre-side penumbra. He found that the displacements were maximal when the slit lay along a line from the spot centre to the centre of the solar disc, and minimal or absent when the slit lay perpendicular to this line. He was cautious about this result because it was unexpected; based on Hale's proposal that the magnetic field of a sunspot is generated by a vortex motion, Evershed expected to see azimuthal rather than radial motions in the

---

[10] Evershed (1909b) points out that these line shifts had actually been noticed many times during the previous three decades, without being interpreted correctly.

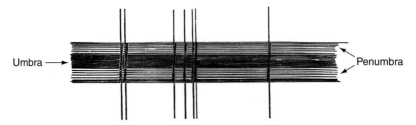

Fig. 2.9. Evershed's original sketch of the characteristic wavelength shift of photospheric absorption lines in a sunspot penumbra. The horizontal axis corresponds to wavelength and the vertical axis corresponds to position along the spectrograph slit. (From Evershed 1909c.)

penumbra. Instead, he concluded, "A hypothesis which seems in harmony with all the facts here stated is one which attributes the displacements to a radial movement outward from the spot centre" and also, "the motion must be essentially horizontal".[11] He also found that the line displacement increases with increasing radial distance outward across the penumbra, and he was already concerned about a problem that has only recently been resolved: "This seems to imply an accelerating movement from the centre of the spot outwards; yet at the limits of the penumbra the motion apparently ceases abruptly."

At Mount Wilson, St. John (1913) measured the Evershed effect in over 500 spectral lines and found the displacement to be roughly proportional to wavelength, thus confirming Evershed's interpretation of the effect as a Doppler shift. St. John also found a systematic decrease in line displacement with increasing line strength, with the displacement becoming negative in the strongest lines (the *inverse Evershed effect*), formed high up in the chromosphere. Abetti (1932) found greater flow velocities (up to $6 \, \mathrm{km \, s^{-1}}$) than Evershed or St. John and also found a substantial azimuthal component of the flow in some cases. Kinman (1952) measured the Evershed effect at many positions within a single sunspot and found the flow velocity to be axisymmetric, radial, and horizontal to within the errors of his measurements. He found that the flow speed increases from about $1 \, \mathrm{km \, s^{-1}}$ at the edge of the umbra to about $2 \, \mathrm{km \, s^{-1}}$ in the middle penumbra before dropping to zero somewhat beyond the outer edge of the penumbra. As we shall see in Chapter 5, more recent high-resolution observations have revealed that the Evershed flow is highly structured on fine scales associated with the penumbral filaments, quite distinct from the smooth flow distribution found by Kinman and others based on observations at moderate spatial resolution.

2.5   Solar granulation and supergranulation

William Herschel (1801) already observed, "the disc of the sun has an appearance which may be called mottled" and he described this "very particular and remarkable unevenness, ruggedness, or asperity" as *corrugations*. Sixty years later there was a controversy over

---

[11] Evershed states, "It is somewhat disappointing, perhaps, that the hypothesis of a radial movement, which is so strongly supported by these observations, seems entirely out of harmony with the splendid discovery of the Zeeman effect in sun-spots, made by Professor Hale. This seems to demand a vortex, or at any rate a circular movement, in sun-spots; and it was only after a considerable amount of evidence had accumulated that the preconceived conviction that the motion must be circular was abandoned."

whether the mottling resembled a pattern of willow-leaves or one of rice-grains, but eventually it was agreed to call the bright mottles *granules* (Bray, Loughhead and Durrant 1984; Spruit, Nordlund and Title 1990).

By 1900 photography had been routinely used to study the solar granulation. Some fine images were obtained by Janssen, Hansky and Chevalier (see Fig. 13 of Kiepenheuer 1953). As resolution improved, it became increasingly clear that the bright granules were enclosed by a network of dark lanes, with spectral lines that were blue- and red-shifted, respectively. By the 1930s it had also been established that, owing to ionization of hydrogen, the outer layers of the Sun must be convectively unstable. It was natural, therefore, to interpret the granulation as convection cells, like those that had been found experimentally by Bénard (Plaskett 1936). It was not obvious, however, that the granules had been adequately resolved. A major advance in the study of granulation was provided by the unmanned Stratoscope balloon, which produced photographs (such as Fig. 2.7) of unprecedented spatial resolution (Schwarzschild 1959; Bahng and Schwarzschild 1961). They showed that the granulation corresponded to coherent cellular structures with an average lifetime of 8.6 minutes and an average diameter – defined as the full width at half-maximum (FWHM) of the correlation function – of around 850 km; the average spacing, i.e. the distance between the centres of adjacent granules, is somewhat larger. More recent observations give similar lifetimes with a FWHM of 1000 km and a spacing of 1400 km in the quiet Sun (Bray, Loughhead and Durrant 1984; Spruit, Nordlund and Title 1990). In magnetic regions the granules are smaller and last longer.

Early spectroheliograms taken in the H and K lines of singly ionized calcium (Ca II) showed a mottled pattern dominated by a bright network, with some resemblance to Bénard convection (Deslandres 1910). Several decades later, Hart (1954, 1956) detected fluctuating horizontal velocities with an amplitude of around $0.5 \, \text{km s}^{-1}$ and a scale of 26 Mm. Simon and Leighton (1964; see also Leighton, Noyes and Simon 1962) confirmed that this motion corresponded to a cellular pattern of *supergranules* with a characteristic diameter of 30 Mm, and radial outflows that fitted into the Ca II network; furthermore, they used a magnetograph to show that the Ca II emission coincided with a magnetic network, consistent with fields being swept aside and concentrated by the horizontal motion.

2.6   Theoretical advances

2.6.1   *The development of magnetohydrodynamics*

Much of our understanding of the physics of sunspots comes from the theory of *magnetohydrodynamics*, the study of the motion of an electrically conducting fluid continuum in the presence of a magnetic field. (Certain detailed aspects of sunspot physics must be studied in the context of *plasma physics*, in which we abandon the continuum approximation and treat the medium as a collection of charged and neutral particles.) The study of solar magnetism originated in early speculations about the Sun and other astronomical objects. Based on the resemblance of coronal plumes seen during a solar eclipse to the lines of force of a uniformly magnetized sphere, Arthur Schuster (1891) and Frank Bigelow (1891) suggested that the Sun is a giant magnet, and Schuster even went on to conjecture that every large celestial body in rotation has an intrinsic magnetic field. These and other speculations prompted Hale to search for the Sun's magnetic field using the Zeeman effect, resulting in his 1908 detection of the kilogauss field in a sunspot. This discovery opened a whole new branch of astrophysics that led eventually to the development of magnetohydrodynamics. Although the basic physical principles involved were already well known even before Maxwell's electromagnetic theory, the important physical phenomena described by magnetohydrodynamics are difficult to produce in the laboratory. However, these phenomena become commonplace in the case of astrophysical bodies with their very large length and time scales.

In 1942 Alfvén showed that in a perfectly conducting fluid the magnetic lines of force are frozen into the fluid, i.e. they move with the fluid. This result, now generally known as *Alfvén's theorem*, is a direct result of Faraday's law applied to a medium of infinite electrical conductivity. Alfvén also showed that, as a consequence of this 'flux freezing', the fluid will support a new kind of wave, in which the magnetic field lines, under their inherent magnetic tension and loaded by the mass of the fluid, will undergo transverse oscillations analogous to those of the classical stretched elastic string. These waves exist even in an incompressible fluid, thereby providing a means of transporting energy through such a fluid without large-scale motions.

Since then, magnetohydrodynamics (MHD) has developed in many directions. A brief outline of key features is provided in Appendix 2. We now know that magnetic fields are ubiquitous, not only in planets like the Earth or stars like the Sun but also in the interplanetary and interstellar plasma, as well as in galaxies and galactic clusters. Moreover, the behaviour of ionized gases in laboratory experiments, notably those related to controlled nuclear fusion, can also be represented by MHD.

2.6.2   *Magnetoconvection*

Hale's discovery that sunspots are the sites of strong magnetic fields led Larmor (1919) to postulate that these fields were maintained by the inductive effect of radial outflows; in 1934, however, Thomas Cowling (1906–1990) proved that such axisymmetric self-excited dynamos could not exist, and that sunspot pairs must be created by subphotospheric fields emerging through the surface. There followed a correspondence between Cowling and Ludwig Biermann (1907–1986) in which they agreed that there had to be a balance between magnetic pressure inside the spot and gas pressure outside (Cowling 1985); then Biermann proposed that the coolness of sunspots could be explained as a consequence of magnetic suppression of convection in the spot (Biermann 1941).[12] Some years later, Hoyle (1949) pointed out that the energy flux at the photosphere would be reduced if convective elements had to follow field lines that splayed out towards the surface of a sunspot.

Schlüter and Temesváry (1958) showed that it is impossible to construct a sunspot model that is in purely radiative equilibrium, and thus established that convection is not completely suppressed by the magnetic field within a spot. The first quantitative estimate of the effect of a magnetic field on convection, due to Walén (1949; Weiss 1991), was taken up by Cowling (1953) who showed that, in a perfect fluid, thermal buoyancy could be overcome by the tension along magnetic field lines if the field strength exceeded a critical value. The effects of adding thermal, magnetic and viscous diffusion were investigated by Thompson (1951) and Chandrasekhar (1952, 1961), who showed that, in a star, linear instability would typically set in as growing (overstable) oscillations rather than as monotonically growing modes. In

---

[12] Biermann's brief but influential statement is reproduced, and translated, by Thomas and Weiss (1992b).

the presence of an inclined magnetic field, rolls oriented along the direction of inclination would be favoured. Hale and Nicholson (1938) had shown that the magnetic field, which is vertical at the centre of a sunspot, becomes increasingly inclined towards its edge and so Danielson (1961b) interpreted penumbral filaments as radially oriented convection rolls.

2.6.3   Dynamo theory

The above considerations led to a picture in which a sunspot was cool (and therefore dark) owing to the presence of a magnetic field whose lines of force were confined to a tight vertical bundle; Cowling (1946) estimated the lifetime of such a static bundle as 300 years and argued that a sunspot pair must be formed by the emergence through the surface of a section of a pre-existing, azimuthally oriented flux tube. This raises the issue of how such magnetic flux, with fields that alternate in direction every 11 years, can be generated. Cowling's (1934) theorem shows that any hydromagnetic dynamo must be non-axisymmetric. Ferraro (1937) had pointed out that a necessary condition for the maintenance of a steady axisymmetric poloidal field is that the angular velocity should be constant along field lines, and it is easy to see how a dipole field can be dragged out by differential rotation to produce an azimuthal (or toroidal) field that is antisymmetric about the equator (Cowling 1953, 1957). The problem is how to generate a reversed poloidal field from the toroidal field. Eugene Parker (1955b) argued that turbulent convective eddies acted on by Coriolis forces would develop into cyclonic motion that could twist the toroidal field and so generate small-scale poloidal fields, which would be smoothed by diffusion to produce a large-scale poloidal field. This process could be parametrized as an electromotive force that is proportional to the toroidal field; the constant of proportionality is now denoted by α and hence this is called the α-effect, while differential rotation is referred to as the ω-effect. Parker demonstrated that this combination could generate dynamo waves with a qualitative similarity to the sunspot cycle. In the same year, Parker (1955a) and Jensen (1955) independently demonstrated that isolated magnetic flux tubes are magnetically buoyant. Six years later, Babcock (1961) put forward a phenomenological model of the solar cycle in which an initial poloidal field is drawn out by differential rotation in the convection zone to form a strong toroidal field; the latter becomes unstable, releasing loops that rise owing to magnetic buoyancy and eventually emerge through the photosphere to form bipolar magnetic regions. These loops are twisted round by Coriolis forces and as they decay the poloidal field is eventually reversed. This model has served as a template for many subsequent calculations.

2.6.4   The solar wind

Eighteenth-century savants were aware of connections between aurorae and magnetic disturbances (manifested as movements of the compass needle); they also knew about the zodiacal light and that comet tails pointed away from the Sun. Already in 1733 de Mairan suggested that there was a material outflow from the Sun that impinged upon the Earth. Sabine, in 1852, established a connection between sunspots and geomagnetic variations, which also show an 11-year periodicity, while magnetic storms recur with a 27-day periodicity that corresponds to the synodic rotation period of the Sun. However, both Herschel and Kelvin remained extremely sceptical. Much more evidence accumulated over the next century and by 1930 Chapman and Ferraro were invoking streams of ionized plasma (now known as coronal mass ejections) to explain magnetic storms. Later, Biermann (1951) argued that the distorted shape of comet tails required a steady corpuscular outflow from the Sun. Then Parker (1958, 1960) realized that a static corona (as recently modelled by Chapman) could not be contained by the pressure of the interstellar medium and predicted the existence of a supersonic solar wind, which was soon confirmed by observations from space. Since magnetic fields are carried outwards by the solar wind, any variations in solar activity are bound to have geomagnetic consequences.

2.7   Recent progress on sunspots

The last 40 years have seen a surge of progress, both in observations and in theory, resulting from technical advances in ground-based telescopes as well as the opportunity of making observations from space, and from the development of powerful digital computers. These new results are the subject of the ensuing chapters and we need only summarize them briefly here.

Perhaps the most striking advance has been in high-resolution observations, which have revealed unexpected fine structure within the penumbrae of individual sunspots. While clarifying some issues, these new observations have raised other theoretical problems that are not yet fully understood. What is clear, however, is that the overall structure and dynamics of a sunspot, with its umbra and penumbra, are determined by fine structure on a scale that is only now beginning to be resolved. In addition to the Dunn Telescope at Sunspot, New Mexico, there is now a whole battery of instruments on the Canary Islands (see Appendix 1 for further details) and the latest generation of telescopes rely on adaptive optics to overcome the effects of seeing. Figure 2.10 shows the new Swedish Solar Telescope, which has produced the remarkable image in Figure 1.2 (see also Fig. 5.1).

Fig. 2.10. The new 1-m Swedish Solar Telescope on La Palma in the Canary Islands. (Courtesy of G. B. Scharmer.)

Meanwhile, important observations of sunspots have been obtained from space, not only in the visible range but also at X-ray and ultraviolet wavelengths which do not reach the ground, by satellites such as Yohkoh, SOHO, TRACE and Hinode. SOHO and Hinode also provide continuous measurements of surface velocities and magnetic fields. Helioseismology relies on precise measurements of the frequencies of acoustic oscillations (p-modes) of the Sun, notably those obtained by SOHO and the ground-based GONG network. From these data the Sun's internal structure and the depth of the convection zone can be precisely calculated, but the great triumph of helioseismology has been to determine how the solar interior rotates – with results that confounded all theoretical expectations (see Fig. 10.6 below).

Gradients in angular velocity, Ω, are a key ingredient of the solar dynamo, producing the strong toroidal fields that emerge in sunspots and in active regions. The other essential process can be represented by the α-effect, which describes the production of poloidal fields from toroidal flux. Mean-field electrodynamics has provided a formal justification for the α-effect in non-mirror symmetric turbulent flows – subject, however, to certain assumptions that do not hold in the Sun. For the last 30 years there has been a steady stream of mean-field dynamo models, both linear and nonlinear, with increasingly reliable distributions of Ω coupled to arbitrary distributions of α. Such models can reproduce the main features of the solar cycle, including grand minima, though they have scarcely any predictive power. Work on direct numerical investigation of the solar dynamo is still in its infancy.

The biggest influence on theory has been the development of large-scale computation. From the mid sixties onwards, it has become possible to tackle nonlinear problems of ever-increasing complexity and two styles of research have developed in parallel. One has focused on idealized models, designed to elucidate individual physical processes and backed up by advances in understanding nonlinear dynamics, while the other has aimed to reproduce properties of the solar atmosphere and interior in full detail. For the moment, however, we can only aspire to a qualitative understanding of the fine structure – particularly that of sunspot penumbrae – that has been revealed by the latest observations. Ever since the time of Galileo, theory has lagged behind observations in the study of sunspots.

2.8 Starspots

The idea that the Sun is a star like other stars goes back to the ancient Greeks (Kuhn 1957).[13] By the mid seventeenth century most astronomers accepted, largely through the influence of Descartes, that the fixed stars were distant suns. It is likely that many astronomers from then onwards speculated about the possible existence of dark spots on stars other than the Sun. In his *Ad Astronomos Monita Duo*, published in 1667, the French astronomer Ismael Boulliau (1605–1694) noted that the brightness of Mira Ceti varied cyclically with a period of 11 months.[14] "Boulliau also offered a physical explanation, by analogy

---

[13] The concept of a universe with many earths and many suns originated with the atomists, Leucippus and Democritus, in the fifth century BC; it became part of the Epicurean tradition and is mentioned by Lucretius. Much later, in the fifteenth century, the idea was revived by Nicholas of Cusa and then, in a Copernican context, by Giordano Bruno in his *De l'Infinito Universo e Mundi* (1584). Kepler, commenting on the difference in appearance between the fixed stars and the planets in his *Dissertatio cum Nuncio Siderio* (1610), said, "What other conclusion shall we draw from this difference, Galileo, than that the fixed stars generate their light from within, whereas the planets, being opaque, are illuminated from without; that is, to use Bruno's terms, the former are suns, the latter, moons or earths?" (Rosen 1965).

[14] Boulliau had previously postulated an inverse square law of attraction, in his *Astronomia Philolaica* (1645), in order to explain Kepler's laws of planetary motion, as later acknowledged by Newton in his *Principia*.

with sunspots: the star had dark patches and its light diminished cyclically when these patches were presented to the observer as the star rotated; but just as sunspots varied, so the dark patches varied, and this resulted in irregularities in the light curve of the star" (Hoskin 2001). This idea was taken up and spread by Bernard de Fontenelle (1657–1757) in his influential book *Entretiens sur la Pluralité des Mondes* (1686), which popularized the opinions of Descartes. Two centuries later, Rudolf Wolf noted the similarity between the irregular behaviour of long-period variable stars and the solar cycle and proposed that this behaviour is caused by rotational modulation of starspots (Tassoul and Tassoul 2004). Pickering (1880) commented similarly that observed luminosity variations in stars might be caused by a pattern of non-uniform surface brightness being carried across the visible hemisphere by the star's rotation. We now know that most variable stars, including the long-period variables, are pulsating, and the detection of starspots had to await the development of photoelectric photometry and spectroscopy.

The earliest direct evidence of solar-like activity on stars other than the Sun was provided in 1900 by Gustav Eberhard's discovery of a central emission core in the calcium K line in the spectrum of Arcturus,[15] analogous to the K-line emission associated with sunspots, discovered by Hale and Deslandres in 1891. Subsequently, Eberhard and Karl Schwarzschild (1913) found similar Ca II emission in the spectra of Aldebaran and strong emission (in both H and K) in the spectrum of σ Geminorum. They concluded, "the same kind of eruptive activity that appears in sun-spots, flocculi, and prominences, we probably have to deal with in Arcturus and Aldebaran and in a very greatly magnified scale in σ Geminorum." They also noted that it remained to be shown whether the emission lines in these stars "have a possible variation in intensity analogous to the sun-spot period."

Some 30 years later, Joy and R. E. Wilson (1949) were able to compile a list of 445 stars whose spectra showed bright emission in the Ca II H and K lines. Then, in 1966, Olin Wilson began a systematic search, using the 100-inch Hooker telescope at Mount Wilson, for long-term variations in Ca II emission from a sample of nearby cool dwarf stars, which finally yielded evidence of cyclic magnetic variability in stars other than the Sun (Wilson 1978).

Following Hale's discovery in 1908 of the magnetic field in a sunspot, and his subsequent detection of an overall solar magnetic field, many astronomers quite naturally expected other stars to possess magnetic fields also. Detecting such fields proved elusive, however, because Zeeman splitting of spectral lines is difficult to see in the integrated light from a star unless the field is very strong and its geometry quite simple. The first direct detection did not come until 1946, with Babcock's discovery of a strong magnetic field in 78 Virginis (Babcock 1947), a typical Ap star. Over the next few years several other stellar magnetic fields were measured, but all of them in non-solar-like stars of types B and A. It was not until the 1980s that new methods of identifying the line broadening due to the unresolved Zeeman effect (Robinson 1980) enabled the discovery of weaker and less organized magnetic fields on cooler stars.

The modern story of starspots begins with the paper of Kron (1947) reporting the detection of patches of varying brightness on the solar-like star AR Lacertae B, a G5 star in an eclipsing binary system. Intrinsic variability of the G5 component had been noticed earlier by Wood (1947), based on observations made in 1938 and 1939. Systematic photoelectric

---

[15] The weak emission core was visible only because the plate was exposed for study of the ultraviolet spectrum and was strongly overexposed in the visible.

observations at Lick Observatory between then and 1947 produced light curves for AR Lacertae with all the usual features of an eclipsing binary system, but with additional small,
sudden steps that could be attributed to the eclipse of patchy areas on the G5 star by the
K0 companion. Subsequently, Kron (1950a) found peculiarities in the dwarf Me star YY
Geminorum that seemed to him likely to be caused by activity similar to that associated with
sunspots.[16] The brightness variations in the two stars studied by Kron were small, however,
corresponding to a relatively small area of spot coverage, so the evidence for starspots was
not conclusive and was not widely accepted.

Several astronomers in the 1960s began calling attention to the possibility of solar-
like activity in late-type stars (e.g. Catalano and Rodonò 1967, 1974; Chugainov 1966;
Godoli 1968). In 1965, Chugainov observed small periodic light and colour variations in
the star HD 234677 (Chugainov 1966). This star, of spectral type K6 V, had shown earlier
evidence of weak flaring (Popper 1953), but it was not known to be a spectroscopic binary
and earlier observations (in 1954 and 1960) had shown no photometric variability, so the
variations were unlikely to be due to eclipses. On the other hand, the smallness of the colour
variations ruled out pulsations as the cause. Chugainov suggested that the variations were
due to a spot on the surface of this star. Since the 1960s, periodic variations in the light output
of stars due to surface patterns of intensity have been used to determine very accurately the
rotation periods of many stars. The method has the advantages over spectrographic methods
(based on the Doppler effect) that it is free of the uncertainty due to the unknown inclination
angle i and it works even for very slowly rotating stars.

Peculiar photometric and spectroscopic variations were observed in the late-type dwarf
binary stars CC Eridani (Evans 1959) and BY Draconis (Krzeminski and Kraft 1967), with
both systems showing sinusoidal intensity variations in the V band with varying amplitude
and phase. Krzeminski (1969) proposed that these variations are caused by starspots a few
hundred degrees cooler than the normal photosphere on these stars. Evans (1971) showed
that the starspot hypothesis might account for the variations in CC Eri but also suggested an
alternative mechanism involving the obscuration caused by circumstellar clouds temporarily
accumulating at the Lagrangian points of the binary system. Subsequently, however, Bopp
and Evans (1973) provided more detailed starspot models for the photometric variations of
CC Eri, and also for BY Dra (see Fig. 2.11). In the case of BY Dra, the spots are large, covering up to 20% of the visible hemisphere, they are some 1500 K cooler than the surrounding
photosphere, they extend up to 60° latitude, and they last for many rotation periods.

A direct analogy between intensity patterns on stars and sunspots was suggested by Hall
(1972) in interpreting the complex photometric behaviour of the binary star RS Canum
Venaticorum. The light curve shows wave-like distortions outside of eclipse that migrate
at a non-uniform rate and vary in amplitude. Hall proposed a model of this system in which

[16] A direct quotation from Kron (1950b) is of interest here: "One day during the recent war, Dr. Olin Wilson of the
Mount Wilson Observatory and I were driving down California Street in Pasadena. We were on war business at
the time, but we could not help noticing that the Sun was veiled by a uniform layer of haze of such properties
that the solar brightness was reduced to the point where one could make direct visual observations, yet the
optical definition of the dimmed solar disc was unimpaired. We stopped the car to look at the Sun, and were
gratified to see in exchange for our efforts a huge centrally-located sunspot. The spot, we learned later, was one
of many record-breaking large ones of the last broad sunspot maximum. Anyone in Pasadena at the time could
have observed it by simply looking at the Sun without optical aid. But what about stars other than the Sun? If
they are at all similar to the Sun, should they not have such spots, too, and, if so, is it possible to learn anything
about their properties?"

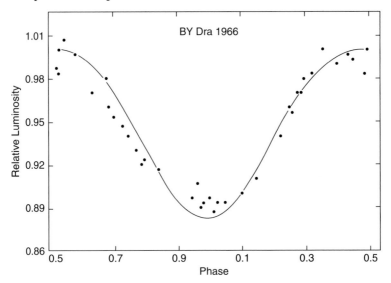

Fig. 2.11. Representation of observed photometric luminosity variations (dots) of BY Draconis by a starspot model (smooth curve). The phases relate to the rotational period of 3.826 days. (From Bopp and Evans 1973.)

the surface of the cooler star had a large region of starspots extending to 30° latitude above and below the equator and over 180° in longitude, with this pattern being periodic with period (between maxima in the spot coverage) equal to 1800 orbital cycles, or 23.5 yr.

Torres and Ferraz Mello (1973) proposed that periodic variations in brightness of several dMe stars are due to spots a few hundred degrees cooler than the star's effective surface temperature, covering a small percentage of the star's surface area. Using a simple spot model, they showed that the light curves for AU Microscopii could be reproduced with two spots, each about 500 K cooler than the normal photosphere, covering a total of about 10% of the surface.

Until the early 1980s there was considerable scepticism about the various claims of starspot detections based solely on photometry. Alternative mechanisms for producing the asymmetric brightness variations were proposed; for example, Kopal (1982) showed that pulsations in one member of an eclipsing binary system can produce an asymmetric light curve. However, beginning in about 1980, new methods based on Doppler imaging and detections of molecular lines provided firm and unambiguous detections of starspots. Today the existence of these spots is no longer questioned, and detailed studies of starspots have revealed not only the properties of the spots themselves but also surface differential rotation and dynamo patterns on several stars.

2.9　Chronology of key developments (1610–1964)

1610–11　The first telescopic observations of sunspots by Harriot, Fabricius, Scheiner and Galileo.

1613　Galileo publishes his *Istoria e Dimostrazioni intorno alle Macchie Solari*.

1630　Scheiner publishes his *Rosa Ursina*.

1647　Hevelius publishes his *Selenographia*.

1769	Alexander Wilson discovers the depression of the visible solar surface within a sunspot.
1801	William Herschel publishes his solar observations, including descriptions of penumbral structure and conjectures that sunspots are holes in the Sun's atmosphere revealing a cooler (and habitable!) surface below.
1826	Capocci and Pastorff detect penumbral filaments in a sunspot.
1843	Schwabe announces his discovery of the sunspot cycle.
1863	Carrington publishes his findings of the equatorward drift of the sunspot belts and the solar surface differential rotation.
1870	Secchi publishes *Le Soleil*, the first monograph on solar physics.
1908	Hale discovers the magnetic field in a sunspot.
1909	Evershed discovers the spectral line shift and asymmetry indicating a radial outflow in the penumbra.
1913	Eberhard and Schwarzschild observe calcium H and K emission in stars and suggest a relation to stellar spots and activity.
1925	Hale's laws of sunspot polarity described by Hale and Nicholson.
1941	Biermann conjectures that a sunspot is cooler than its surroundings because of the inhibition of convection by the magnetic field.
1942	Alfvén originates the concepts of magnetic flux freezing and magnetohydrodynamic waves.
1946	Cowling investigates time scales for growth and decay of sunspot magnetic fields.
1946	First direct detection of a stellar magnetic field by Babcock, in the Ap star 78 Virginis.
1947	First, tentative detection of starspots by Kron, on the solar-like star AR Lacertae B.
1951	First theoretical investigations of magnetoconvection by Thompson and Chandrasekhar.
1955	Parker and Jensen independently demonstrate the mechanism of magnetic buoyancy. Parker also introduces the concept of a solar dynamo based on differential rotation and helical turbulence, later to become known as the $\alpha\omega$-dynamo.
1957	Cowling discusses the darkness of sunspots in his influential textbook *Magnetohydrodynamics*.
1958	Parker predicts the existence of a supersonic solar wind.
1959	The balloon-borne Stratoscope provides images of unprecedented spatial resolution, revealing new details of penumbral filaments, umbral dots and photospheric granulation.
1961	Publication of the Stratoscope results; Danielson interprets penumbral filaments as convection rolls. Babcock presents a phenomenological model of the solar cycle based on differential rotation and latitudinal flux transport by near-surface motions.
1964	Bray and Loughhead publish their monograph *Sunspots*.

3

Overall structure of a sunspot

In this chapter we describe the gross features of sunspots, including their shapes and sizes and their overall thermal and magnetic structure. We also discuss those models of a sunspot that ignore its fine structure and make the simplifying assumption that the spot can be treated as an axisymmetric magnetic flux tube. This is a reasonable approximation, for although most spots are irregularly shaped, as can be seen from Figure 1.2, there are still many examples that are approximately circular, like that in Figure 3.1.

Sunspots are dark because they contain strong magnetic fields that partially inhibit the normal transport of energy by convection at, or just below, the solar photosphere. A well-developed spot may have a radius of 10 000–20 000 km, with a dark central nucleus (the *umbra*), surrounded by a less dark, filamentary *penumbra*. Such sunspots have approximately similar structures. The umbra occupies about 18% of the area of the spot, corresponding to an umbral radius that is about 40% of that of the spot. The umbra radiates energy at only 20% of the normal photospheric rate, corresponding to a temperature deficit of 2000 K, while the average penumbral intensity is about 75% of that outside the spot, corresponding to a deficit of only 400 K (Bray and Loughhead 1964; Thomas and Weiss 1992a; Stix 2002), so the total 'missing energy' is about 35% of that from a corresponding field-free area. The magnetic field is almost vertical at the centre of the spot (where its strength is typically around 2800 G, or 0.28 T) but its inclination to the vertical increases with increasing radius, reaching an average value of $70°$ at the edge of the spot, where the field strength has dropped to less than 1000 G.

3.1 Morphology of sunspots

Sunspots come in a wide range of sizes. The largest spots have diameters of 60 000 km or more, while the smallest have diameters of only about 3500 km, less than the diameters of the largest pores (about 7000 km). The area of a sunspot is typically given in units of millionths of the surface area of a solar hemisphere: $1.0 \times 10^{-6} A_{\odot/2} = 3.044 \times 10^6 \text{ km}^2$, where $A_{\odot/2} = 2\pi R_\odot^2$ is the area of the visible hemisphere. The largest recorded sunspot (in March 1947) had an area of $4300 \times 10^{-6} A_{\odot/2}$ (Newton 1955; Abetti 1957), corresponding to a mean diameter of 130 000 km (almost 10% of the solar diameter and 10 times that of the Earth).

Bogdan *et al.* (1988) analysed the distribution of umbral areas in over 24 000 individual sunspots from the Mount Wilson white-light plates covering the period 1917–1982. They found this distribution to be lognormal for umbral areas in the range $1.5–141 \times 10^{-6} A_{\odot/2}$.

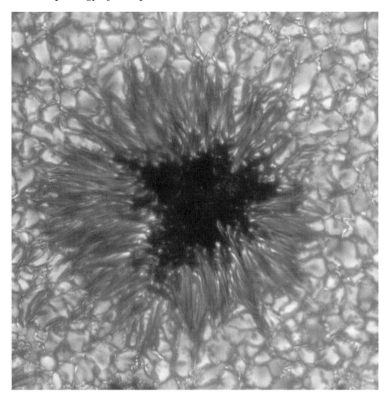

Fig. 3.1. High-resolution G-band image of a symmetrical sunspot, obtained with the SST on La Palma. Penumbral filaments are clearly visible, as are a few umbral dots. The bright points nestling between granules indicate the presence of small-scale magnetic fields. (Courtesy of L. H. M. Rouppe van der Voort and the Royal Swedish Academy of Sciences.)

They obtained the same distribution for each of the solar cycles in the period 1917–1982, and for all phases of an individual cycle (minimum, ascending phase, maximum, descending phase). Since the total area of a sunspot scales nearly linearly with umbral area (see below), we can infer that the total areas of sunspots have roughly the same lognormal distribution.

The question of the relative size of the umbra and penumbra has been investigated by a number of observers, beginning with Nicholson (1933) and Waldmeier (1939). Unfortunately, the results of these studies are largely inconsistent, because of different measurement techniques, different selection procedures and the lack of a precise and consistent definition of the boundary between the umbra and penumbra. Early micrometer measurements of diameters (assuming spot symmetry; e.g. Waldmeier 1939) have been replaced by photometric determinations of the umbral (A_u) and penumbral (A_p) areas based on continuum intensity. For example, Brandt, Schmidt and Steinegger (1990) define the umbra–penumbra and penumbra–photosphere boundaries to be at 59% and 85% of the quiet photospheric intensity. Their study of 126 sunspots produced the plot of $\log(A_u)$ versus $\log(A_p)$ shown in Figure 3.2, which shows large scatter for small sunspots but a fairly tight fit given by

$$\log(A_u) = -(0.79 \pm 0.35) + (1.10 \pm 0.17)\log(A_p) \qquad (3.1)$$

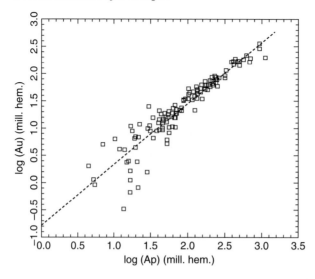

Fig. 3.2. Umbral area versus penumbral area (in units of $10^{-6} A_{\odot/2}$) plotted on a log-log scale, for 126 sunspots and spot groups observed during the period 19 August – 4 September 1980. (From Brandt, Schmidt and Steinegger 1990.)

for medium-size and large spots. Hence they find a weak tendency for the area ratio A_u/A_p to increase with spot size, being about 0.24 for the smallest spots (of total area around $50 \times 10^{-6} A_{\odot/2}$) and 0.32 for the largest spots in the sample; the corresponding values of the ratio of A_u to the total spot area A are 0.19 and 0.24, respectively. Jensen, Nordø and Ringnes (1955) measured over 600 large, regular, single spots (from the Greenwich Photoheliographic Survey) and obtained a ratio $A_u/A_p = 0.23$, corresponding to $A_u/A = 0.19$ and a ratio $R_u/R = 0.43$ for the umbral to the total radius. In practice, it is commonly assumed that the umbra occupies about 40% by radius of the spot.

The relative sizes of the umbra and penumbra may well also depend on the atmospheric height at which they are measured. Indeed, Wilson and Cannon (1968) and Wilson and McIntosh (1969) have argued that the size of the penumbra increases with height while that of the umbra decreases. In support of this conclusion, Collados, del Toro Iniesta and Vázquez (1987) find that the width of the umbra parallel to the limb decreases as a spot approaches the limb. They also confirm the finding of earlier studies that the width of the penumbra on the western side of a sunspot is on average less (by about 8%) than that on the eastern side, a fact which alone can explain an 'inverse' Wilson effect observed in spots moving within a heliocentric angle of 40°.

Jensen, Nordø and Ringnes (1955) and Ringnes (1964) found a dependence of R_u/R on the phase of the sunspot cycle in the sense that the average value of R_u/R is smaller at sunspot maximum than at sunspot minimum. Since they also showed that the average value of R itself varies in phase with solar activity, this is consistent with the Bogdan *et al.* (1988) result that the distribution of umbral area is the same at all phases of the cycle. The finding of Jensen *et al.* suggests that penumbrae (but not umbrae) are relatively larger at solar maximum, though this has not been tested by more modern methods.

3.2   Thermal properties of sunspots

Here we discuss the distribution of temperature and other thermodynamic variables in a sunspot, beginning with the fundamental question of why sunspots are significantly cooler and darker than the surrounding quiet photosphere.

3.2.1   *The cooling of sunspots*

In a typical sunspot the umbra radiates only 20 to 30% of the flux (integrated over wavelength) of the quiet Sun and the penumbra radiates some 75 to 85% of the quiet Sun flux. This implies that, on average, the umbra is 1000 to 1900 K cooler and the penumbra is 250 to 400 K cooler than the quiet Sun at the photosphere. Of course, the brightness is not uniform across either the umbra or the penumbra: in the umbra there are large-scale intensity variations and small-scale bright features (the umbral dots), while in the penumbra there are radial variations and conspicuous azimuthal variations associated with the filamentary structure.

The low surface temperature of pores and sunspots is caused by the influence of the magnetic field on the transport of energy in the optically thick layers below the surface. There are two important ways in which the heat flux within these large magnetic flux concentrations can be reduced: inhibition of convection by the magnetic field, and dilution of the heat flux as the cross-sectional area of the magnetic flux concentration increases with height.

Biermann (1941) suggested that a sunspot is cooler than its surroundings because convection within the spot is suppressed by the strong magnetic field acting to prevent the motions of convective eddies (see Section 2.6.2). We now know that this idea is basically correct, although convection is not fully suppressed because a fairly significant convective flux of energy is needed to maintain the brightness of the umbra. In modelling the thermal structure of a sunspot, the partial suppression of convection can be represented most simply by employing mixing-length theory with a mixing length that is reduced within the spot compared to that in the surrounding convection zone (Chitre 1963; Deinzer 1965; Chitre and Shaviv 1967; Yun 1970; Jahn 1989).

Another effect of the sunspot's magnetic field was first pointed out by Hoyle (1949), who argued that convective motions within the spot are largely channelled along the magnetic field lines, so that convective heat transport can occur along the field lines but is very inefficient across them. This helps to insulate the sunspot's flux tube from its surroundings, and spreading of the cross-sectional area of the flux tube with height dilutes the heat flux and hence reduces the temperature of the spot at the surface.

A third possible way in which the magnetic field might in principle affect the energy transport in a sunspot is by carrying away a flux of mechanical energy in the form of MHD waves (Alfvén waves or magneto-acoustic waves). The idea that a sunspot is cooled by a flux of such waves was put forward in several papers in the 1960s (Danielson 1965; Zwaan 1965; Musman 1967; Savage 1969) and was later revived by Parker (1974; see also Boruta 1977). He argued that if the heat flux were diverted owing to inhibition of convection there should be a conspicuous bright ring around the spot, which is not observed, whereas an enhanced transport of energy due to Alfvén waves would produce a cool sunspot without a bright ring. Cowling (1976b) refuted this proposal on the basis of both magnetohydrodynamic and thermodynamic considerations (see also Schmidt 1991). Observations (Beckers and Schneeberger 1977) then showed that the upward flux of wave energy in a sunspot is insufficient to produce significant cooling, and theory (Thomas 1978) confirmed that a significant upward flux of wave energy is prevented by strong downward reflection of Alfvén waves in the umbral atmosphere. Meanwhile, it was shown that a wave-energy flux is not needed to explain the absence of a bright ring: instead, the absence is explained naturally within the context of magnetic inhibition of convection in terms of the spatial and temporal properties of convective heat transport and storage in the surrounding convection zone (see Section 3.2.3).

3.2.2   Brightness of the umbra and penumbra

Measurements of sunspot brightness are important for determining the thermal structure of the sunspot atmosphere and the effect of sunspots on the Sun's total and spectral irradiance. Estimates of the continuum intensity of radiation emerging from a sunspot, especially from the umbra, go back over 150 years to the work of Henry and Alexander (1846) and are the subject of several reviews (Bray and Loughhead 1964; Zwaan 1965, 1968; Obridko 1985; Maltby 1992). The development of photoelectric photometry substantially increased the accuracy of these measurements. They are influenced significantly by the presence of stray light in the sunspot image, produced by the Earth's atmosphere (due to 'seeing') or by scattering within the instrument, and some method of correction for this stray light is required. It has only been in the past 40 years or so that improved instruments and accurate methods for correcting for stray light have produced reliable results. Here we will restrict the discussion to measurements made in integrated light or in selected bands in the continuous spectrum.

Sunspot brightness is usually measured in relative terms, as the *brightness ratio* I_s/I_{qs} between the wavelength-integrated continuum intensities I_s in the sunspot and I_{qs} in the surrounding quiet photosphere, or as the *contrast* $\alpha = (I_{qs} - I_s)/I_{qs}$. The brightness is usually determined separately for the umbra and the penumbra; typical values of the umbral contrast α_u are in the range 0.5 to 0.8, and typical values of the penumbral contrast α_p are in the range 0.15 to 0.2. For modelling variations in the Sun's total and spectral irradiance due to solar activity (e.g. Foukal 1981b, 2004; Hudson *et al.* 1982; Foukal *et al.* 2006), a single value of the overall effective contrast is often used, a typical value being $\alpha = 0.32$. The umbral and penumbral contrasts do not vary significantly from centre to limb (Maltby *et al.* 1986), but because the umbra is partly obscured by the penumbra as a spot approaches the limb (due to the Wilson depression; see Section 3.2.4), the overall contrast decreases near the limb.

A simple question concerning sunspot brightness is whether it varies with the size of the sunspot. This question is closely coupled to that of whether the magnetic field strength in a sunspot varies with the size, for theoretical models of sunspots based on magnetohydrostatic equilibrium and inhibition of convective heat transport (e.g. Deinzer 1965; Yun 1970) have lower temperatures for stronger magnetic fields. Until the 1960s, it was generally accepted that large sunspots are darker than small sunspots (see Bray and Loughhead 1964), but the measurements on which this was based often lacked sufficient correction for stray light. Zwaan (1965) pointed out that the amount of stray light affecting the measured umbral intensity increases rapidly with decreasing umbral radius and that this spurious effect could account for most of the measured dependence of umbral intensity on umbral size. Some subsequent observations indicated no dependence of umbral intensity on umbral size for umbral diameters greater than about 8″ (Rossbach and Schröter 1970; Albregtsen and Maltby 1981). However, more recent observations, beginning with those of Stellmacher and Wiehr (1988) and Sobotka (1988), have established that umbral intensity does decrease with increasing umbral radius. The observations of Kopp and Rabin (1992) in the near infrared at 1.56 μm, where stray light is less of a problem than in the visible, showed a nearly linear decrease in umbral brightness with umbral radius for six sunspots. A similar result was obtained for seven spots at visible wavelengths by Martínez Pillet and Vázquez (1993; see also Collados *et al.* 1994), who also found that the umbral contrast is well correlated with the umbral field strength, allowing one to predict the peak field strength of a sunspot from its brightness to an accuracy of about 100 G (Norton and Gilman 2004). Recently, in a study of continuum images of more than 160 sunspots taken during solar Cycle 23 with the MDI instrument aboard SOHO, Mathew *et al.* (2007) found a strong and clear dependence of umbral brightness (and a weak dependence of penumbral brightness) on umbral radius.

Observations of large samples of sunspots allow the dependence of the overall sunspot contrast α on sunspot area to be estimated: Chapman, Cookson and Dobias (1994) found $\alpha = 0.276 + 3.22 \times 10^{-5} A_s$ and Brandt, Stix and Weinhardt (1994) found $\alpha = 0.2231 + 0.0244 \log(A_s)$, where A_s is the sunspot area in millionths of the solar hemisphere.

The size dependence of umbral brightness has implications for the issue of whether a sunspot is essentially a monolithic flux tube or a cluster of individual flux tubes (see Section 3.5). As support for a cluster model, Parker (1979b) pointed out that one would expect the umbral brightness to decrease with size for a monolithic model but not necessarily for a cluster model. At that time the best observational evidence suggested that there was no significant size–brightness relation, but since then such a relation has been firmly established and this objection to the monolithic model no longer stands (cf. Martínez Pillet and Vázquez 1993).

A more detailed question concerns the local relation between brightness and magnetic field strength within a sunspot. Such a relationship is expected on theoretical grounds; for example, simulations of magnetoconvection within a sunspot umbra show hot rising plumes that tend to expel the magnetic field and concentrate it into cooler regions (Nordlund and Stein 1990; Weiss *et al.* 1990; Weiss, Proctor and Brownjohn 2002; Schüssler and Vögler 2006). Detailed observations have established a general inverse relation between magnetic field strength and continuum intensity at different positions within a sunspot, although the precise form of this relation has been subject to debate (von Klüber 1955; Gurman and House 1981; Lites *et al.* 1991; Kopp and Rabin 1992; Martínez Pillet and Vázquez 1993; Lites *et al.* 1993; Hofmann *et al.* 1994; Stanchfield, Thomas and Lites 1997: Westendorp Plaza *et al.* 2001; Penn *et al.* 2003). Figure 3.3 shows the relation in the form of a scatter plot for all the pixels in high-resolution observations of a single large sunspot. Note that the relation is distinctly nonlinear in the umbra (points to the left of the dotted vertical line).

A surprising result concerning sunspot brightness was the discovery by Albregtsen and Maltby (1978, 1981) that the umbra–photosphere brightness ratio of large sunspots varies over the solar cycle. They found that sunspots were darkest at the beginning of sunspot Cycle 20 and that spots appearing later in the cycle were progressively brighter (with a nearly linear dependence on the phase of the cycle), up until the new Cycle 21 spots appeared which were again darkest. Subsequent observations showed the same behaviour occurred in Cycle 21 (Maltby *et al.* 1986). Recently, Penn and Livingston (2006) found similar behaviour during Cycle 23: from observations of more than 900 sunspots during 1998–2005, they found an increase in the normalized umbral intensity (from 0.60 to 0.75) accompanied by

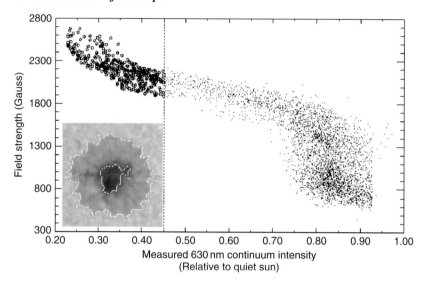

Fig. 3.3. Scatter plot of magnetic field strength versus relative continuum intensity for individual points within a sunspot. Open circles denote points within the umbra and dots denote points in the penumbra. The vertical dotted line corresponds to the continuum intensity level that defines the umbra–penumbra boundary shown in the inset continuum image of the sunspot. (From Stanchfield, Thomas and Lites 1997.)

a decrease in the maximum umbral magnetic field strength (at a rate of about $52\,\mathrm{G\,yr^{-1}}$). However, the results of Penn and Livingston are directly contradicted by the more recent results of Mathew *et al.* (2007) in their study of more than 160 spots mentioned above; they find no significant change in umbral brightness over Cycle 23. From an observational viewpoint, then, any solar-cycle dependence of umbral brightness seems to be quite uncertain.

Several complicating factors might influence the observational results. The behaviour might be related to the known dependence of umbral brightness on latitude and the fact that spots tend to appear at lower latitudes as the cycle progresses, although Albregtsen, Jorås and Maltby (1984) claim to have corrected for this effect and still find the cycle dependence. The established dependence of umbral brightness on umbral size may influence the results: although there is apparently no systematic variation in the distribution of umbral sizes over the solar cycle (Bogdan *et al.* 1988), there may be such a variation in a particular sample of spots. If there were a variation of the quiet photospheric brightness over the cycle, it would cause a variation in the umbral brightness ratio, but Maltby (1992) claims to have ruled this out. From a theoretical viewpoint, possible mechanisms for producing the cycle dependence in the context of dynamo theory have been proposed. Schüssler (1980) suggested that the umbral brightness might depend on the age of the toroidal flux tube that formed a sunspot, while Yoshimura (1983) suggested that it might depend on the depth at which the flux tube forms. Nordlund and Stein (1990) suggested that solar-cycle changes in the relative pressure between a magnetic flux tube and its surroundings deep in the convection zone might be transmitted to the surface and affect the efficiency of magnetoconvection within a sunspot.

3.2.3   The weak bright ring around a sunspot

Observations of the bright ring

The presence and strength of a bright ring surrounding a sunspot is a long-standing issue (see references in Rast *et al.* 1999, 2001). Early photographic studies found bright rings with an intensity excess of 2–3% above the normal photospheric intensity (see Bray and Loughhead 1964). More recent photoelectric studies, however, find a much lower contrast of only 0.1–0.3% (Fowler, Foukal and Duvall 1983; Rast *et al.* 1999, 2001). In either case, the observed intensity excess in the bright ring is far too small to account for the deficit in the sunspot. (Rast *et al.* 2001, for example, estimate that the bright ring accounts for only about 10% of the heat flux deficit.) The radiative intensity in the umbra is roughly only a quarter, and in the penumbra three quarters, of that in the quiet photosphere. If the missing surface heat flux due to a sunspot were balanced at the same time by an enhanced flux in the immediate vicinity of the spot, there would be a very conspicuous brightening of the photosphere around the spot, which is not observed. This point was quantified by Parker (1974) on the basis of a simple steady-state thermal model of the heat flux blocked by a shallow sunspot (due to inhibition of convection).

What then is the fate of the missing heat flux in a sunspot? As we shall discuss next, the missing flux is apparently redistributed so widely, both spatially and temporally, as to be almost imperceptible.

Thermal models of sunspots

Horizontal temperature gradients, however small, drive motions in a fluid layer, and the resulting convection is extremely efficient at redistributing thermal energy (Sweet 1955). This nonlinear process can be parametrized in terms of an (isotropic or anisotropic) thermal conductivity (Parker 1974; Spruit 1977, 1982a,b, 1992; Eschrich and Krause 1977; Clark 1979; Foukal, Fowler and Livshits 1983; Stix 2002). A sunspot can then be regarded as a non-conducting cylindrical plug inserted into a conducting layer, and the resulting temperature distribution can be calculated. There are two aspects of this problem: first of all, a steady-state solution can be obtained, but then the more general time-dependent problem, modelling the emergence of a sunspot at the solar surface, has to be tackled. Spruit (1992) has provided a very apt and familiar analogy. Consider an insulating disc (e.g. a ceramic tile) placed on top of a thick slab of highly conducting copper that is heated electrically from within (like an electric hob on a stove). In a steady state the disc will be cooler than its surroundings and radiate less; underneath it, the temperature will be marginally higher, so that the otherwise vertical heat flux is diverted laterally. To be sure, the rest of the surface will be slightly hotter and radiate slightly more – but all horizontal gradients of temperature will be small and any bright ring will be very faint. Now consider what happens just after the insulating tile is applied to the surface. There is an immediate reduction in the total energy emitted but the missing energy is rapidly distributed throughout the copper block, on a time scale determined by its thermal conductivity. It is only on a much longer time scale, depending on the thermal capacity of the block, that it reaches an equilibrium, radiating as much energy as the wattage supplies.

In a more realistic situation it is essential to distinguish carefully between the different time scales that are involved. Spots appear (and usually disappear) on a time scale of days, though a large spot may survive for several months; if we assume a turbulent diffusivity of 10^8 m^2 s^{-1}, the thermal diffusion time τ_d for the convection zone is about a month (comparable to the turnover time of the largest eddies); this is far less than the time taken for the entire convection zone to respond to changes at its surface, its Kelvin–Helmholtz time τ_{KH} (the ratio of its total thermal capacity to the solar luminosity), which is about 10^5 years; this is, in turn, much less than the Kelvin–Helmholtz time for the whole Sun, which is of order 10^7 years. The immediate response of the Sun to the appearance of a sunspot is therefore a slight reduction in its luminosity; on a time scale of order τ_d (or less, if the adjustment applies in a shallow layer only) a weak bright ring appears but the 'missing energy' is simply absorbed by the whole convection zone. On the longer time scale τ_{KH} the convection zone adjusts its structure so that the average rate at which energy is radiated from the solar surface equals the rate of supply from the solar interior. Since we know from proxy data that solar activity has been maintained at a similar level for the last 100 000 years, we may assume that there is a balance between input and output of energy over times much longer than the intervals between grand minima, i.e. over times of order 1000 years.

More detailed model calculations, for a stratified atmosphere with a depth-dependent thermal diffusivity, show that the proportion, α_{br}, of the missing energy that is radiated from the bright ring depends on the aspect ratio of the cylindrical plug (Spruit 1982a,b; Foukal, Fowler and Livshits 1983). Taking a plug whose depth is equal to its radius R (a reasonable assumption for the blocking effect of a spot), Spruit (1992) estimates that α_{br} rises from zero to a maximum value given by

$$\alpha_{br} \approx \frac{1}{2} \left(1 + \frac{R}{3H_0} \right)^{-2}, \tag{3.2}$$

where H_0 is the pressure scale height at the surface, which is small compared with R. This is consistent with what is observed.

Having said all this, we should note that, while the measured solar irradiance does indeed fall when a new sunspot appears, there is a further compensating effect. Active regions contain not only dark sunspots but also bright faculae, and the excess radiation from faculae actually exceeds the deficit from sunspots (Foukal 2004; Foukal *et al.* 2006). Consequently the total solar irradiance *increases* when activity is highest, varying by about 0.1% over the solar cycle. We shall return to this topic in Section 12.1.

3.2.4 The Wilson depression

Observations dating back to those of Alexander Wilson in 1769 reveal that, as a large sunspot approaches the limb, the width of the penumbra on the disc-centre side of the spot decreases more rapidly than the width on the limbward side, an effect that can be explained by foreshortening if the spot is a saucer-shaped depression of the solar surface. We now understand that this *Wilson depression* is essentially due to the decreased opacity of the sunspot atmosphere (owing to its lower temperature and gas pressure), allowing one to see deeper into a sunspot than into the quiet photosphere. The depression can be specified by giving the geometric depth z_W as a function of the optical depth τ_{500} and horizontal position within the spot. The value of z_W for unit continuum optical depth ($\tau_{500} = 1$) varies across a spot, with a maximum value of about 600 km in the centre of the umbra.

A strictly observational determination of the Wilson depression z_W is complicated by evolutionary changes in the shape of the spot. Older observations, summarized by Bray and Loughhead (1964) and Gokhale and Zwaan (1972), give maximum values in the range of

400–800 km for mature sunspots. Balthasar and Wöhl (1983) used an indirect method, comparing the rotation rate determined by the disc passage of a spot (which is affected by the Wilson depression) with that determined by successive passages of a spot across the central meridian (which is not), and found values of the depression in the range 500–1000 km. There are other possible complications. If the relation between geometric and optical depth is different in the umbra, penumbra and photosphere, z_W will then depend on the heliocentric angle θ. The determination of z_W will also be affected by variations in the relative size of the umbra and penumbra with height (Wilson and Cannon 1968; Wilson and McIntosh 1969) and by the raggedness of the umbral boundary revealed in high-resolution observations (Solanki and Montavon 1994).

The reduced opacity in a sunspot, and the consequent depression of the $\tau_{500} = 1$ level, arise mainly from two effects: the reduced temperature in the spot atmosphere causes a marked decrease in the H^- bound–free opacity, and the radial force balance including magnetic pressure and curvature forces demands a lower gas pressure within the spot, further reducing the net opacity. A purely theoretical determination of the Wilson depression z_W requires a complete magnetohydrostatic model of a sunspot in which magnetic pressure and curvature forces as well as the temperature distribution are specified. One could instead use observed values of the vector magnetic field in the equation for the radial force balance in order to determine z_W. An alternative approach has been developed by Martínez Pillet and Vázquez (1990, 1993); they assumed a linear relation between magnetic pressure and temperature in a spot, as indicated by observations, in which case the radial force balance yields a simple relationship between the net magnetic curvature force and the Wilson depression. They found that for typical values of the observed Wilson depression, the radial curvature force must be comparable to the radial pressure force within the spot. Solanki, Walther and Livingston (1993) used this approach to determine the variation of z_W with radius across a sunspot; they found values of 50–100 km in the penumbra and 400–500 km in the umbra, with a fairly sharp transition at the umbra–penumbra boundary.

3.3 Spectroscopy and atmospheric models

3.3.1 Spectral analysis of the umbra and penumbra

The vertical temperature structure in the photosphere of a sunspot is revealed by measurements of the variation of continuum intensity with wavelength within the spot at a fixed position on the solar disc, or by the variation of intensity at a fixed wavelength as the spot moves from disc centre to the limb (Maltby 1992). Measuring the centre-to-limb variation of continuum intensity ('limb darkening') is a standard technique for determining the temperature stratification in the solar atmosphere (see, for example, Mihalas 1978), and the possibility of using similar measurements as a means of determining the temperature stratification in a sunspot was first pointed out by Minnaert and Wanders (1932). A number of early studies found that the umbral limb darkening was either absent or significantly less than that for the quiet photosphere. However, Albregtsen, Jorås and Maltby (1984) found a significant decrease in the umbral intensity ratio towards the limb, arguing that previous studies had made insufficient corrections for stray light, and later studies have generally confirmed their results.

Photometric measurements of continuum intensities in sunspots are done either in fairly narrow, clean 'continuum windows' lying between spectral lines or in broader spectral bands that include spectral lines. In the latter case, the calculations must be corrected for the effect of the spectral line blanketing on the total absorption coefficient. Modern measurements usually include a band in the near infrared around the opacity minimum at 1.6 μm, which corresponds to the deepest visible layers of the umbra, and for which the limb darkening is relatively greater and line blanketing is less pronounced than in the visible.

The temperature structure in the sunspot atmosphere can also be deduced from high spectral resolution measurements of the profiles of absorption lines. For the sunspot photosphere, the profiles of weak spectral lines provide a check on the continuum measurements, and the profiles of molecular lines formed only in the cool umbra are especially useful. For levels in the atmosphere above the photosphere, line profiles provide the best diagnostic. For the chromosphere, the profiles of strong absorption lines, such as the Ca II H and K lines, are used, and for the transition region and corona emission lines in the UV and EUV are used.

3.3.2   *Semi-empirical models of umbral and penumbral atmospheres*

A number of one-dimensional, semi-empirical models of an umbral or penumbral atmosphere have been constructed, giving the variation of thermodynamic variables with optical depth (and geometric height) based on empirical data and theoretical considerations of mechanical equilibrium and radiative transfer. There are both one-component models, meant to represent a horizontal average over the umbra or penumbra, and two-component models meant to represent bright and dark components separately. In most cases, a single model atmosphere is meant to represent spots of all sizes, or at least all large spots; this approach was supported by observational evidence that the brightness of the dark cores of large umbrae is independent of umbral size, but as we have mentioned (in Section 3.2.2) recent observations contradict this evidence. Although the models represent only a mean atmosphere, averaged in some sense over small-scale horizontal inhomogeneities, they are nevertheless useful in constraining certain physical processes that determine the structure of a real sunspot atmosphere, and also in providing a background model for studies of element abundances, wave propagation, and other behaviour in a sunspot. Here we shall present a brief general discussion of atmospheric models with only a few examples; for more details, see the detailed review of early models in Bray and Loughhead (1964) and the comprehensive review by Solanki (2003).

In calculating a semi-empirical model, the atmosphere is assumed to be in magneto-hydrostatic equilibrium, often including a turbulent pressure. For the lower photospheric layers, measurements of the centre-to-limb variation of continuum intensity, the spectral distribution of continuum intensity, and the profiles of weak spectral lines (or the wings of strong lines) are used. The radiative transfer calculations usually assume local thermodynamic equilibrium (LTE), and the temperature–optical depth relation is adjusted so that the model intensities agree with the observed values. The deepest layers of the photosphere are detected at infrared wavelengths near 1.64 μm where the H^- continuum opacity is a minimum. For the chromospheric layers and the transition region, profiles of strong spectral lines in the visible and UV ranges are used, and the radiative transfer is necessarily treated as non-LTE. The general procedure in constructing a model atmosphere is first to determine a temperature–optical depth relation, $T(\tau)$, as a best fit to the empirical data, and then to determine the gas and electron pressures by integrating the equation of hydrostatic equilibrium.

Umbral models

Most semi-empirical models have been for the umbra. The umbral magnetic field is usually assumed to be uniform and vertical, in which case it exerts no vertical force and the stratification is purely hydrostatic. There are both one-component models, meant to represent a mean umbra or perhaps a dark umbral core free of umbral dots, and two-component models meant to represent separately the dark main component and a bright component corresponding to the umbral dots. The simplest one-component models of the umbral core are based on a model of the quiet-Sun atmosphere scaled to the lower effective temperature of the umbra. Such models are usually expressed in terms of the parameters $\Theta = 5040/T$, where the temperature T is measured in K, and

$$\Delta\Theta = 5040\left(\frac{1}{T} - \frac{1}{T_{\mathrm{ph}}}\right) \tag{3.3}$$

where T_{ph} is the temperature in the quiet-Sun model.

Several umbral models were constructed in the 1950s and 1960s; these are discussed in the reviews by Bray and Loughhead (1964) and Solanki (2003). Two important umbral models were constructed in the early 1980s based on large amounts of empirical data and the efforts of several modellers: the model of Avrett (1981), produced at the 1981 Sacramento Peak workshop held at Sunspot, NM, and hence called the 'Sunspot sunspot model', and the model of Staude (1981). Improvements to the deepest layers of the Sunspot sunspot model were made later by Maltby *et al.* (1986); Figure 3.4 shows the variation of temperature with geometric height in their model M, meant to represent a dark umbral core within a large sunspot, and, for comparison, in a model of the quiet solar atmosphere. In each temperature profile, the height is measured relative to the corresponding level where the optical depth is $\tau_{500} = 1$; because of the Wilson depression, these two zero levels are offset by several hundred kilometres and hence the umbral temperature curve needs to be shifted to the right by this amount in order to compare values at the same geometric height inside and outside the

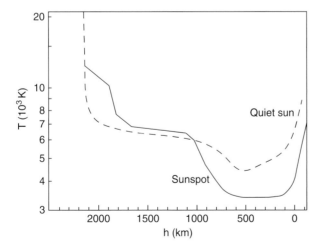

Fig. 3.4. Temperature as a function of height in the sunspot umbral model of Maltby *et al.* (1986) and in the quiet Sun (model C of Vernazza, Avrett and Loeser 1981). (From Stix 2002, courtesy of Springer Science and Business Media.)

sunspot. (Note that a value for the Wilson depression is not an inherent part of the umbral model but instead must be determined from additional considerations.) The profiles show that the umbra is significantly cooler than the quiet Sun in the photospheric layers but slightly hotter in the chromosphere, and that the transition to coronal temperatures begins at a lower height in the umbra. Modifications of the Maltby *et al.* (1986) models have been proposed by Lites *et al.* (1987), Severino, Gomez and Caccin (1994), and Ayres (1996).

Penumbral models

The penumbral atmosphere is more inhomogeneous and complicated than the umbral atmosphere, and hence models with one and even two components are of more limited applicability than umbral models and have received less attention. One-component models were constructed by Makita and Morimoto (1960), Kjeldseth Moe and Maltby (1969), Yun, Beebe and Baggett (1984) and Ding and Fang (1989). At least at photospheric heights, a two-component model is needed to account for the different thermal structure of the bright and dark filaments. Because the penumbra is much closer in temperature to the quiet photosphere than the umbra, a simple scaling of a quiet-Sun model with a fixed value of $\Delta\Theta$ will give a reasonable first approximation. An example is the two-component model of Kjeldseth Moe and Maltby (1974), which has different constant values $\Delta\Theta_b = 0.010$ and $\Delta\Theta_d = 0.093$ for the bright and dark components and assumes that the density at a given height is the same in both components.

More recent work has focused on models with more components, to better represent penumbral fine structure. Rouppe van der Voort (2002) constructed three atmospheric models meant to represent cool, intermediate, and hot features within the penumbra, with temperature differences of order 300 K between them. A different approach was taken by del Toro Iniesta, Tarbell and Ruiz Cobo (1994), who constructed models of the vertical temperature profile at many different spatial positions in the penumbra, and also average temperature profiles at different radial distances from the spot centre and an overall average temperature profile.

3.3.3   The chromospheric superpenumbra

In the chromosphere, particularly as seen in the centre of the Hα line (formed at a height of about 1500 km), a large, isolated sunspot often displays a *superpenumbra*, a distinctive pattern of dark, nearly radial fibrils similar to that seen in white light but extending outward well beyond the edge of the white-light penumbra (Howard and Harvey 1964; Bray and Loughhead 1974). The fibrils often show a spiral-like pattern with individual filaments being slightly curved. An example of a superpenumbra is shown in Figure 3.5. Dark superpenumbral filaments typically begin near the outer edge of the penumbra (as defined in white light), although about a third of them begin well within the penumbra (sometimes as far in as the umbral boundary) and some begin outside the penumbra. Some of the fibrils appear to branch or coalesce. The spacing between adjacent fibrils is as little as $1''$ in the inner penumbra and increases to $2''$–$3''$ or more at the outer edge of the superpenumbra, which can lie more than a spot diameter beyond the white-light penumbra. The overall fibril pattern of the superpenumbra changes slowly over time periods of several hours to a day or more, although individual fibrils show small changes in brightness or shape over a few minutes. The dark superpenumbral fibrils are the locations of the strongest reverse (inward) Evershed flow in the sunspot chromosphere (see Section 5.4).

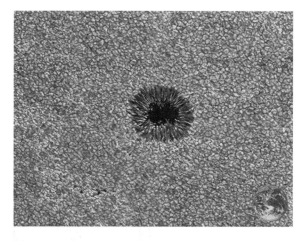

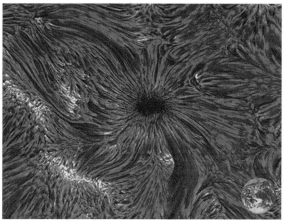

Fig. 3.5. White-light and Hα images of a large sunspot. The Hα image shows the dark filaments of the chromospheric superpenumbra extending radially outward well beyond the outer edge of the penumbra seen in the white-light image. Images of the Earth are superimposed to indicate the scale. (Courtesy of NASA GSFC.)

Low-resolution chromospheric images also show the superpenumbra, and although they may not resolve the individual dark fibrils they do reveal their sometimes spiral-like appearance. The spiral configuration of Hα sunspot filaments was studied intensively for many years, particularly by Hale (1908a; Hale *et al.* 1919) and Abetti (1957), as evidence that sunspots correspond to cyclonic storms on the Sun, an idea that was abandoned when it was learned that the filaments are aligned with magnetic field lines. Hale (1925) determined the spiral pattern of 51 sunspots during two successive sunspot cycles and found that about 80% of the spots in the northern hemisphere had counter-clockwise twist while about 80% of those in the southern hemisphere had clockwise twist (consistent with the sense of twist of terrestrial cyclones). He found no correlation between the sense of twist and the magnetic polarity of the sunspot; instead, he suggested that the twist was due to the Coriolis force acting on the reverse Evershed flow. A subsequent study of 141 sunspots by Richardson (1941), also at Mount Wilson, confirmed Hale's findings, but a later study by Nakagawa *et al.* (1971) of 240 spots failed to confirm Hale's hemisphere rule for the direction of twist. More recently, Balasubramaniam, Pevtsov and Rogers (2004) studied 897 individual super-penumbral fibrils in 139 sunspots. They found that both clockwise and counter-clockwise twisted fibrils can be present in the same spot and that the geometry of the fibrils is influenced by the magnetic field distribution outside the spot; nevertheless, they found that on average the hemisphere rule of Hale is obeyed.

Various attempts to construct static, force-free magnetic field configurations to match the spiral pattern of the superpenumbra have been made (e.g. Nakagawa *et al.* 1971; Nakagawa and Raadu 1972; Schmieder *et al.* 1989). Peter (1996) has presented a simplified MHD model of the superpenumbra that includes an Evershed-like inflow and the associated Coriolis force; this model can reproduce the essential features of the superpenumbra and the statistical properties of the observed spiral patterns.

3.3.4   The transition region and corona above a sunspot

Although sunspots are generally thought of as dark features in the solar atmosphere, in the transition region and corona they are often brighter than their surroundings. Measurements of EUV emission in active regions, from rockets (Brueckner and Bartoe 1974) and from the Apollo Telescope Mount (ATM) aboard the Skylab mission (Foukal *et al.* 1974; Noyes *et al.* 1985), revealed that the areas directly above sunspot umbrae are often the brightest features in the transition region. These bright regions, known as *sunspot plumes*, show relative enhancements (compared to the quiet Sun) of up to a factor of 40 or more in the intensities of emission lines formed at temperatures in the range of 10^5 to 10^6 K (see Fig. 3.6). Analysis of the ATM data indicated that the plume emission comes from extended regions of relatively cool, dense plasma within large magnetic loops, in which the temperature is one or two orders of magnitude lower than in the surroundings (Foukal 1976, 1981a; Raymond and Foukal 1982).

In a study of the ATM data for 22 large sunspots, Foukal (1976) found that a sunspot plume is essentially a steady-state feature within a cool coronal loop emanating from the sunspot umbra, with enhanced EUV emission extending up into the low corona to heights as great as 10^5 km (several density scale heights). He pointed out that the plume material cannot be supported hydrostatically (even with turbulent pressure included) or magneto-hydrostatically; instead, plasma must be falling down along the loop. Subsequent spectroscopic measurements did indeed reveal the presence of strong downflows in the transition region above sunspots, with speeds typically in the range 20–40 km s^{-1} (Brueckner 1981; Dere 1982; Nicolas *et al.* 1982; Gurman 1993) but occasionally reaching highly supersonic speeds of up to 200 km s^{-1} (Brekke *et al.* 1987). More recent observations have found downflows located precisely within the plumes, with speeds greater than 25 km s^{-1} (Maltby *et al.* 1999; Brynildsen *et al.* 2001; Brosius 2005; Brosius and Landi 2005). The downflow, which seems to be an essential feature of a plume, is fed by an inflow of plasma at transition-region temperatures from locations well outside the sunspot, maintaining the enhanced emission of the plume in a quasi-steady state.

A major study of sunspot plumes, in 42 sunspot regions, was carried out by the Oslo group (Brynildsen *et al.* 2001) using data from the Coronal Diagnostic Spectrometer and the SUMER instrument aboard SOHO. They found that the plumes are almost invariably present in spots where one magnetic polarity dominates throughout a region extending well away ($50''$ or more) from the centre of the spot. The plumes show their maximum spatial

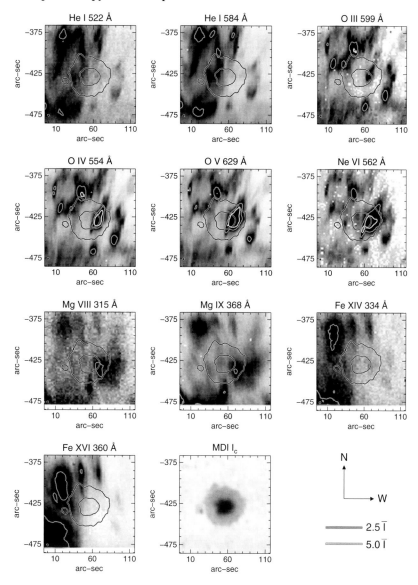

Fig. 3.6. Images of peak EUV line intensities associated with sunspot plumes. Regions with enhanced intensity are shown as dark areas. The images are ordered (starting at the upper left) with increasing line-formation temperature. The contours of the umbra and penumbra are from continuum intensity observations. The scales in arcseconds are in a reference system with its origin at the centre of the solar disc. (From Brynildsen *et al.* 2001.)

extent in lines formed at temperatures near 40 000 K and are essentially invisible in lines formed at temperatures less than 10 000 K or greater than 100 000 K. Observations of a few spots at the limb indicate a plume height of no more than 15 000 km, considerably less than the heights estimated from the ATM data. Transition-region lines are almost invariably red-shifted in sunspot plumes, with the line-of-sight velocity peaking in lines formed at temperatures near $32\,000$ K and decreasing rapidly for higher temperatures. The line-of-sight velocity in the corona above a plume is too small to provide the material to sustain the downflow in the plume; instead, the downflow is maintained by a more horizontal inflow of material from the surroundings along individual flow channels, of widths ranging from $2''$ to nearly $20''$ (see also Brynildsen *et al.* 1998; Maltby *et al.* 1999). A mean flow in these channels persists for times up to about a day, while fluctuations about the mean flow occur on a time scale of about 10 minutes. As Brynildsen *et al.* (2001) point out, these inflows are almost certainly field-aligned and the flow channels correspond to magnetic flux tubes. The flows are most likely produced by the 'siphon-flow' mechanism, which we shall discuss in some detail in Chapter 5 in connection with the Evershed effect. The characteristic 3-minute chromospheric umbral oscillations are also present in sunspot plumes; these oscillations are discussed in Chapter 6.

3.4   Observations of the magnetic field in sunspots

Measurements of magnetic fields on the Sun rely on the Zeeman effect, which produces splitting and polarization of spectral lines. Here we outline how this effect can be used not only to measure strong longitudinal fields directly but also to detect weaker fields by measuring circular polarization and, more recently, to measure the full vector magnetic field by determining the four Stokes parameters across a spectral line. Then we go on to describe the average magnetic structure, first of pores and then of sunspots. For this purpose, we regard them as axisymmetric configurations, leaving the fine structure of the umbral and penumbral magnetic fields to be described in Chapters 4 and 5.

3.4.1   The Zeeman effect

Here we give only a brief discussion of the Zeeman effect and its use in measuring solar magnetic fields. A fuller account can be found in the books by Stenflo (1994) and Stix (2002).

Individual atoms precess in the presence of a magnetic field and the precession frequency combines with the atomic transition frequency. As a result, spectral lines formed in a magnetized layer of the solar atmosphere get split up into different components, separated in wavelength. The pattern of this splitting depends on the quantum numbers of the atomic transition and on the strength and direction of the magnetic field. The simplest form of splitting is the so-called *normal* Zeeman triplet, consisting of an unshifted π component and two σ components shifted symmetrically to either side of the central π component. The wavelength shift can be represented in terms of the Landé factor g, defined as

$$g = 1 + \frac{J(J+1) - L(L+1) + S(S+1)}{2J(J+1)}, \tag{3.4}$$

where L, S and J are the quantum numbers characterizing the orbital, spin, and total angular momentum vectors, respectively. If we let M denote the additional 'magnetic' quantum number that determines the component of total angular momentum in the direction of the magnetic field, then the wavelength displacement of a spectral line from its original wavelength λ_0 is given by

$$\lambda - \lambda_0 = \frac{e}{4\pi m_e c} g^* \lambda^2 B, \tag{3.5}$$

where e and m_e are the charge and mass of an electron, c is the speed of light, B is the magnetic field strength, and g^* is the Landé factor for the transition, given by

$$g^* = g_u M_u - g_l M_l, \tag{3.6}$$

where the subscripts u and l denote the upper and lower levels of the transition.

More generally, there is *anomalous* splitting producing a Zeeman multiplet with several π and σ components. Because solar spectral lines are considerably broadened, these individual components are usually not distinguishable. If the magnetic field is not too strong, however, the multiplet can be considered as a triplet with an *effective* Landé factor g_{eff} calculated from the individual g^* values of the components, each weighted by the corresponding intensity.

In observing a Zeeman triplet of a solar absorption line, if the line of sight is along the magnetic field then one sees only the two σ components, which are circularly polarized in opposite senses; this is known as the *longitudinal* Zeeman effect. On the other hand, if the magnetic field is perpendicular to the line of sight then one sees all three components, with the π component being linearly polarized perpendicular to **B** and the σ components being linearly polarized parallel to **B**; this is known as the *transverse* Zeeman effect.[1] In the more general case, when the magnetic field is neither along nor perpendicular to the line of sight, all of these components are superimposed, but they can be singled out by taking advantage of their different polarization states (polarimetry).

Direct measurement of Zeeman splitting is possible only if $\Delta\lambda$ is greater than the half-width of the broadened spectral line. A typical half-width of a Zeeman-sensitive spectral line is 0.1 Å, and typically $\Delta\lambda$ will exceed this width only for field strengths greater than about 1500 G. Hence Zeeman splitting is directly measurable in a sunspot umbra but not in most local magnetic fields outside of spots.

If the magnetic field is strong enough, as in a sunspot (see Fig. 3.7), the σ components are clearly separated and the longitudinal magnetic field strength can be measured directly in integrated light. (This is the basis for Hale's original discovery of the sunspot magnetic field.) For weaker fields, the σ components overlap but they can be resolved by taking advantage of their opposite circular polarization produced by the longitudinal (line-of-sight) component of the magnetic field; this is the basis for the *magnetograph*, developed by the Babcocks at Mount Wilson in the 1950s, which measures the longitudinal magnetic field. More recently, polarimeters have been developed that measure the full polarization state (circular and linear) of a spectral line and thus are capable of determining the full vector magnetic field **B**. These instruments, known as *vector magnetographs* or *Stokes polarimeters*, measure the four *Stokes parameters* I, Q, U and V which together uniquely determine the polarization state of the light. For a monochromatic, completely polarized light wave propagating in the z-direction with its electric field lying in the xy-plane, with components

$$E_x = \xi_x \cos(\omega t - kz), \quad E_y = \xi_y \cos(\omega t - kz + \phi), \tag{3.7}$$

where ξ_x and ξ_y are constant amplitudes and ϕ is the phase difference between E_x and E_y, the Stokes parameters are defined by

$$I = \xi_x{}^2 + \xi_y{}^2, \quad Q = \xi_x{}^2 - \xi_y{}^2, \quad U = 2\xi_x\xi_y\cos\phi, \quad V = 2\xi_x\xi_y\sin\phi, \tag{3.8}$$

---

[1] For an emission line, the senses of the circular polarization are reversed in the longitudinal Zeeman effect and the directions of the linear polarization are exchanged in the transverse Zeeman effect; this is simply because the emission lines have their own intensity, whereas in an absorption line we see only the residual intensity.

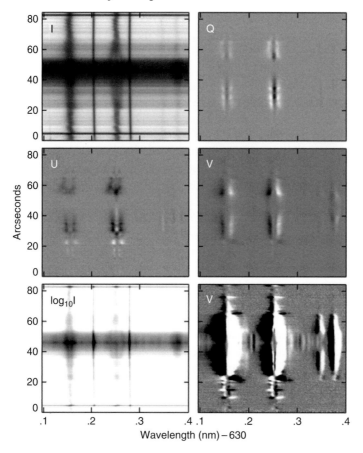

Fig. 3.7. Zeeman splitting of spectral lines in a sunspot. Shown are sample Stokes I, Q, U, and V profiles of the spectrum near 630 nm along the spectrograph slit, which lies across the image of a sunspot in the focal plane. In each panel the vertical direction corresponds to distance along the spectrograph slit (in arcseconds) and the horizontal direction corresponds to wavelength (in nm). The spectral lines visible in total intensity I (upper left panel) are Fe I 630.15, telluric O_2 630.20, Fe I 630.20, telluric O_2 630.30, Fe I 630.35 and Ti I 630.40 nm. (The dark horizontal lines near the top and bottom of this panel are crosshairs used to aid in registration of consecutive frames.) In the bottom two panels, the Stokes I and V profiles have been scaled to bring forward the weak molecular lines of TiO and CaH that occur in the umbra between the stronger lines. (From Lites *et al.* 1998: data taken with the Advanced Stokes Polarimeter at the NSO Dunn Telescope.)

and they obey the relation $I^2 = Q^2 + U^2 + V^2$. For a beam of light of finite bandwidth, consisting of a superposition of many independent wave trains of different amplitudes and phases, we replace the right-hand sides of Equations (3.8) by suitable time averages. In this case, the beam need not be fully polarized. If it is completely unpolarized, then all transverse directions are equivalent and all phases ϕ between 0 and 2π occur with equal probability, so Q, U and V are all zero. If the beam is partially polarized, then Q and U describe the net linear polarization, V describes the net circular polarization, and the degree of polarization is given by $P = [(Q^2 + U^2 + V^2)/I^2]^{1/2}$.

An example of a measurement of the four Stokes parameters as functions of wavelength (Stokes profiles) for spectral lines formed in a sunspot and its immediate surroundings is shown in Figure 3.7. Such measurements of Stokes profiles must be transformed into information about the distribution of the velocity and magnetic fields in the solar atmosphere through some sort of inversion technique. All such inversion techniques are based on some model of the physical state of the solar atmosphere, and this model dependence is usually the largest source of uncertainty in the measured vector magnetic field. The various inversion methods are based on the radiative transfer equations applied to the model atmosphere. A non-uniform magnetic field within a resolution element typically produces abnormal Stokes profiles with more than two lobes. The simplest model atmosphere that can reproduce these profiles consists of two different components sitting side by side within the resolution element. In some cases, one component is magnetic and the other is non-magnetic (and hence may represent stray light from non-magnetic surroundings). Computed synthetic Stokes profiles for the model atmosphere are fitted iteratively to the observed profiles using response functions and an algorithm that measures the goodness of fit, through variations in a set of free parameters for the model atmosphere such as magnetic field strength and inclination, line-of-sight velocity, and the filling factor of the magnetic component. An alternative approach is based on a single-component model atmosphere with vertical gradients. Further information on Stokes inversion methods may be found in several recent reviews (e.g. Socas-Navarro 2001; Bellot Rubio 2003; del Toro Iniesta 2003).

Beckers (1969b) provides a useful table of solar Zeeman multiplets with their effective Landé factors g_{eff}, while Harvey (1973) gives a selected list of spectral lines with large Zeeman splitting ($g_{\text{eff}} \geq 2.5$).

According to Equation (3.5), Zeeman splitting increases as the square of the wavelength, but the width of spectral lines increases only linearly with wavelength. Thus, the separation of the Zeeman components becomes more easily detectable at larger wavelengths, which makes magnetic field measurements in infrared lines highly desirable. Work in recent years has concentrated on spectral lines near 1.6 μm and 12 μm.

3.4.2 The magnetic field in a pore

Figure 1.2 shows a continuous hierarchy of magnetic features at the solar surface, ranging from tiny intergranular magnetic elements with kilogauss fields, which appear as bright points in the continuum, through dark pores, with typical diameters of 1500–4000 km, to sunspots with fully fledged penumbrae. Magnetic fields in pores are typically measured in Fe I lines, for instance those at 630.2 nm formed about 350 km above the level where $\tau_{500} = 1$. Owing to the Wilson depression, that surface of constant optical depth dips within a pore, lying about 300 km deeper at its centre (Sütterlin 1998). The measured field strength at the centre ranges from 1800 G to 2300 G, depending on the size of the pore, and drops to about 1000 G at the edge, after which the field falls abruptly. The field's strength decreases with height at a rate of about $4 \, \text{G km}^{-1}$; it is vertical at the centre, and estimates of its inclination at the edge range from 35° to 60° (Brants and Zwaan 1982; Muglach, Solanki and Livingston 1994; Sütterlin, Schröter and Muglach 1996; Keppens and Martínez Pillet 1996; Martínez Pillet 1997; Sütterlin 1998). Apparently the field splays out rapidly above the photosphere, for the diameter of the magnetic structure in the Fe I 630.2 nm lines is up to 25% higher than that of the pore radius in the continuum (Martínez Pillet 1997).

Magnetohydrostatic equilibrium of a pore's flux tube in the highly stratified photosphere indeed requires that the magnetic field should fan out with height.

A pore is surrounded by granules with cool downflows at their edges. It is not surprising therefore that Doppler velocity measurements show the existence of an annular region of downflow immediately surrounding most pores (Keil *et al.* 1999; Tritschler, Schmidt and Rimmele 2002). These annuli have widths of $1''$–$2''$ and flow speeds (measured in photospheric spectral lines) of order 300–500 m s^{-1}, occasionally reaching 1 km s^{-1}. The associated dynamic pressure helps to confine the magnetic field.

3.4.3   *Magnetic fields in sunspots*

Hale (1908b) demonstrated the presence of longitudinal fields with strengths of 2500–3000 G in sunspots by measuring the separations of the circularly polarized σ components in various Zeeman doublets. Subsequent measurements at Mount Wilson revealed fields of up to 4600 G (Hale and Nicholson 1938; Livingston *et al.* 2006); the record is held by a field of 6100 G in a large sunspot group that appeared in February 1942 (Mulders 1943; Livingston *et al.* 2006). Nicholson (1933) found that large spots (with radii of more than about 12 000 km) typically had central fields of around 2600 G. Systematic observations of spots as they were carried across the solar disc by the Sun's rotation were used to locate the positions where the longitudinal component vanished and hence to determine the variation of the field's inclination with radius within a spot, yielding a roughly linear increase from zero, at the centre, to a maximum of 70–80° at the edge of the spot (Hale *et al.* 1919; Hale and Nicholson 1938). Broxon (1942) proposed that the variation of the field strength with radius followed a parabolic law of the form $B = B_0(1 - \hat{r}^2)$, where B_0 is the central field and $\hat{r} = r/R$ is the normalized radius, so that the field actually dropped to zero at the outer boundary of a spot.

The acceleration of subsequent progress can be followed through a sequence of reviews (Kiepenheuer 1953; Bray and Loughhead 1964; Skumanich 1992; Martínez Pillet 1997; Solanki 2003). The gradual development of spectropolarimetry has allowed transverse as well as longitudinal components to be determined. Here we will only describe some of the key developments. Beckers and Schröter (1969) obtained the Stokes V parameter from Zeeman spectra with opposite circular polarizations as a chosen spot traversed the Sun's disc. Hence (using a theoretical model of line formation) they calculated both the field strength B and its angle of inclination γ as functions of the normalized radius $\hat{r} = r/R$. They found that the radial variation of B could best be represented by the expression

$$B(\hat{r}) = B_0/(1 + \hat{r}^2), \tag{3.9}$$

so that the field fell to half its central value at the edge of the spot. More specifically, in the spot they studied, with $B_0 = 2600$ G, the field reached a value of 1300 G at the outer boundary of the penumbra and then dropped abruptly to a very small value. Meanwhile, the apparent angle of inclination increased until the field became horizontal at $\hat{r} = 1$. Similar results were obtained by Kawakami (1983), based on the first polarimetric measurement of all four Stokes parameters, and hence of the vector field.

A reliable measurement of the vector field was obtained by Adam (1990). She measured interference fringes produced by a Babinet compensator in order to calculate the elliptical polarization of the σ components at different positions in a sunspot. The corresponding field strength fell to about half its central value at $\hat{r} = 0.7$, while the inclination γ approached a maximum value of about 70° at the edge of the spot. This limiting inclination of the mean field has been confirmed by subsequent observations.

Since 1990, such measurements have been transformed by the availability both of spectropolarimeters and of high-resolution observations. The development first of the Stokes II Polarimeter (Lites and Skumanich 1990; Arena, Landi degl'Innocenti and Noci 1990), and then of the more powerful Advanced Stokes Polarimeter at Sacramento Peak (Keppens and Martínez Pillet 1996; Stanchfield, Thomas and Lites 1997; Westendorp Plaza *et al.* 2001), followed by the Tenerife Infrared Polarimeter (Bellot Rubio *et al.* 2003; Mathew *et al.* 2003; Bellot Rubio, Balthasar and Collados 2004) has made it possible to make precise measurements of all four Stokes components, and hence to determine the vector magnetic field. These results all concur in showing azimuthally averaged mesoscale fields that drop to 700–900 G at the visible edge of the penumbra, where the field reaches an inclination of 70–80°. These averaged fields lie in meridional planes and the azimuthal component of the field is always weak. Figure 3.8 shows the two-dimensional distribution of B and γ in a nearly axisymmetric, medium-sized sunspot (diameter about 20 000 km). Superimposed on the mesoscale structure – which extends beyond the visible edge of the penumbra – are significant variations, notably in the orientation of the field but also in its magnitude, which are apparent in this figure. This fine structure, which will receive a detailed discussion in Chapter 5, is clearly seen in the high-resolution images of the line-of-sight magnetic field obtained first on the Swedish Vacuum Solar Telescope (Title *et al.* 1992, 1993) and, more recently, on the 1-m Swedish Solar Telescope (Langhans *et al.* 2005) on La Palma. Magnetograms made from filtergrams taken in circularly polarized light, using the Lockheed Solar Optical Universal Polarimeter, show the line-of-sight component of the magnetic field at different orientations around a sunspot. Provided the spot is off-centre on the solar disc, the mesoscale field and the fluctuating fine-scale field can be determined. The radial variation of the azimuthally averaged field in the penumbra, in both magnitude and inclination, is included in Figure 5.7 below. The inclinations of 70–75° at the outer boundary of the spot are consistent with those obtained by spectropolarimetry but the corresponding field strengths are higher (Title *et al.* 1993).

Information on the vertical structure of the magnetic field can be obtained either by comparing measurements made with absorption lines that are formed at different levels in the photosphere or by inverting Stokes profiles obtained across a given line. Westendorp Plaza *et al.* (2001) used the Fe I 630.15 and 630.25 nm lines to investigate the variation of the mesoscale vector field at continuum optical depths $0.0 \geq \log \tau_{500} \geq -2.8$; in the outer penumbra they found that B decreased, while γ increased, with increasing optical depth. Mathew *et al.* (2003) used the Fe I 1564.8 and 1565.3 nm infrared lines, which are formed low in the atmosphere (Solanki, Rüedi and Livingston 1992), and obtained similar results. These changes probably reflect the depth-dependent fine structure discussed by Bellot Rubio, Balthasar and Collados (2004). In all these measurements it is apparent that the magnetic field does not vanish outside the visible penumbra. Instead, the field persists in an elevated canopy that extends into the chromospheric superpenumbra, with its lower boundary rising gradually to reach a height of about 300 km above $\tau_{500} = 1$ at a distance of twice the spot radius (Solanki *et al.* 1999; Solanki 2002, 2003).

The field strength at the outer edge of the penumbra varies somewhat from spot to spot but the central field in the umbra of a sunspot depends on its size. Various estimates, going back to Nicholson (1933; see Bray and Loughhead 1964), agree that the central field B_0

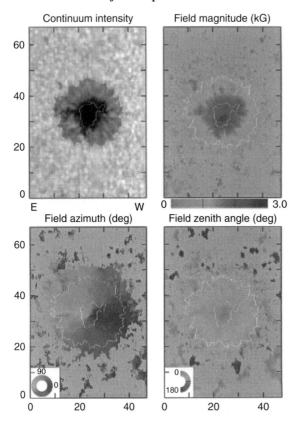

Fig. 3.8. The vector magnetic field in a sunspot, as determined by spectropolarimetry from the Stokes profiles of the Fe I 630.15 and 630.25 nm lines, formed about 160 km above the $\tau_{500} = 1$ level in the photosphere. The four panels show a continuum spectroheliogram (at 630 nm), the field magnitude, B, the azimuth of the horizontal component and the angle of inclination, γ. Note that the magnetic field extends beyond the visible boundary of the sunspot at $\hat{r} = 1$. Although the overall structure corresponds to a meridional field whose strength drops to around 800 G at $\hat{r} = 1$, where $\gamma \approx 70°$, there is clear evidence of fine structure associated with penumbral filaments. Furthermore, there are isolated regions, both inside and outside the penumbra, where the field plunges downwards, with $\gamma > 90°$. (From Stanchfield, Thomas and Lites 1997.)

increases with increasing umbral radius (e.g. Brants and Zwaan 1982; Collados *et al.* 1994; Martínez Pillet 1997; Solanki 2002), as might be expected if larger spots have deeper Wilson depressions. On the other hand, the average field strength of 1200–1700 G apparently shows much less variation from one spot to another (Solanki 2002).

3.5   Modelling the overall magnetic structure of a sunspot

We turn now to a discussion of theoretical models of the gross magnetic structure of a sunspot. Two alternative pictures were originally proposed (see Fig. 4 of Thomas and Weiss 1992b). Cowling (1946, 1976a) had favoured a single, coherent, monolithic tube of magnetic flux both above and below the solar surface. Ignoring fine structure, this assumption is justified for the visible layers of the spot, where the plasma beta is relatively low and the field must be essentially space-filling. For the layers beneath the surface, the issue is more controversial. Parker (1975) imagined that the field split into many independent flux tubes which spread out like the tentacles of a jellyfish immediately below the photosphere; such a configuration, with a tight throat, would be liable to interchange instabilities and unlikely to survive.

We now know from helioseismology that a sunspot is sufficiently coherent to support magneto-acoustic modes of oscillation down to a depth of at least 10 Mm (Kosovichev 2002, 2006; Zharkov, Nicholas and Thompson 2007). Figure 3.9 shows the fluctuations in sound speed beneath a spot; the sound speed decreases below the surface (where the temperature is reduced) and increases again at depths greater than about 4 Mm; this pattern is consistent with the thermal models described in Section 3.2.3 above. In the modern form of Parker's model, the isolated flux tubes are supposed to be gathered into a tight cluster, within which they are separated by regions of field-free plasma where heat transport by convection is locally unimpeded (Parker 1979b; Spruit 1981b; Choudhuri 1992; Spruit and Scharmer 2006). The alternative model is of a coherent but inhomogeneous flux tube, with convection transporting energy upwards and consequent variations of field strength within

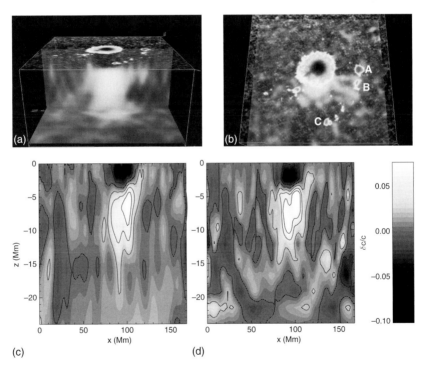

Fig. 3.9. Helioseismic measurement of the sound speed beneath a sunspot, determined from measurements with the MDI instrument on SOHO. The variation with radius and depth of the sound speed c is shown (a) projected onto vertical and horizontal planes, with surface intensity represented on the top, and (b) in a horizontal plane at a depth of 4 Mm. The lower images show vertical cuts through the spot with sound speeds determined by two different techniques, using (c) Fresnel-zone and (d) ray-approximation kernels. In each case, the thermal disturbance extends to depths of at least 14 Mm beneath the spot. Note also the connection to the two pores labelled A and B. (From Kosovichev 2006, courtesy of Elsevier.)

 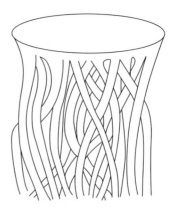

Fig. 3.10. Monolith and cluster models of a sunspot. (From Thomas and Weiss 1992b, courtesy of Springer Science and Business Media.)

the magnetic column. In either case, local convective plumes penetrate into the visible layers of the umbra to form umbral dots (see Section 4.1 below). These versions of the monolith and cluster models are illustrated schematically in Figure 3.10.

3.5.1   *Static axisymmetric models*

The simplest static models of a pore or a sunspot ignore azimuthal variations and treat it as an axisymmetric, meridional magnetic field (with no azimuthal component) confined to a homogeneous flux tube of circular cross-section, and maintained by an azimuthal current sheet (the magnetopause) separating the tube from its non-magnetic surroundings. If $\mathbf{B}$ is referred to cylindrical polar co-ordinates (r, ϕ, z), the assumed axial symmetry implies that $B_\phi = 0$ and $\partial B_r/\partial\phi = \partial B_z/\partial\phi = 0$.

The equation describing the magnetohydrostatic equilibrium of the flux tube is

$$0 = -\nabla p + \rho\mathbf{g} + \frac{1}{\mu_0}(\nabla \times \mathbf{B}) \times \mathbf{B}. \tag{3.10}$$

Because the surrounding atmosphere is strongly stratified, with the pressure dropping nearly exponentially with height, the flux tube must expand radially with height in order to maintain a total magnetohydrostatic pressure balance across the boundary.

Potential-field models

In the simplest approach, the magnetic field within the sunspot flux tube is assumed to be force-free, in which case the atmosphere is horizontally stratified, with variations only in the vertical direction. The force-free field must satisfy the condition

$$(\nabla \times \mathbf{B}) \times \mathbf{B} = 0, \tag{3.11}$$

but the axisymmetry in fact requires that the field satisfy the more stringent current-free condition $\nabla \times \mathbf{B} = 0$, and hence $\mathbf{B}$ is a potential field $\mathbf{B} = -\nabla\Phi$ where Φ satisfies Laplace's equation, $\nabla^2\Phi = 0$.

Simon and Weiss (1970) constructed models of a pore based on the simple Bessel function solution $\Phi = AJ_0(kr)\exp(-kz)$ of Laplace's equation with a single value of k. For a given total magnetic flux, exact pressure balance with the surrounding atmosphere is then possible only at two distinct values of the height z, but fixing these values still gives a fairly satisfactory model at other heights. For models with total magnetic flux in the range 0.5–3.0×10^{20} Mx the surface diameter is in the range 2200–5400 km. The inclination (to the vertical) of the outermost field lines at the surface increases, becoming nearly horizontal for the highest value of the total flux. Presumably the non-axisymmetric penumbra forms at some critical value of the total flux (or field inclination); we shall discuss this further in Section 5.6. More elaborate versions of this model were subsequently developed by Spruit (1976) and by Simon, Weiss and Nye (1983), who represented the field in a pore by a potential field such that, at the photosphere, B_z was uniform over a disc with a prescribed radius and zero outside it.

A more satisfactory potential field model can be derived by constructing a field contained within a flux tube of radius $R(z)$, whose boundary is a field line, such that the difference between the internal and external gas pressures is balanced by the magnetic pressure at the magnetopause (Jahn 1989, 1992; Pizzo 1990). Schmidt and Wegmann (1983; see also Schmidt 1991) developed an elegant procedure for calculating such a field, by reducing the associated free boundary problem to an integral equation. The stratifications inside and outside the flux tube can be computed in the normal manner, using the standard mixing-length description of convective transport but with a reduced mixing length inside the flux tube (to simulate magnetic suppression of convection). This approach can provide acceptable models of the magnetic field in pores but it obviously fails for sunspots, where there is a distinction between the umbra and penumbra.

Self-similar models

The visible surface of the umbra is lower than that of the surrounding photosphere (the Wilson depression) and it would require a field of about 5000 G to maintain a lateral pressure balance between the centre of a sunspot and the field-free plasma outside the magnetopause if the poloidal field were force-free. Since the peak fields that are observed are significantly weaker, it becomes necessary to include azimuthal volume currents in the penumbra. The Lorentz force then has a horizontal component that can balance the horizontal gradient in the gas pressure. There is, however, considerable arbitrariness in assigning the distribution of this volume current and the corresponding radial structure in the sunspot atmosphere. As first shown by Schlüter and Temesváry (1958), the problem can be greatly simplified by assuming a self-similar form for the magnetic field, regarded as a function of (r/R).

At the visible surface of the umbra the atmosphere is stably stratified and energy is transported by radiation. Schlüter and Temesváry (1958) showed, however, that it is impossible to construct a purely radiative model of a sunspot and that energy transport must be primarily convective just below the surface. The energy emitted from a spot is reduced both by the expanding cross-sectional area of the flux tube and by magnetic inhibition of convection. The latter leads to a greater superadiabatic temperature gradient in the tube, with the result that the sunspot appears cooler and darker, and its visible surface is depressed. A number of sunspot models have been constructed, some with self-similar fields and others adopting equivalent assumptions but all relying on various estimates of the partial inhibition of convective transport by the magnetic field (Chitre 1963; Deinzer 1965; Jakimiec 1965; Jakimiec and Zabza 1966; Chitre and Shaviv 1967; Yun 1970; Landman and Finn 1979; Low 1980).

The role of the penumbra

As we have seen, there must be volume currents within a sunspot in order to produce a curvature force that opposes the radial gradient in gas pressure. Since the umbral field is fairly uniform, these currents are likely to flow in the penumbra. Jahn (1989) explored the effects of introducing a volume current distributed beneath the visible surface of the outer penumbra; subsequently, Jahn and Schmidt (1994) introduced a convenient approximation whereby these volume currents were replaced by a second azimuthal current sheet at the umbral–penumbral boundary.[2] As shown in Figure 3.11, their sunspot model contains three distinct regions, separated by two azimuthal current sheets: an inner region with a current-free potential magnetic field (the umbra), an outer region, also with a potential magnetic field (the penumbra), and a field-free exterior region. These regions are separated by two current sheets: one at the umbra–penumbra interface, representing all of the volume current within the flux tube, and one at the outer penumbral boundary (the magnetopause) which tapers inward with increasing depth. They furthermore assumed that the umbra is thermally insulated from the penumbra, while some of the energy radiated from the penumbra itself is supplied by convective processes that transfer energy across the magnetopause. They conjectured that this transport is accomplished by small flux tubes that are heated at the magnetopause and driven radially inward by buoyancy forces as far as the inner current sheet. This approach leads to a family of models, with different total magnetic fluxes, that provide a reasonable description of the overall global (azimuthally averaged) structure of sunspots below the solar surface. For a large spot with a flux of 2×10^{22} Mx (200 TWb) the

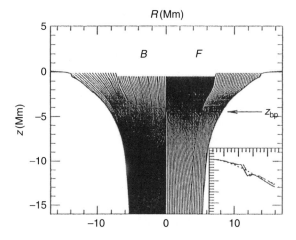

Fig. 3.11. Schematic diagram of the axisymmetric static sunspot model of Jahn and Schmidt (1994), for a spot containing a magnetic flux of 10^{22} Mx. The left half shows the magnetic field configuration and the right half shows the pattern of heat flow; the base of the penumbra is at a depth given by z_{bp}. The intensities of shading are proportional to the field strength and inversely proportional to the energy flux, which is augmented by transport across the magnetopause. The inset shows the surface profile of the magnetic field, as compared with that obtained by Beckers and Schröter (1969). (After Jahn and Schmidt 1994.)

---

[2] In their picture, peripatetic flux tubes transport energy inwards from the magnetopause; hence Schmidt (1991) dubbed this boundary the 'peripatopause'.

radius of the magnetopause drops smoothly from 20 Mm at the surface to 12 Mm at a depth of 5 Mm and then to 9 Mm at 10 Mm depth. This corresponds to a fourfold decrease in the cross-sectional area of the flux tube, with a corresponding increase in field strength.

3.5.2   *Stability of static models*

It is well known that attempts to contain a plasma by an external magnetic field that is concave toward the plasma are doomed owing to interchange instabilities that lead to fluting of the field. Since a static, vertical flux tube that fans out near the solar surface is concave toward its field-free surroundings, it too might appear susceptible to a fluting instability. There is, however, an additional stabilizing effect: in a stratified atmosphere, a configuration with less dense fluid lying above denser fluid is Rayleigh–Taylor stable, and the gas within the flux tube is indeed less dense than the external plasma. Meyer, Schmidt and Weiss (1977) used the energy principle of Bernstein *et al.* (1958) to show that a flux tube in an adiabatically stratified atmosphere is stabilized by the buoyancy force provided the radial component of the magnetic field at the surface of the tube decreases with height. The field at the magnetopause becomes more nearly horizontal, while it decreases in strength, and stability depends on the competition between these two effects. Simple potential field models of sunspots, as well as the models of Jahn and Schmidt (1994), turn out to be stable near the solar surface. With increasing depth, however, the field at the magnetopause becomes less steeply inclined and the configuration is only marginally stable. If the field is nearly vertical it becomes unstable to interchanges (although it could be stabilized by a sufficient twist of the flux tube about its axis). If the flux tube beneath a sunspot had a throat below which it expanded, it would certainly be liable to fluting instabilities that would rapidly destroy the spot.

It appears therefore that some further dynamical effect is needed to stabilize the entire configuration, and it is suspected that the required containment is provided by some sort of supergranule-scale 'collar' flow in the surrounding gas. Such a collar might be provided by the moat cell that surrounds a fully developed sunspot, with an outflow at the photosphere, for mass conservation requires that there should be an inflow at greater depth and this inflow might act as the collar that preserves the spot.

3.6   The moat flow

Once a stable sunspot has formed, a characteristic pattern of surface motion is quickly established around it, consisting of a persistent radial outflow in an annular region, or *moat*, around the spot (Shine and Title 2001; Solanki 2003). The diameter of this cell ranges from about 15 Mm for very small spots to 100 Mm or more for large ones. The moat flow was first detected by Sheeley (1969) in the form of a continual radial outflow of bright points from sunspots in CN spectroheliograms. Before the moat is formed, the typical photospheric network pattern adjoins the outer penumbral boundary. As the moat flow develops, it sweeps magnetic flux to its periphery, leaving the moat free of magnetic field except for small magnetic features (of both polarities) that move outward across the moat at speeds of order $1 \, \mathrm{km \, s^{-1}}$ (Sheeley 1969, 1972; Vrabec 1971, 1974; Harvey and Harvey 1973). Granules within the moat are swept outward too (Muller and Mena 1987; Shine *et al.* 1987). Moats are 10 to 20 Mm wide, and the outer radius of the moat scales with the size of the enclosed sunspot, being about twice the radius of the spot itself (Brickhouse and LaBonte

1988). The moat flow begins immediately after the formation of a penumbra; if a spot possesses only a partial penumbra, then the moat develops in sectors where penumbral filaments are present (Vargas Domínguez *et al.* 2007). The flow usually persists over the rest of the spot's life, while the spot's area and magnetic flux decrease at a roughly constant rate.

Sheeley (1969) suggested that the moat flow is similar to that in a supergranule, but with the sunspot at the centre. Many others have noted that the size, shape, and velocity pattern in the moat are similar to those in a supergranule, although the surface flow speed in the moat is on average about twice that in a typical supergranule (e.g. Brickhouse and LaBonte 1988). Although the moat flow was first seen as a motion of individual features, it can also be detected as a Doppler shift (Sheeley and Bhatnagar 1971; Sheeley 1972). Doppler maps show a systematic decrease of the flow speed with radius across the moat, as in a supergranule. The moat flow has also been detected through local helioseismology using f-mode surface gravity waves (Gizon, Duvall and Larsen 2000, 2001). Somewhat surprisingly, p-mode measurements show an *in*flow near the photosphere (Zhao, Kosovichev and Duvall 2001). These conflicting results could only be reconciled if the flow changed direction within 2 Mm of the surface. That seems unlikely, although it has been suggested that the moat flow is merely a superficial extension of the Evershed outflow (Vargas Domínguez *et al.* 2007).

The moat cell is not a simple axially symmetric outflow. Local correlation tracking of the proper motions of granules, G-band bright points, and magnetic features shows azimuthal variations in the outward motion within the moat (Shine and Title 2001; Hagenaar and Shine 2005; Bonet *et al.* 2005). The azimuthal motion appears to diverge from certain orientations and to converge on spoke-like channels that extend radially across the moat. Correspondingly, magnetic features tend to accumulate along these channels as they are transported outwards.

Meyer *et al.* (1974) modelled the moat flow as essentially an annular supergranule, anchored around the sunspot, with an inflow at depth and an outflow at the surface. The somewhat larger size and longer lifetime of the moat, compared to a typical supergranule, can then be ascribed to the stabilizing influence of the central sunspot. They suggested that the upflow around a sunspot is caused by the spreading penumbral magnetic field, which blocks small-scale convective transport and causes local heating.

One way to investigate this possibility is to simulate the formation of the sunspot flux tube by magnetoconvection in an idealized geometry. Hurlburt and Rucklidge (2000; see also Botha, Rucklidge and Hurlburt 2006) conducted numerical experiments designed to model the formation of axisymmetric magnetic flux concentrations (pores and sunspots) in a convecting compressible atmosphere. In all cases they found that the preferred mode of convection is an annular cell with a downflow along the flux concentration, driven by local cooling there, and an inflow (toward the central flux concentration) at the surface. This flow pattern corresponds to the formation of a pore, but contradicts the observed moat flow in a fully developed sunspot. For larger flux concentrations, however, they found that the convective cell has an annular countercell outside it, with outflow at the surface. They conjectured that in a real sunspot the inner 'collar' cell that confines the flux concentration is hidden beneath the penumbra, leaving only the outer moat cell visible at the surface. This is an attractive possibility, for it reconciles the expected presence of a downflow along a cooled sunspot flux tube and the observed outflow in the moat; furthermore, there are suggestions from helioseismology that such an inflow is indeed present. It must be borne in mind, however, that in reality both the inside and the outside of the flux tube are superadiabatically stratified and unstable to convection, and that the resulting motion will be fully three-dimensional.

Further calculations, no longer restricted to axisymmetric configurations, reveal that the axisymmetric flow outside the central flux tube is unstable to non-axisymmetric perturbations that are essentially hydrodynamically (rather than magnetically) driven (Botha, Rucklidge and Hurlburt 2007). In the nonlinear domain, these instabilities develop into a pattern of sectors, separated by spoke-like regions of convergence, that are reminiscent of the channels in moat cells around sunspots (Hurlburt, Matthews and Rucklidge 2000; Hurlburt and Alexander 2002).

4

Fine structure of the umbra

We turn now to a discussion of the fine structure of a sunspot, beginning here with features in the umbra and continuing in the next chapter with features in the penumbra. Our knowledge of this fine structure has been transformed in recent years due to remarkable improvements in high-resolution observations. We review the results of these observations and theoretical interpretations of them. Our enhanced knowledge of the fine structure of sunspots has not only provided us with a far larger collection of details; it has also stimulated new insights that allow us to start assembling a coherent picture of the formation and maintenance of a sunspot, with its dark umbra and its puzzling filamentary penumbra.

4.1   Umbral dots

In many images, sunspot umbrae – like pores, which are just isolated umbrae – appear uniformly dark. When such images are appropriately exposed, however, as in Figure 4.1, it becomes apparent that there is an intensity pattern in sunspot umbrae, composed of many small, isolated, bright features embedded in a darker, smoothly varying background. These features are called *umbral dots* and they are found in essentially all sunspots and also in pores (Sobotka 1997, 2002). Earlier observations of an intensity pattern in umbrae, with a resolution of about 1″, had failed to resolve the umbral dots and instead showed a pattern that looked more like a weaker version of the photospheric granulation (Chevalier 1916; Bray and Loughhead 1964). Umbral dots themselves were first reported by Thiessen (1950), who resolved the granulation pattern into bright points as small as the diffraction limit (0.3″) of his 60-cm refractor. They were later rediscovered in photographs from the Stratoscope balloon-borne telescope and named by Danielson (1964). As Beckers and Schröter (1968) explained, "these features do not form a kind of closed pattern as observed in the photosphere. They appear rather as isolated emission points . . ."

In fact, the umbral background is not uniformly dark. The intensity is lowest in *dark nuclei*, which cover 10–20% of the umbral area and are almost free of umbral dots (Sobotka, Bonet and Vázquez 1993), so much so that Livingston (1991) termed them 'voids'. They are comparable in size to photospheric granules, with diameters of around 1000 km. The relative intensity I_{min} of these dark nuclei, expressed as a fraction of the intensity I_{phot} of the undisturbed photosphere, lies in the range $0.05 \leq I_{min} \leq 0.33$ with typical values around $I_{min} \approx 0.15$ (Sobotka, Bonet and Vázquez 1993; Sobotka and Hanslmeier 2005).

Umbral dots cover only 3 to 10% of the umbral area but contribute 10 to 20% of the total umbral brightness (Sobotka, Bonet and Vázquez 1993). Although they are distributed

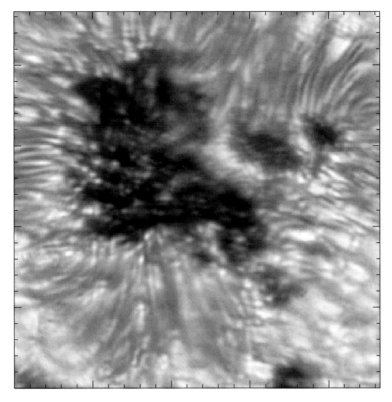

Fig. 4.1. Sunspot image exposed to show umbral dots. This image has been corrected using a phase-diversity reconstruction. Tick marks are at intervals of 1″. (From Tritschler and Schmidt 2002a.)

throughout the umbra, their distribution is not uniform: they can occur in clusters and alignments (Rimmele 1997), and no large dots are found in dark nuclei. Grossmann-Doerth, Schmidt and Schröter (1986) introduced a distinction between peripheral and central umbral dots. The former are typically brighter, as is to be expected, since the intensity of umbral dots is positively correlated with that of the local umbral background (Sobotka, Bonet and Vázquez 1993; Tritschler and Schmidt 2002b). The peripheral umbral dots are also associated with bright grains that flow inwards across the penumbral–umbral boundary (Sobotka, Brandt and Simon 1997b). The average relative intensity of central umbral dots lies in the range 0.38–0.64, with peak values as high as 1.24 (Sobotka and Hanslmeier 2005). Thus individual umbral dots may exceed the average photospheric intensity I_{phot}, though not its maximum value within an individual granule. On average, umbral dots are 500–1000 K cooler than the photosphere outside a spot, but about 1000 K hotter than the coolest parts of the umbra itself (Sobotka and Hanslmeier 2005; Kitai *et al.* 2007).

Accurate measurements of the diameters of umbral dots require observations with sufficient resolution. Early measurements (e.g. Beckers and Schröter 1968; Koutchmy and Adjabshirzadeh 1981) suggested that the true diameters might be as low as 150–200 km; observations with the Swedish Vacuum Solar Telescope at La Palma confirmed this estimate (Lites *et al.* 1991; Sobotka, Bonet and Vázquez 1993; Socas-Navarro *et al.* 2004) but Sobotka, Brandt and Simon (1997a; see also Tritschler and Schmidt 2002b) found that the probability distribution function kept on rising down to the smallest dots that could be resolved, with diameters of about 200 km. Finally, Sobotka and Hanslmeier (2005) used the new Swedish Solar Telescope to measure effective diameters (for dots defined as having intensities more than 5% above the local umbral background) down to 100 km: they found that the distribution peaked at diameters of 175 km – and that the observed diameters were not correlated with intensities. They also investigated the spacing between umbral dots, and found that the mean distance between nearest neighbours was about 300 km, while the average peak-to-peak distance was about twice that. Comparable results have also been obtained with the Solar Optical Telescope on Hinode (Kitai *et al.* 2007).

Umbral dots are dynamic features. Sobotka, Brandt and Simon (1997a) found that larger and brighter dots were also long-lived. A few survived for longer than two hours but most lifetimes were much shorter. In their sample, lifetimes appear to decrease with decreasing size, down to the limit set by resolution; they found an average lifetime of about 14 minutes and a median of 6 minutes. Similar values were obtained by Kitai *et al.* (2007). Long-lived individual dots show variations in intensity and Sobotka, Brandt and Simon (1997b) found peaks in power spectra at periods from 3 to 32 minutes. The bright features also travel horizontally, with typical velocities of several hundred metres per second. Rapidly and slowly moving dots are distributed throughout the umbra but those with the longest lifetimes are most nearly stationary. Interestingly, there are examples of fast-moving penumbral grains or umbral dots that disappear after colliding with one side of a dark nucleus but stimulate brightenings of umbral dots on the far side of the same nucleus (Sobotka *et al.* 1995; Sobotka, Brandt and Simon 1997b).

The physical properties of umbral dots – brightness, temperature, diameter, vertical motion and magnetic field strength – all vary rapidly with height in the photosphere and measurements are therefore sensitive to the level at which they are made. Some observations with limited resolution did show upward velocities of no more than a few hundred $m\,s^{-1}$ (Lites *et al.* 1991; Rimmele 1997). More recent Doppler measurements, with high resolution, show upflows exceeding $1\,km\,s^{-1}$ in the lower photosphere (using a C I line) but fail to detect any significant motion in the upper photosphere (Rimmele 2004); velocities of around $100\,m\,s^{-1}$ at unit optical depth were reported by Socas-Navarro *et al.* (2004). There has been a general consensus that the magnetic field in umbral dots is weaker than that in the umbral background (Sobotka 1997). Socas-Navarro *et al.* (2004) found differences of several hundred gauss and deduced that the fields were more inclined to the vertical, by about $10°$.

The umbral dots can naturally be interpreted as hot convective plumes that overshoot to a level where the photosphere is stably stratified (Weiss *et al.* 1990, 1996; Degenhardt and Lites 1993a,b). The plumes decelerate as they rise and the magnetic field is reduced, and swept aside, as they expand. These features owe their origin to some form of magnetoconvection, which will be discussed in the next section.

4.2   Convection in the umbra

Below the visible surface of the umbra the predominant mode of energy transport has to be by convection (as explained in Section 3.5.1) and umbral dots provide some clues as to its subphotospheric pattern. Sunspots have provided the principal motivation for studying convection in an imposed magnetic field, and magnetoconvection is a fascinating topic in its own right. In what follows, we first outline what can be learnt from idealized model calculations and then go on to describe recent, more realistic simulations of umbral convection; finally, we discuss the deeper, subphotospheric structure of the umbra in the light of these results.

4.2.1   Idealized model calculations

In idealized models the umbra is represented by a plane layer, containing an electrically conducting fluid and heated from below, with an imposed vertical magnetic field. The simplest approach is to assume that the fluid is incompressible; the main properties of magnetoconvection in this Boussinesq approximation are summarized in Appendix 2, and the subsequent development of the subject can be followed in a series of reviews (Weiss 1991, 2003; Schüssler 2001; Proctor 2005). Two robust features of Boussinesq convection in a strong magnetic field are that it occurs in narrow, vertically elongated cells, and that its nature depends on the ratio, ζ, of the magnetic to the thermal diffusivity: if $\zeta > 1$ convection sets in as steady overturning motion but if $\zeta < 1$ oscillatory convection is preferred.

These properties hold also for a stratified, compressible layer. In the absence of any diffusion, convection would set in as a steady overturning mode (Moreno-Insertis and Spruit

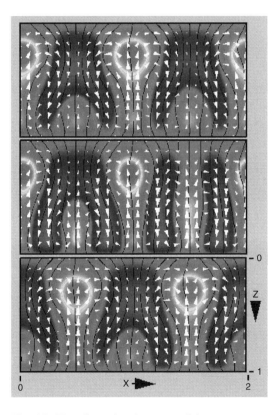

Fig. 4.2. Two-dimensional compressible magnetoconvection: spatially modulated oscillations over a half-period. Shown are field lines, arrows that indicate the fluid velocity, and shading corresponding to the relative temperature fluctuations. (After Hurlburt, Matthews and Rucklidge 2000.)

1989), but if magnetic and (radiative) thermal diffusion are included and $\zeta \ll 1$, as is typically the case in a star, then oscillatory convection appears in the form of thermally destabilized slow magneto-acoustic modes (Cowling 1976b). In the Sun, however, the radiative diffusivity decreases owing to the effects of ionization, so that $\zeta > 1$ at depths between 2000 and 20 000 km below the surface (Meyer *et al.* 1974). The effects of this variation can most simply be modelled by considering two-dimensional magnetoconvection in a polytropic atmosphere with $\zeta < 1$ at the top and $\zeta > 1$ at the bottom. Then convection first sets in as a steady mode, which becomes unstable to oscillatory perturbations as the superadiabatic gradient, measured by a Rayleigh number Ra, is increased (Weiss *et al.* 1990). In this highly idealized configuration, solutions then take the form of spatially modulated oscillations, with slender rising plumes that are anchored at their bases but wax and wane alternately in vigour, as illustrated in Figure 4.2 (Hurlburt, Matthews and Proctor 1996; Hurlburt, Matthews and Rucklidge 2000). Three-dimensional behaviour is naturally richer. Convection sets in as a hexagonal array of steady, isolated plumes; as Ra is increased (or, equivalently, the imposed field is reduced) an irregular time-dependent pattern develops, as shown in Figure 4.3 (Weiss, Proctor and Brownjohn 2002). In all these examples, the slender plumes expand as they rise and spread as they impinge upon the upper boundary. Thus magnetic flux is swept aside and, at the top of the layer, the field is weakest where the temperature is highest.

Fig. 4.3. Three-dimensional compressible magnetoconvection: spatially modulated oscillations. (From Weiss, Proctor and Brownjohn 2002.)

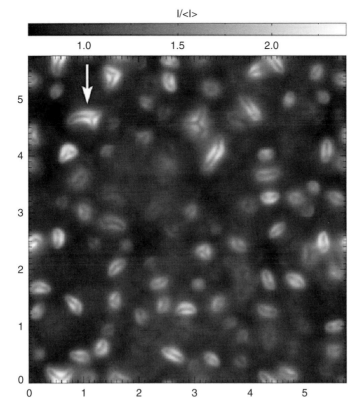

Fig. 4.4. Realistic simulation of umbral magnetoconvection: the pattern of vertically emerging intensity. (From Schüssler and Vögler 2006.)

4.2.2 *Realistic simulations*

Schüssler and Vögler (2006) have adopted a much more ambitious approach. They represent a realistic umbral atmosphere, including both radiative transfer and partial ionization, and describe three-dimensional convection in a region 1600 km deep and approximately 5800 km wide, extending to an 'open' lower boundary at a depth of 1200 km below the level of average optical depth $\tau_{500} = 1$ (Vögler *et al.* 2005). Figure 4.4 shows a snapshot of the temperature at this level. As expected, there is an irregular pattern of slender, short-lived plumes, with a typical size of 200 to 300 km and a lifetime of around 30 min. They achieve a peak upward velocity of about $3 \, \mathrm{km \, s^{-1}}$ and spread laterally, so that the local field strength is drastically reduced. Interestingly, the plumes are not circular but oval, with dark streaks along their major axes, giving them a 'coffee-bean' appearance, as shown in Figure 4.4. Moreover, the horizontal outflow is anisotropic and focused along the dark streaks, as though the plumes had undergone a non-axisymmetric $m = 2$ instability. Figure 4.5 shows the velocity and field strength in a detailed cross-section perpendicular to a dark streak. Note that the surfaces of constant optical depth are elevated above the central axis of the plume, implying that the weakest fields could not be observed, while upward and downward motions would scarcely be resolved. (This explains the disparities in observations made in lines formed at different levels in the atmosphere.) As the rising plume encounters

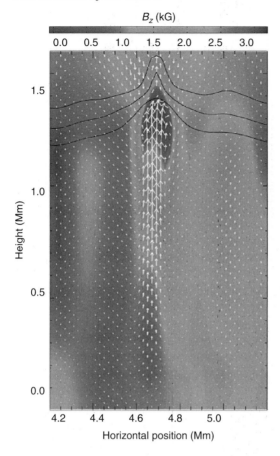

Fig. 4.5. Realistic simulation of umbral magnetoconvection: vertical cut across a rising plume. The shading indicates the strength of the magnetic field, which is reduced at the head of the plume, and the arrows represent the projected velocity. The dark lines indicate surfaces of constant optical depth, which are elevated above the rising plume. (From Schüssler and Vögler 2006.)

the stably stratified layer of the atmosphere it is decelerated by buoyancy braking. It follows that both the pressure and the density are locally high, and it is this local density maximum that is responsible for the dark streaks as absorption features. This remarkable numerical model demonstrates how convection can occur in the umbra of a sunspot and convincingly reproduces the principal features of umbral dots.

4.2.3   *Magnetic structure of the umbra*

As already mentioned in Section 3.5, there are two competing theoretical pictures of the magnetic field beneath the visible surface of the umbra. In the cluster model (Parker 1979b; Spruit 1981b; Choudhuri 1992; Spruit and Scharmer 2006) it is supposed that magnetic flux separates out into isolated tubes that are surrounded by field-free plasma. Where the field is strong, convection is effectively suppressed but energy transport is unimpeded in the field-free regions, which extend to just below the visible surface and form umbral dots. In that case we should, however, expect to see a bright network enclosing dark features – the opposite of what is actually observed. Furthermore, although numerical experiments do reveal examples of flux separation, with patches of almost field-free convection surrounded by strong fields with smaller-scale motion (Tao, Proctor and Weiss 1998; Weiss, Proctor and Brownjohn 2002; Schüssler and Vögler 2006), the converse does not occur: an isolated clump of magnetic flux always escapes along the interstices between convection cells and disperses.

An inhomogeneous monolithic model, on the other hand, is compatible not only with the observations but also with numerical experiments and simulations. The calculations of Schüssler and Vögler (2006) indicate that umbral dots do indeed correspond to convective plumes that are capable of supplying the reduced energy that is emitted from the umbra itself. This energy is drawn from below, either by transport along the underlying flux concentration or by draining its internal energy. We conclude, therefore, that the field below the umbra is contained in an inhomogeneous column that extends at least for several tens of megametres below the surface. Whether such a column continues right down to the base of the convection zone is a different issue, to which we will return in Chapter 11.

Although an ideal umbral model has a field that is uniform (or varies only with distance from the centre of the spot) when averaged over a region large compared with umbral dots, it is not to be expected that such uniformity will be found in real sunspots. Observations show that there are dark nuclei, which are patches with relatively stronger and more vertical fields (Stanchfield, Thomas and Lites 1997) and significantly weaker convection. Schüssler and Vögler (2006) found that the averaged energy output was reduced when the imposed field strength was increased from 2500 to 3000 G. It has been conjectured that these dark nuclei may be examples of flux separation (e.g. Blanchflower, Rucklidge and Weiss 1998) but they may well be relics of a sunspot's earlier history and formation.

4.3   Light bridges

The umbrae of most sunspots are at some time, especially late in their lives, crossed by narrow, bright features known as *light bridges*. These features come in a variety of shapes, sizes, and brightnesses, with the largest of them extending all the way across the umbra and covering an appreciable fraction of its area. Most light bridges are segmented, with bright segments separated by narrow dark lanes lying perpendicular to a long, narrow, central dark lane running along the length of the bridge (Berger and Berdyugina 2003). Some light bridges show little or no segmentation, however; many of these extend well into the penumbra where they resemble the other elongated bright penumbral filaments. Indeed, these features might better be thought of as penumbral filaments extending into the umbra.

A light bridge of each of these types may be seen in the umbral image shown in Figure 4.6. In this blue-continuum image (from Lites *et al.* 2004), taken near the solar limb, we have a perspective view in which we can see that the light bridge at the upper left is an elongated, raised, tent-like structure, segmented along its length. The dark central lane runs along the elevated ridge of the tent. The elevation and the dark lane are explained by a combination of the increased opacity of the plasma due to the increased temperature and the increased gas pressure due to the reduced magnetic field strength in the light bridge. The light bridge along the bottom of the image shows less segmentation and extends into the penumbra.

The term 'light bridge' was originally coined because it was thought that these structures consisted of elevated, bright facular material lying above the dark umbral photosphere. For some time now it has been known that they are instead low-lying features of the umbral

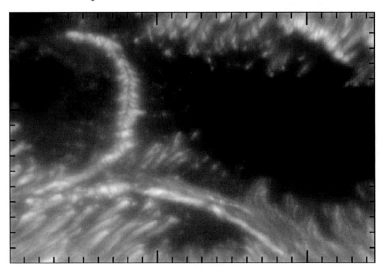

Fig. 4.6. Blue-continuum image of a large umbra near the solar limb, with two conspicuous light bridges. The interval between tick marks is $1''$, and the image is oriented with its vertical direction along a solar radius and the closest point on the limb toward the top. (From Lites *et al.* 2004.)

photosphere itself, but now the recent discovery of the slightly elevated nature of the light bridges, as seen in Figure 4.6, again makes the term appropriate.

Several earlier attempts were made to classify light bridges into different types, but these classifications have not been particularly helpful in understanding them (a point emphasized by Leka 1997). Some of the widest and brightest features that have been called light bridges are actually strips of nearly normal granulation separating two umbrae lacking penumbrae. True light bridges are narrow features (less than about 10 Mm across) of roughly penumbral (not photospheric) intensity that lie within an umbra of single magnetic polarity. In complex δ-sunspots there are usually bright lanes separating umbral cores of opposite magnetic polarity, but these bright lanes are filamentary and are more closely akin to penumbrae than to light bridges.

Light bridges often follow sutures or fissures in the umbra, outlining individual pores that assembled to form the umbra or the segments into which the umbra will split during the decay phase of the spot. Narrow, faint light bridges come and go during the lifetime of a sunspot, but a few might remain throughout, outlining a pore that retains its identity. One of the first indications of the imminent breakup of a sunspot is brightening of existing light bridges or the appearance of new bright light bridges in the umbra. These light bridges grow in intensity and width, reaching photospheric intensity and displaying a nearly normal granulation pattern as the sunspot fragments and expands during its decay phase (Vázquez 1973). The formation of a new light bridge occurs as a number of umbral dots emerge sequentially from the inner end of a penumbral filament and move rapidly inward into the umbra, where they collect to form the light bridge (Katsukawa *et al.* 2007b).

The magnetic field strength at photospheric heights in a light bridge is significantly less than that in the surrounding umbra (Beckers and Schröter 1969; Abdussamatov 1971; Lites *et al.* 1991; Rüedi, Solanki and Livingston 1995; Leka 1997; Katsukawa *et al.* 2007b) and it decreases with depth. The magnetic field configuration varies considerably from one light bridge to another, but the field is generally more inclined to the vertical than in the surrounding umbra (Leka 1997) and in some cases may be locally horizontal, or even of reverse polarity (Lites *et al.* 2004). On the basis of a study of many light bridges, Leka (1997) suggested that light bridges consist of intrusions of field-free gas from below, with the magnetic field diverted around the intrusion and merging again above the intrusion. Jurčák, Martínez Pillet and Sobotka (2006) find a field configuration consistent with such a field-free intrusion at the deepest visible level but with magnetic canopies spreading from either side of the light bridge and merging above the bridge, where the associated intense electric currents heat the base of coronal loops having a footpoint above the light bridge (as seen in TRACE images).

It seems then that a light bridge corresponds to a fissure that separates two distinct components of a spot, which only merge as their expanding magnetic fields meet on either side of a separatrix surface above the photosphere. Beneath the visible surface the plasma is apparently more or less field-free, implying that the subphotospheric flux concentrations remain separate to some indeterminate depth. Rimmele (1997) found a correlation between vertical velocity and continuum intensity in a light bridge that is consistent with a pattern of segmented, time-dependent convection in a narrow slot. This pattern resembles the spatially modulated oscillations illustrated in Figure 4.2. As in the model of umbral convection described in Section 4.2.2 above, the surface of unit optical depth will be raised above the row of plumes and buoyancy braking will lead to an enhancement of density along the axis of the slot, just as in Figure 4.5, which helps to explain the occurrence of a dark central lane.

5

Fine structure of the penumbra

The most striking recent development in the study of sunspots has been the revelation of fine structure in intensity, magnetic fields and velocity patterns in the penumbra. Within the past 15 years new observations, notably those made first with the Swedish Vacuum Solar Telescope (SVST) on La Palma, then with the aid of adaptive optics on the Dunn Telescope at Sacramento Peak and the 1-m Swedish Solar Telescope (SST) and, most recently, with the Solar Optical Telescope on the Hinode spacecraft, have resolved delicate features with unprecedented clarity – and, in so doing, have posed major problems for theory to explain (Thomas and Weiss 2004). The filamentary penumbra appears clearly in Figures 1.2 and 3.1 and is shown here in greater detail in Figure 5.1.

In this chapter we discuss this fine-scale filamentary structure and the associated interlocking-comb configuration of the penumbral magnetic field. We begin with observations and describe first the two-dimensional intensity pattern in the penumbral photosphere. Then we go on to discuss the complex three-dimensional structure that is revealed by measuring two vector fields, the magnetic field $\mathbf{B}$ and the velocity $\mathbf{u}$. This rich and intricate magnetic geometry results from interactions between convection and the inclined magnetic fields in the outer part of a sunspot. We next outline the current theoretical understanding of this form of magnetoconvection, and attempt to interpret the observed penumbral structure in the light of available theoretical models. Then we consider in turn the outward photospheric Evershed flow, which is organized on fine scales that are closely associated with the filamentary structure, and the outward-moving magnetic features in the moat surrounding a sunspot. Finally, we present some theoretical ideas on how the penumbra is formed and maintained.

5.1   Penumbral filaments

At moderate spatial resolution ($1''$–$2''$) the penumbra is seen to consist of alternating bright and dark elongated filaments. These filaments have a predominantly radial alignment, which is most apparent in a single isolated spot, like that in Figure 3.1. In a minority of such spots, the pattern shows a distinct vortical structure (which is much more prominent in Hα spectroheliograms; see Section 3.3.3), with a sense that corresponds (statistically) to that of terrestrial cyclones and is antisymmetric about the equator.[1] In many cases the pattern leaves the impression of narrow bright filaments superimposed on

---

[1] It was this pattern that prompted Hale to search for a magnetic field, and later stimulated Bjerknes (1926) to introduce a model of the solar cycle that relied on vortex tubes.

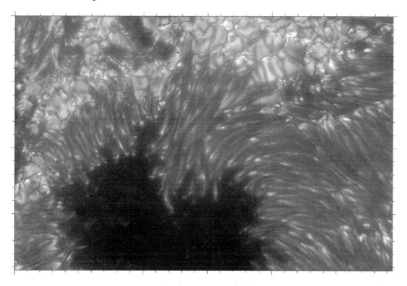

Fig. 5.1. Detailed structure of penumbral filaments in Figure 1.2, as observed at high resolution with the Swedish Solar Telescope. (From Scharmer *et al.* 2002.)

a dark background. However, the terms 'bright' and 'dark' are here only relative; there are also larger-scale intensity variations within the penumbra, and some bright filaments may actually have lower intensity than a dark filament elsewhere.

The width of the narrowest penumbral filaments is apparently near or below even the current limit of resolution (0.1″), so it is not surprising that reported widths have decreased as spatial resolution has improved. Early studies determined the width directly from photometric profiles (e.g. Muller 1973b; Bonet, Ponz and Vázquez 1982), and Scharmer *et al.* (2002) measured widths of 150–180 km for well-resolved bright filaments extending into the umbra. Some recent studies have tried to make more objective measurements based on the spatial power spectrum of intensity. For example, Sánchez Almeida and Bonet (1998) found a flat power spectrum and hence concluded that the actual widths are typically well below their resolution limit of about 0.2″. Sütterlin (2001), on the other hand, found an enhancement of spatial power at around 0.35″ (250 km) and suggested that this is the preferred width of filaments. Rouppe van der Voort *et al.* (2004), using observations with spatial resolution of 0.12″ (80 km), found a power spectrum that drops off roughly as k^{-4} (where k is the horizontal wavenumber) and has no distinct peak corresponding to a preferred width. This implies that, although there certainly are bright filaments that have been resolved, there are also many unresolved features with widths less than 80 km.

Bright filaments appear most prominently near the inner penumbral boundary, where they protrude into the dark umbra. Some filaments may extend across the entire penumbra, while others peter out within it. There are, moreover, many filaments that originate within the penumbra itself and fan out with increasing distance from the umbra. Individual filaments may extend for lengths of 3500–7000 km (Rouppe van der Voort *et al.* 2004; Langhans *et al.* 2005). The filamentary pattern becomes less distinct in the outer penumbra, where brighter and darker features are apparently intermingled.

5.1.1   *Bright grains in penumbral filaments*

At sub-arcsecond resolution, a bright penumbral filament is seen to consist of a number of elongated bright features known as *penumbral grains* (Muller 1973a,b, 1992). The widths of these grains are typically $0.5''$ or less, and their lengths range from about $0.5''$ to $3.5''$. The intensity of the grains ranges from about $0.85I_{phot}$ to $1.10I_{phot}$, with the brightest grains being hotter than the brightest granules outside the sunspot by some 150 K (Tritschler and Schmidt 2002b). Images at the highest available resolution (near $0.1''$) show that some parts of the bright filaments consist of several narrower, long ($5''$–$9''$), truly filamentary features, while other parts (especially at the ends nearest the umbra) consist of more grain-like, segmented features made up of yet smaller bright features separated by narrow dark bands (see Fig. 5.2). These transverse bands appear as dark streaks, apparently inclined

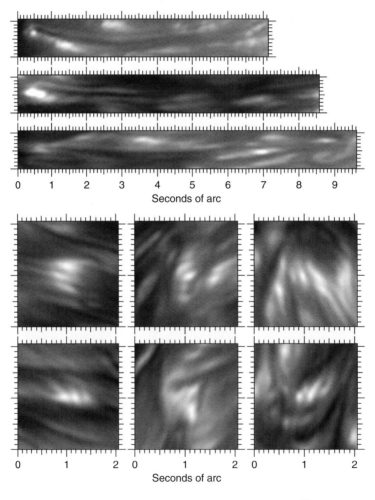

Fig. 5.2. Images of bright penumbral filaments at nearly $0.1''$ resolution, taken with the Swedish Solar Telescope. The bright filaments are seen to divide into long, even narrower filaments and (especially nearest the umbra) segmented penumbral grains. (From Rouppe van der Voort *et al.* 2004.)

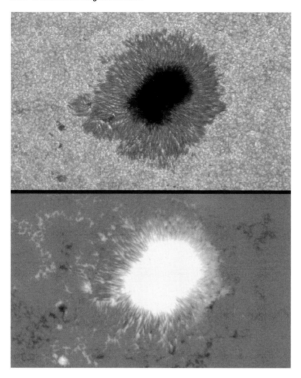

Fig. 5.3. Images of a sunspot observed with the Solar Optical Telescope on the Hinode satellite, with exceptionally clear resolution. Upper panel: G-band image, showing hyperfine structure within bright filaments. Lower panel: magnetogram, demonstrating the variations in the line-of-sight field strength caused by the interlocking-comb structure of the magnetic field. (Courtesy of NAOJ/LMSAL/JAXA/NASA.)

at about 45° to the filaments, and give the illusion of a twisted structure; this pattern migrates inward across the grain and may even continue over an adjacent filament (Scharmer *et al.* 2002; Rouppe van der Voort *et al.* 2004). The G-band image in Figure 5.3 shows a number of these features. Careful observations with Hinode reveal, however, that although the orientation of the dark streaks is consistent in any quadrant of the solar surface, their inclinations are opposite in the E and W hemispheres (Ichimoto *et al.* 2007a). This mirror symmetry implies the presence of a three-dimensional structure; the streaks appear then as two-dimensional projections of transverse cuts across the elevated filaments, in planes that are inclined to the local vertical.

Penumbral grains show systematic proper motions in the radial direction (see Fig. 5.4). Early observations found only inward radial motion, toward the umbra. More recent observations, however, have revealed a pattern of both inward and outward motion: the grains move inward in the inner penumbra and outward in the outer penumbra (Wang and Zirin 1992; Denker 1998; Sobotka, Brandt and Simon 1999; Sobotka and Sütterlin 2001; Márquez, Sánchez Almeida and Bonet 2006). There seems to be a dividing circle located at about 60% of the radial distance from the inner to the outer edge of the penumbra, inside of which the grains move radially inward at speeds of about 0.5 km s^{-1}, and outside of which the grains move outward at about the same speed (Wang and Zirin 1992) or perhaps

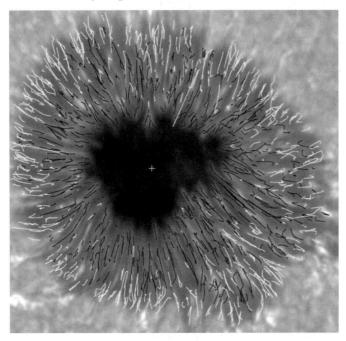

Fig. 5.4. Paths of proper motions of penumbral grains, showing inward-moving grains (black curves) in the inner penumbra and outward-moving grains (white curves) in the outer penumbra. The image size is 29.8×28.4 Mm, and the white cross marks the centre of the umbra. (From Sobotka and Sütterlin 2001.)

slightly higher speeds of about $0.75 \, \mathrm{km \, s}^{-1}$ (Sobotka and Sütterlin 2001). The inward-moving penumbral grains often penetrate into the umbra, where they become peripheral umbral dots (see Section 4.1), and may continue to move inward for a while at speeds of up to $0.5 \, \mathrm{km \, s}^{-1}$ (Ewell 1992). The majority of the outward-moving grains disappear before reaching the outer penumbral boundary, but roughly a third of them cross this boundary and evolve into either a small bright feature (of diameter less than $0.5''$) or a normal photospheric granule, and continue to move radially outward away from the sunspot (Bonet *et al.* 2004). New penumbral grains appear and tend to follow trajectories identical to those of the grains that preceded them at that location (Sobotka, Brandt and Simon 1999). White-light images of the penumbra averaged over 2–4 hours still show the filamentary structure (Balthasar *et al.* 1996; Sobotka, Brandt and Simon 1999), indicating a long-term stability of this pattern of motion, and indeed of the magnetic structure of the penumbra too.

Márquez, Sánchez Almeida and Bonet (2006) applied the technique of local correlation tracking (November and Simon 1988) to the sequence of high-resolution G-band images obtained by Scharmer *et al.* (2002) in order to determine proper motions of fine-scale features within the penumbra. As expected, they found a predominantly radial velocity field, corresponding to inward motion; but they also detected an apparent transverse flow out of the bright filaments that converged on the intervening dark filaments. This pattern of proper motions was demonstrated by following passive test particles ('corks'), which moved rapidly out of adjacent bright filaments and accumulated in a linear array, cospatial with the inter-mediate dark filament. The fundamental question regarding these observed proper motions, as well as those of the penumbral grains, is whether they correspond to material motions of the gas itself (along with the magnetic field) or to a travelling pattern of magnetoconvection. We shall return to this issue in Section 5.3, after we have discussed the arrangement of the magnetic field in the penumbra.

5.1.2   *Dark cores within bright filaments*

High-resolution observations have also revealed hyperfine structure within the bright filaments. The slender dark cores that can be seen clearly in Figures 5.1, 5.2 and 5.3 were first reported by Scharmer *et al.* (2002).[2] Many – but not all – bright filaments contain dark cores, which may extend right across the penumbra. The width of a typical dark core is around 0.2″ (Bellot Rubio, Langhans and Schlichenmaier 2005; Langhans *et al.* 2007). They are most apparent near the umbra–penumbra boundary, where they often split to give a Y-shaped structure that penetrates into the umbra. The dark cores themselves seem to be elevated features, which are more apparent on the centre side than on the limb side of a spot, indicating that they are shallow structures perched upon bright filaments (Sütterlin, Bellot Rubio and Schlichenmaier 2004).

5.2   The intricate structure of the penumbral magnetic field

Early measurements of azimuthally averaged velocities and magnetic fields had already raised a contradiction: in a steady state the flow should be parallel to the magnetic field in a highly conducting plasma, yet the persistent Evershed flow was horizontal while the magnetic field reached an inclination of only 70° (with respect to the local vertical) at the edge of the spot (e.g. Adam and Petford 1991), as shown schematically in Figure 5.5a. This paradox could only be resolved by assuming an inhomogeneous magnetic structure. We now know that, like the intensity pattern, the magnetic field in the penumbra is not axi-ally symmetric. The inclination of the magnetic field varies azimuthally, being more nearly horizontal in the dark filaments, as first observed by Beckers and Schröter (1969). More recent high-resolution observations, beginning in about 1990, have gradually revealed the complex *interlocking-comb*, or interlocking-sheet, configuration[3] of the penumbral mag-netic field (Degenhardt and Wiehr 1991; Title *et al.* 1992, 1993; Schmidt *et al.* 1992; Lites *et al.* 1993; Solanki and Montavon 1993; Hofmann *et al.* 1994; Stanchfield, Thomas and Lites 1997; Westendorp Plaza *et al.* 1997), which is sketched in Figure 5.5b. These mea-surements showed that the field inclination in the bright filaments increases from about 40° at the umbra–penumbra boundary to around 60° at the outer edge of the spot. Although the field in the dark filaments is approximately aligned with that in bright filaments at the inner boundary of the penumbra, it becomes almost horizontal at the outer edge. Thus the line-of-sight component of the magnetic field yields the spiny pattern that is apparent in Figure 5.3.

---

[2] In fact, the dark cores can also be discerned in an earlier image, reproduced by Thomas and Weiss (1992b, Fig. 7), obtained with the 50 cm SVST, the predecessor of the 1-m SST on La Palma. At the time, it was not clear whether these features were genuine or an artefact of the image restoration process.

[3] The configuration has been given various other names: 'spines' (the more vertical fields) and 'inter-spines' (Lites *et al.* 1993); 'fluted' (Title *et al.* 1993); 'uncombed' (Solanki and Montavon 1993); and even 'interdigitated'. We prefer to stick with 'interlocking-comb' (Thomas and Weiss 1992b), which accords with manual demonstrations.

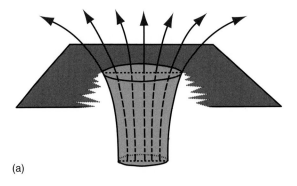

(a)

(b)

Fig. 5.5. Sketches showing (a) the axisymmetric field configuration in an idealized model of a sunspot and (b) the interlocking-comb configuration of individual flux tubes in a sunspot penumbra. Note that these flux tubes combine with their vertically displaced neighbours to form interlocking sheets. (From Weiss *et al.* 2004.)

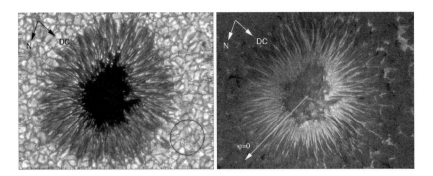

Fig. 5.6. The fluted structure of the penumbral magnetic field in a sunspot. Left panel: a broadband continuum image of the sunspot. Right panel: a magnetogram showing the line-of-sight magnetic field. The arrows indicate the directions of the disc centre and north on the Sun; the spot was 16° off disc centre. Images obtained with the SST. (From Langhans *et al.* 2005.)

5.2.1   *The interlocking-comb magnetic structure*

Figure 5.6 shows two images of a sunspot, obtained with the SST (Langhans *et al.* 2005). It is apparent that the line-of-sight magnetic field varies rapidly in the azimuthal direction and is stronger in the bright filaments than in the intervening darker filaments. This line-of-sight component is also stronger on the centre-side of the spot, where the field points towards the observer, than on the limb-side, where it is strongly inclined to the line of sight and points away from the observer in the dark filaments. Langhans *et al.* (2005) exploited this magnetic geometry (which is more pronounced for spots that are further off disc centre) in order to compute the radial variation of field inclinations in the bright and dark filaments, and their results are summarized in Figure 5.7.

Their analysis clearly shows that the relative field strength is well correlated with variations in intensity, and that fields in bright filaments are stronger than those in dark filaments. Nevertheless, the measured field inclinations differ depending on whether line-of-sight field

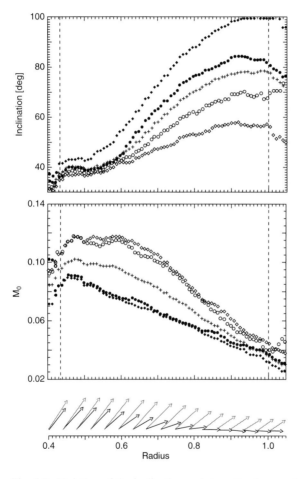

Fig. 5.7. Variation of the inclination and strength of magnetic fields in bright and dark filaments across the penumbra. Top panel: field inclinations; middle panel: relative strengths of the magnetic field; bottom panel: vectors showing average strengths and inclinations in strong (grey) and weak (black) field components of the penumbra. The abscissa runs from just inside $(0.4R)$ to just outside $(1.05R)$ the penumbra, and the dashed lines indicate its inner and outer boundaries. In the upper panels, filled symbols refer to dark (or weaker field) components, hollow symbols refer to bright (or stronger field) components; diamonds and circles indicate components distinguished by magnetic and intensity measurements, respectively; $+$ symbols denote azimuthal averages. (From Langhans *et al.* 2005.)

strength or relative intensity is used as a criterion. Taking the former as more reliable, we see that the tilt in the less inclined component increases monotonically from around 40° in the inner penumbra to about 60° at the edge of the spot. The more inclined component has a slightly greater inclination near the umbra but its inclination increases rapidly to reach 90° at 0.8R and goes on to around 100° for $r > 0.9R$. Thus the field in this component is almost horizontal in the outer 20% by radius of the penumbra, and field lines actually reverse their direction and plunge below the surface in the outermost 10%. This configuration agrees with that found by Bellot Rubio, Balthasar and Collados (2004; see also Bellot Rubio *et al.* 2003; Mathew *et al.* 2003), in a spectropolarimetric investigation. Despite being unable to resolve the fine structure of the field, they were able to invert the Stokes profiles of three infrared Fe I lines and to demonstrate the presence of two field components with different inclinations. In a related treatment, Borrero *et al.* (2006) found that the steeply inclined field component has a limited vertical extent, reaching no higher than an optical depth $\tau_{500} \approx 3 \times 10^{-2}$, a few hundred km above the photosphere.[4] Hinode measurements of the vector field (Jurčák *et al.* 2007) indicate that the more steeply inclined field component only appears as the optical depth τ_{500} approaches unity; they also suggest that the field strength above the bright filaments decreases at this level. The vertical extent of this field component remains undetermined. These results confirm earlier, less well resolved measurements, which revealed patches in the outer penumbra where the vertical component of the field reversed (Stanchfield, Thomas and Lites 1997; Westendorp Plaza *et al.* 1997).

It is possible therefore to distinguish between three different components that make up the penumbral magnetic field. First, the field lines that are less steeply inclined, and typically associated with bright filaments, rise up to form loops that extend for great distances across the solar surface, connecting either to other sunspots or to distant footpoints. These loops appear both in X-ray images of the corona (Sams, Golub and Weiss 1992) and, more strikingly, in the extreme ultraviolet, as shown in Figure 5.8 (Winebarger, DeLuca and Golub 2001; Winebarger *et al.* 2002). Then there are steeply inclined field lines that emerge from darker regions of the penumbra and rise to form a shallow canopy, elevated above the photosphere, which extends well beyond the visible boundary of the spot (e.g. Giovanelli and Jones 1982; Solanki, Rüedi and Livingston 1992; Solanki, Montavon and Livingston 1994; Solanki 2002, 2003; Rezaei *et al.* 2006). This magnetic canopy has a diameter that may be more than twice that of the spot itself, while its base rises gradually upwards to a height of around 300 km above the $\tau_{500} = 1$ level in the photosphere (Solanki 2002). Finally, there is the third component, with strongly inclined field lines that emerge in the outer penumbra and then bend over to return below the solar surface, either within the penumbra or just outside it. It seems clear that there is little scope for interchanges between the first component and the other two, and so they must remain essentially distinct. Nevertheless, since the magnetic structure cannot be entirely current-free, and is also dynamic and constantly evolving, some localized magnetic reconnection is inevitable. Evidence for this comes from Hinode observations of fine-scale jets in Ca II H emission, aligned with the less inclined magnetic field component (Katsukawa *et al.* 2007a).

---

[4] They also identify a lower boundary at $\tau_{500} \approx 1$, in accordance with the 'uncombed' model of Solanki and Montavon (1993).

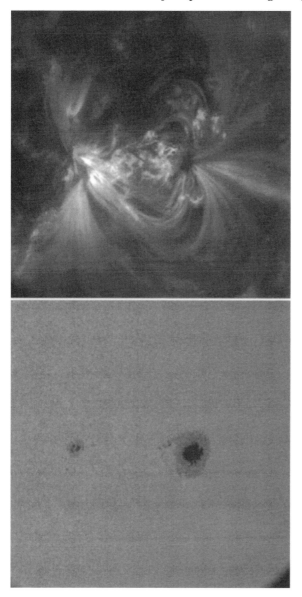

Fig. 5.8. Coronal loops connecting a sunspot pair. The lower image, in white light, shows the two spots, while the upper image, obtained in the extreme ultraviolet by the TRACE spacecraft, reveals fine loops, following magnetic field lines in the corona, that link the two spots or extend to distant footpoints on the solar surface. (Courtesy of Lockheed-Martin Solar and Astrophysics Research Laboratory.)

5.2.2 *Velocity structure*

Although the radial outflow in the outer penumbra was discovered by Evershed (1909a) almost a century ago, its detailed velocity structure was not established until observations could be made with much higher resolution. Recent observations have confirmed that the Evershed flow is associated with dark filaments, and therefore with more inclined magnetic fields. Correlations between upflows and bright features, and downflows and dark features, first pointed out by Beckers and Schröter (1969), have also been firmly established (Schmidt and Schlichenmaier 2000; Bellot Rubio, Schlichenmaier and Tritschler 2006). Spectropolarimetric inversions indicate that the velocity is everywhere parallel to the magnetic field (Bellot Rubio *et al.* 2003; Bellot Rubio, Balthasar and Collados 2004), as expected for a highly conducting plasma, and that the Evershed flow is carried along nearly horizontal fields (Borrero *et al.* 2004, 2005; Bello González *et al.* 2005), with a velocity that increases with increasing optical depth (Bellot Rubio, Schlichenmaier and Tritschler 2006). (We will discuss the high-resolution observations of the Evershed flow in more detail in Section 5.4.)

The relationship between the velocity $\mathbf{u}$, the magnetic field $\mathbf{B}$ and filamentary structure has been clearly demonstrated by Langhans *et al.* (2005), using measurements with a resolution of 0.2″ on the SST. They show that $\mathbf{u}$ is roughly parallel to $\mathbf{B}$ in both bright and dark filaments, with the velocity always directed outwards in the penumbra and hence with a much stronger upward component in bright filaments, although the greatest speeds are attained in the most nearly horizontal fields. The strong outflow masks any indications of local convective motion. The strongest upflows are associated with small bright grains (around 0.2″ in size) in the inner penumbra, and preferentially at the umbra–penumbra boundary (Rimmele 2004; Rimmele and Marino 2006; see also Ichimoto *et al.* 2007a). These upflows (with speeds of up to $0.5\,\mathrm{km\,s^{-1}}$) move inward and are closely associated with outflows along dark filaments, where the fields are more inclined. In the outer penumbra the Evershed flow is predominantly, though not exclusively, associated with dark filaments (Rouppe van der Voort 2003), while the downflows necessarily coincide with downward-pointing fields.

5.2.3   *Hyperfine structure and dark cores*

Dark cores within bright filaments were originally identified in continuum and G-band images (Scharmer *et al.* 2002; Rouppe van der Voort *et al.* 2004; Sütterlin, Bellot Rubio and Schlichenmaier 2004) but they are even more obvious in magnetograms (Langhans *et al.* 2005, 2007; Bellot Rubio, Langhans and Schlichenmaier 2005; Bellot Rubio *et al.* 2007). Ground-based measurements have shown that the line-of-sight magnetic field is weaker in the dark cores than in the parallel bright features that enclose them, and also more steeply inclined to the vertical (Langhans *et al.* 2005; Bellot Rubio, Langhans and Schlichenmaier 2005). Langhans *et al.* (2007) estimate that the actual field strength $|\mathbf{B}|$ drops by 30–40% in the dark cores, while the difference in inclination rises rapidly, from zero at the inner end of the filament to about 10°, within 1″. Correspondingly, the tilt of the field in a dark core increases from 40° to 60° within a few arcseconds. Spectropolarimetric measurements from the Hinode satellite have made it possible to determine the profiles of the four Stokes parameters across two Fe I lines in dark cores, and hence to calculate the vector field $\mathbf{B}$ (Bellot Rubio *et al.* 2007). Dark cores within filaments are clearly visible in maps of total polarization, which yield field strengths only 150–200 G less than those in the lateral brightenings around them and (rather surprisingly) a difference of only 4° in inclination.

Doppler shifts indicate larger line-of-sight velocities in dark cores, corresponding to upflows with a stronger horizontal component than that in the neighbouring bright features (Bellot Rubio, Langhans and Schlichenmaier 2005; Langhans *et al.* 2007). Rimmele and Marino (2006) studied the flow patterns in dark cores and found an abrupt transition from upflows in bright features at the edge of the umbra to outflows that are more horizontal.

Since this relationship is so similar to that between upflows and dark filaments, it raises a fundamental question: could the Evershed flow emanate entirely from dark cores? This will be answered by future observations at yet higher resolution; for the moment, it seems that the flows in dark cores are distinct from those in dark filaments, and that it is the latter that are linked to Evershed flows in the outer penumbra of a sunspot.

5.3   Convection in the penumbra

Having discussed the observations, we now turn to theoretical models of the processes that result in the complicated structure of the penumbra. It is already clear that the various features that we have described are intimately connected to different patterns of convection. In his study of linearized Boussinesq magnetoconvection, Chandrasekhar (1952, 1961) pointed out that in an inclined magnetic field convection first sets in as rolls oriented parallel to the horizontal component of the field; in an infinite layer with a horizontal field such rolls would be unimpeded by magnetic forces. Danielson (1961b) interpreted penumbral filaments as convection rolls in a strongly inclined magnetic field – and it now seems intuitively obvious that the penumbra's filamentary structure results from the interactions between its inclined meridional field and convection.

In any model of an isolated flux tube embedded in a strongly stratified layer, the magnetic field fans out with height. As the enclosed flux increases, so does the inclination of the field lines at the edge of the flux tube. If we consider a simplified model of a pore that gradually accrues more magnetic flux, the field at its boundary with the external photosphere will become increasingly tilted, until the total flux reaches a critical value (Simon and Weiss 1970). Prior to that, we expect to see a tesselated pattern of convection in the pore; once the critical tilt is exceeded, a filamentary penumbra will appear. This transition is nicely illustrated in Figure 5.9, which shows results obtained for a highly idealized two-dimensional model system, governed by an extended Swift–Hohenberg equation and relying

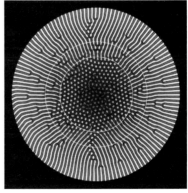

Fig. 5.9. Schematic illustration of the transition from a pore, with an isolated umbra containing a tesselated pattern of convection, to a sunspot, with an umbra and a filamentary penumbra, as the total magnetic flux is increased. Patterns generated by an extended Swift–Hohenberg equation, including terms modelling the tilt of the field. Only hexagonal patterns are stable within the inner circle, and only rolls are stable outside the outer circle. Dislocations appear because the rolls have fixed widths. (Courtesy of S. D. Thompson.)

solely on symmetry constraints (Thompson 2006b). In this nonlinear model system, a hexagonal pattern is favoured when the imposed 'field' is vertical or only slightly inclined, but rolls take over at steeper inclinations; analogous effects appear in a numerical experiment on convection in an arched field configuration. A more realistic simulation of magnetoconvection with an arched field, in Cartesian geometry, shows a rudimentary penumbra, with hints of dark-cored structures (Heinemann *et al.* 2007).

Here we start by describing what is known about the behaviour of convection in both weak and strong inclined magnetic fields. Then we attempt to apply these results to convection in the penumbra itself. It is helpful to distinguish behaviour in the inner penumbra, whose structure is dominated by bright filaments, from that in the outer penumbra, where the dark component is more prevalent, and we shall consider these two regions separately. For convenience we choose as the boundary between them the line separating inward and outward moving grains in bright filaments; this line, at about $0.75R$, or 60% of the radial distance from the inner to the outer edge of the penumbra (see Section 5.1.1) divides the penumbra into two roughly equal areas. We first try to explain the patterns of convection in bright and dark filaments in the inner penumbra and then go on to discuss dark cores. Next, we consider convection in the outer penumbra, where the spiny magnetic structure is most apparent. Finally, we comment on an alternative model that uses the motions of individual flux tubes to represent convective processes. It is clear that the available theoretical models can only be regarded as tentative and preliminary descriptions of the actual penumbral structure.

5.3.1 *Travelling patterns in inclined magnetic fields*

In the presence of an imposed vertical magnetic field, all horizontal directions are equivalent, but once the field is tilted this degeneracy is removed. If the field is vertical and convection is steady at onset, then hexagonal cells are preferred, both in the Boussinesq regime (Clune and Knobloch 1994) and for a compressible layer (Rucklidge *et al.* 2000). This pattern survives even for a tilted field if the up–down symmetry of a Boussinesq layer is maintained. Once that symmetry is broken, as it is in a stratified fluid layer, then stationary solutions cease to exist for inclined fields (Matthews *et al.* 1992). It is easy to see, for instance, that, while rolls with vertical boundaries need not travel, rolls with tilted boundaries are bound to do so. It is convenient to consider an imposed field $\mathbf{B}_0$ that lies in the xz-plane, referred to Cartesian co-ordinates, and is tilted at an angle ϕ to the upward vertical, so that $\mathbf{B}_0 = B_0(\sin\phi, 0, \cos\phi)$. We note that for ϕ sufficiently small the critical Rayleigh number Ra_c for the onset of convection in parallel rolls (with axes in the x-direction) cannot depend on the sign of ϕ, so that $Ra_c = Ra_0 + \mathcal{O}(\phi^2)$; for transverse rolls, on the other hand, either left-travelling or right-travelling rolls (as viewed along the y-axis) will be preferred and hence $Ra_c = Ra_0 + \mathcal{O}(\phi)$. It follows therefore that one or other family of transverse rolls will set in first, and a similar argument holds for growth rates when $Ra > Ra_c$ (Matthews *et al.* 1992; Thompson 2005). Linearized theory confirms the prediction of this simple argument, which is based on symmetry alone. Transverse rolls are indeed preferred at onset for small values of ϕ but as ϕ is increased there is a transition first to oblique rolls (which grow progressively more oblique) and then another jump to parallel rolls as ϕ approaches $90°$.

Whether left-going or right-going rolls are preferred is a delicate matter that depends on the details of the configuration studied. Two-dimensional numerical studies of nonlinear transverse waves in a stratified compressible layer (Hurlburt, Matthews and Proctor 1996) revealed tilted cells, with clockwise motion predominating over anticlockwise, so that the mean surface velocity was in the direction of tilt; the direction and speed of travel of the waves were, however, sensitive to the parameters of the problem – the degree of nonlinearity, the field strength, the angle of tilt and the choice of boundary conditions – making it hard to draw any general conclusions. Hurlburt, Matthews and Rucklidge (2000) extended these results to three dimensions, for a relatively shallow atmosphere which allowed stationary, spatially modulated oscillations when the field was vertical. For a mildly tilted field ($22°$) they found a hexagonal pattern of leftward-travelling oscillations, but there was a gradual transition to roll-like solutions as the tilt was increased. For $\phi = 45°$ there were larger plumes in parallel rows, elongated in the direction of tilt, and for $\phi = 67°$ there was an almost roll-like, modulated, leftward-travelling wave (see Fig. 6 of Thomas and Weiss 2004). These model solutions provide a hint of what might be expected in the penumbra of a sunspot.

Julien, Knobloch and Tobias (1999, 2000, 2003) have developed an asymptotic treatment of high Ra magnetoconvection in a very strong magnetic field, which relies on assuming very small horizontal scales for the motion. Within this framework, they consider both transverse and parallel rolls and find, interestingly, that as the tilt increases there is in each case an abrupt transition to an extremely inefficient 'horizontal' mode of convection, which they associate with the formation of a penumbra. It is not clear, however, that such a transition exists when wider horizontal scales are admitted.

5.3.2   Convection in the inner penumbra

At the inner edge of the penumbra, the inclinations of the magnetic fields in bright and dark filaments are not significantly different, though the difference increases to 15–30° at the boundary with the outer penumbra. The overall convective pattern seems to be one of parallel rolls, with hot gas rising in the bright filaments and cooler gas sinking within the darker gaps between them. The proper motions detected by local correlation tracking (Márquez, Sánchez Almeida and Bonet 2006) provide evidence of transverse outflows from bright filaments, which converge in the adjacent dark lanes, as expected for such rolls. The inward-moving bright grains are presumably travelling waves that penetrate into the umbra, where their contrast is highest and where there is a strong upward velocity (Rimmele and Marino 2006). These grains, in turn, are apparently modulated by a pattern of transverse waves that passes radially inward across them.

Thus the observed patterns in the inner penumbra can be explained as a combination of three different scales of motion. The tilted field forces a roll-like structure, with regular overturning motion at least in the innermost penumbra. This pattern is then modulated by the appearance of bright grains which move radially inward and eventually enter the umbra; their behaviour is consistent with that of the time-dependent umbral dots, except that they are now constrained to migrate radially inwards. Morover, the grains exhibit a fine-scale segmented structure, apparently caused by transverse waves that travel inwards at a faster rate. Within bright filaments the plasma velocity itself is predominantly upwards and outwards along the field lines, although these may themselves be laterally displaced by interchanges with adjacent darker lanes. It is difficult to conceive that interchanges between bright and dark filaments can persist when the magnetic field inclinations differ significantly between them. It seems more likely that bright filaments are enclosed by darker lanes with sinking plasma and that, while the bright and dark filaments merge in the innermost penumbra, further outward they separate progressively as their field inclinations come to differ significantly. Before the boundary with the outer penumbra, the two families of field lines are distinct, and their convective structures have to be considered separately.

5.3.3   Buoyancy braking and the origin of dark cores

The dark cores within bright filaments have been convincingly explained by Spruit and Scharmer (2006) as absorption features caused by a density excess above the rising two-dimensional plume. There is a close analogy with the dark streaks found by Schüssler and Vögler (2006) in their simulations of umbral convection (see Section 4.2.2). As they point out, in any convecting system there has to be a pressure excess around the stagnation point (or line) where a rising plume is brought to a halt; this leads to a density excess that results in buoyancy braking of the rising gas (e.g. Spruit, Nordlund and Title 1990) and also acts as an absorption feature. Hence the dark cores can be regarded as slender absorption features perched on top of the upwelling plasma in a bright filament. (Similar features appear in light bridges for the same reason – see Section 4.3.) In an umbral dot, the enhanced pressure drives an outward flow in both directions along the dark streak (Schüssler and Vögler 2006), but the tilted fields in the penumbra inevitably force a radial outflow in dark cores (where the field is only slightly more inclined than in the rest of the bright filament). Thus dark cores can be seen as an inevitable consequence of roll-like convection. The flow along them is driven by a persistent pressure excess, as in the flux tube models of Schlichenmaier (2002). As in all manifestations of magnetoconvection, a rising and expanding plume sweeps the magnetic field aside and the field strength immediately above it is reduced. Given the narrow width of the bright filaments, such a local reduction will fall off very rapidly with height. In summary, therefore, we may regard the dark cores as a remarkable – but very natural – feature of penumbral convection.[5]

5.3.4   Dark filaments and convection in the outer penumbra

It is apparent that convective transport is more efficient in the bright filaments, which allow some form of oscillation in bright grains, than it is in the dark filaments, where the field is more inclined. This difference becomes most acute in the outermost part of the penumbra ($r > 0.8R$, about 35% of the total area of the spot) where the magnetic field is almost horizontal over a large fraction of the area. Theoretical models indicate that in such a field convection should take the form of horizontal rolls whose axes lie in vertical planes containing the magnetic field, as originally suggested by Danielson (1961b). These rolls are presumably confined to long, narrow slots of limited vertical extent. Within these slots there must be some form of time-dependent interchange convection that transfers heat both upward, across the average field, and inward, from the surrounding field-free plasma toward the umbra (Schmidt 1991).

Bright grains in the outer penumbra resemble those in the inner penumbra, except that they move outwards rather than inwards, and at greater average speeds (Sobotka, Brandt and Simon 1999; Sobotka and Sütterlin 2001). We may presume that they too represent a travelling wave pattern and that the transition from inward to outward proper motion results

---

[5] In particular, there is no need to invoke field-free 'gaps' (Spruit and Scharmer 2006; Scharmer and Spruit 2006) beneath the inner penumbra in order to explain their presence. It should be noted also that the three-dimensional magnetic configuration in Fig. 4 of Spruit and Scharmer (2006), where the inclination of the field in meridional planes is greater above bright filaments than above dark filaments, directly contradicts the observations.

from the increase in the angle of inclination of the magnetic field in bright filaments, as in some theoretical models. The darker regions have a somewhat mottled appearance; the Evershed outflow appears in both bright and dark patches and may even link one with the other at different radii (Schlichenmaier, Bellot Rubio and Tritschler 2005). There are also prevalent 'dark clouds' that move outwards into the surrounding granulation; they appear also in the Doppler signal and may extend over several filaments. All this suggests that there is an irregular, confused pattern of overturning convection in the almost horizontal fields of the outer penumbra.

The downward extent of these highly inclined fields in the outer penumbra cannot be directly observed, though the field geometry itself implies a depth of not less than several hundred kilometres (cf. Borrero *et al.* 2006). Moreover, since these fields lie in regions that are relatively dark, they must extend sufficiently deep for convective energy transport to be seriously inhibited: that in turn indicates a depth of at least 1000 km. Some rough estimates suggest depths of up to 5 Mm (Weiss *et al.* 2004; Brummell *et al.* 2008). If so, then a significant fraction of the spot's magnetic flux extends outward below the photosphere and into the surrounding moat region.

A further issue is the extent to which convection in these darker regions is related to that in the external field-free plasma. It is clear from observations (see, for example, Fig. 5.1) that the penumbra has a very ragged outer boundary and that shallow dark features overlie bright granules and eventually disappear into the cracks between them. Conversely, it seems highly probable that external convection will enter into the outward-flaring flux tube below the outer part of the visible penumbra, providing an inward flux of thermal energy, as envisaged by Jahn and Schmidt (1994). While it is not clear how far such field-free tongues can penetrate, it seems unlikely that they will extend farther than the outermost 10–20% of the spot radius, and extremely unlikely that they will reach the inner penumbra. In any case, they must be overlain by a horizontal field that is thick enough to impede heat transport upwards.

5.3.5 *Thin flux tubes in the penumbra*

An alternative approach to modelling penumbral convection is to consider the motion of thin flux tubes within an unstably stratified background atmosphere (Schlichenmaier, Jahn and Schmidt 1998a,b; Schlichenmaier 2002). It is envisaged that an individual thin flux tube within a sunspot is initially located at the outer edge of the spot configuration, in thermal contact with the external field-free plasma, and then gradually migrates inwards, thereby transporting energy from outside into the penumbra. Although this is a helpful representation of the convection process, it must be borne in mind that individual flux tubes will not be able to maintain their identities in a turbulent background flow. An interesting feature of this approach is the resulting behaviour of the free upper end of the flux tube. Initially it rises above the photosphere, but subsequently it falls down towards the surface, squirting an outward flow along its length.[6] We shall return to this outflow in the next section.

5.4 The Evershed flow

The horizontal Evershed outflow at photospheric heights in the penumbra is an inherent feature of sunspots. The existence of this flow is inferred from the Evershed effect,

---

[6] As previously suggested by Wentzel (1992).

which consists of a wavelength shift and an asymmetry of spectral lines formed in penumbra. The effect is seen in essentially all fully developed sunspots and appears immediately after the penumbra first forms. Beginning with Evershed's (1909a) first report, the cause of the Evershed effect has been generally assumed to be a radial, nearly horizontal outflow across the penumbra. (Alternative interpretations invoking small-scale, unresolved wave motions have been proposed but are now considered untenable; see Thomas 1994 for a discussion.)

The *normal* Evershed effect, in weak lines formed at photospheric heights, is consistent with a radial outflow of gas, while the *reverse* Evershed effect, seen in strong lines formed at chromospheric heights, is consistent with a radial inflow. Early observations at moderate spatial resolution show a rather smooth radial flow field with flow speed decreasing with height (i.e. with increasing line strength) and reversing direction in the low chromosphere (St. John 1913; Kinman 1952). This reverse flow follows field lines that emanate from the umbra.

At a fixed height in the penumbral photosphere, the speed of the normal Evershed flow increases outward across the penumbra, reaching a maximum in the outer penumbra before disappearing rather abruptly at or near the outer penumbral boundary (e.g. Brekke and Maltby 1963; Maltby 1964; Beckers 1969a; Wiehr *et al.* 1986; Wiehr and Degenhardt 1992; Wiehr 1996), although there is evidence that a small fraction of the flow continues outward along the elevated magnetic canopy (Solanki, Montavon and Livingston 1994; Rezaei *et al.* 2006). At moderate resolution, a typical peak outflow speed in the photosphere is 1–$2 \, \mathrm{km \, s^{-1}}$.

5.4.1   *Fine-scale organization of the Evershed flow*

High-resolution observations have revealed that the Evershed flow is structured on fine scales and is episodic. Beckers (1968) and Beckers and Schröter (1969) first established that the flow is mostly concentrated in the dark penumbral filaments, and subsequent observations have generally confirmed this (e.g. Title *et al.* 1993; Rimmele 1995a; Stanchfield, Thomas and Lites 1997), although the most recent observations show that the flow often originates within a bright feature in the inner penumbra but continues along a dark feature the rest of the way outward (Rimmele 2004; Rimmele and Marino 2006; Ichimoto *et al.* 2007b). There is a strong spatial correlation between the Evershed flow and the most horizontal magnetic fields in the penumbra. Stokes polarimetry reveals that the flow is confined to thin, loop-like channels elevated above the surface (Rimmele 1995a,b), and that many of these flow channels (and their associated magnetic fields) actually arch back downward and dive below the surface somewhere in the outermost penumbra or just outside the spot (Börner and Kneer 1992; Rimmele 1995b; Stanchfield, Thomas and Lites 1997; Westendorp Plaza *et al.* 1997; Schlichenmaier and Schmidt 1999, 2000; del Toro Iniesta, Bellot Rubio and Collados 2001; Bellot Rubio *et al.* 2003). Detailed inversions of Stokes profiles (Bellot Rubio *et al.* 2003; Bellot Rubio, Balthasar and Collados 2004) show the flow and magnetic field to be very well aligned everywhere across the penumbra, as one would of course expect on the basis of magnetohydrodynamic theory.

The Evershed flows along individual flux tubes are time dependent (Shine *et al.* 1994; Rimmele 1994; Rouppe van der Voort 2003): the flow waxes and wanes along these channels with a time scale of 10 to 20 minutes. The flow often apparently repeats along the same channel, and the episodes of flow in different channels seem to be uncorrelated (Rimmele 1994);

outward-moving coherent 'clouds' of Evershed flow, extending over several penumbral filaments, do occur (Shine *et al.* 1994; Cabrera Solana *et al.* 2007), but these may be associated with a large-scale wave motion superimposed on the flow.

5.4.2   Theoretical models of the Evershed flow

The observations discussed above generally support the idea that the Evershed flow consists of many individual flows along arched magnetic flux tubes in the penumbra.[7] These individual flows must be driven by pressure gradients along the flux tubes. Meyer and Schmidt (1968) first proposed that the Evershed flow (both normal and reverse) consists of 'siphon flows' along individual arched magnetic flux tubes, driven by a pressure difference between the two footpoints of each tube. For footpoints on the same gravitational equipotential surface, the total pressure (gas plus magnetic) will be the same, but the gas pressure will be lower at a footpoint where the magnetic pressure is higher. Thus, they suggested that for an arched flux tube with a footpoint in the umbra, where the magnetic field strength is high, the flow is likely to be inward, whereas for an arched tube originating in the penumbra, where the field strength is much lower, the flow is likely to be outward, thus explaining the normal and reverse Evershed flow. This picture was clarified by the later discovery that most of the photospheric magnetic flux outside of sunspots is concentrated into small, intense elements with field strengths of 1200 to 1500 G, intermediate between the typical values for an umbra and a penumbra; hence, if the outer footpoints of the arched flux tubes are such elements, the normal and reversed flows follow naturally (Spruit 1981c).

The model of Meyer and Schmidt (1968) was based on the limit of small plasma beta, in which an individual magnetic flux tube (embedded in the space-filling field) is effectively rigid, with its geometry unaffected by a flow within it. This approximation is valid in the chromosphere and corona but not in the photosphere and below, where the magnetic pressure is comparable to the gas pressure and the siphon flow will affect the equilibrium path of the tube and its cross-sectional area (Thomas 1984b). The siphon-flow model has been reformulated to include these 'flexible tube' effects (Thomas 1988; Montesinos and Thomas 1989, 1993, 1997; Degenhardt 1989, 1991; Thomas and Montesinos 1990, 1991, 1993). In this case, the critical speed for the flow is not the sound speed c_s but rather the 'tube speed' $c_t = [c_s^2 v_A^2/(c_s^2 + v_A^2)]^{1/2}$, where v_A is the Alfvén speed. (Note that c_t is always less than both the sound speed and the Alfvén speed.) Figure 5.10 shows two examples of computed steady siphon flows for penumbral flux tubes (Montesinos and Thomas 1997). One flow is subcritical everywhere. The other flow passes through the critical speed near the top of the arch and continues to accelerate downstream (up to a speed of about 8 km s^{-1}) until it is slowed suddenly to subcritical speed at a standing 'tube shock' in the downstream leg of the arch. This supercritical flow corresponds well with the supersonic Evershed downflows that are regularly observed in the outer penumbra (e.g. del Toro Iniesta, Bellot Rubio and Collados 2001). So far, all of the siphon-flow models have assumed a steady state and hence do not explain the episodic nature of the Evershed flow, but in principle time-dependent siphon flows could also be computed.

A variant of the siphon-flow model, also based on the thin-flux-tube approximation, is the 'moving-tube' model of Schlichenmaier, Jahn and Schmidt (1998a,b). In this model, the flux

---

[7] On the other hand, the detailed structure of the Evershed flow is not consistent with a simple convectively driven outward flow (Galloway 1975; Busse 1987).

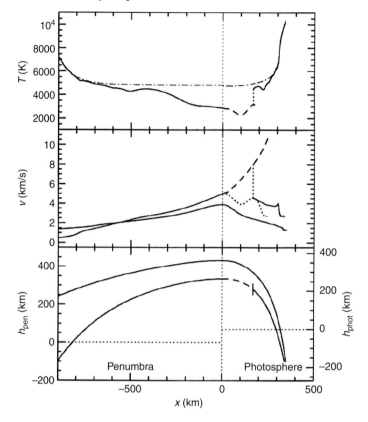

Fig. 5.10. Examples of steady Evershed siphon flows along penumbral flux tubes. The bottom two panels show the flow velocity and the equilibrium path of the flux tube for a purely subcritical flow (solid line) and for a critical flow with supercritical flow (dashed line) and a standing tube shock (vertical dotted line) in the descending part of the arch. The upper panel shows the temperature within the flux tube for the critical flow and the temperature of the external atmosphere (dot-dash line). (From Montesinos and Thomas 1997.)

tube lies initially along the outer edge of the sunspot's overall magnetic field and extends radially outward along the magnetic canopy, where at some point the tube is truncated and an open boundary condition is applied. The flux tube is heated in its lower parts where it is in contact with the hotter surroundings: this creates pressure and buoyancy forces that drive an outward flow along the tube and cause the footpoint of the tube (where it crosses the visible surface) to move inward (see Fig. 5.11). The flow may be associated with the Evershed flow, and the inward-moving footpoint may be associated with a bright penumbral grain. This model has the advantage of being time-dependent and hence capable, in principle, of explaining the episodic nature of the Evershed flow. On the other hand, it has the distinct disadvantage of not having an arched form with a downstream footpoint where the flow dives back below the surface, as observed for much of the Evershed flow.[8] All of the

---

[8] High-speed, super-Alfvénic flows in the moving-tube model produce a serpentine configuration of the flux tube (Schlichenmaier 2002) which does dive back below the surface, but this configuration has been shown to be gravitationally unstable and hence will not occur (Thomas 2005).

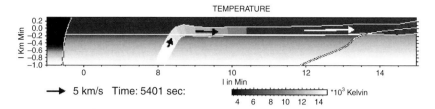

Fig. 5.11. Snapshot of the evolution of a penumbral flux tube in the moving-tube model for the Evershed flow, showing temperature according to a grey-scale and flow velocity as scaled arrows. (From Schlichenmaier, Jahn and Schmidt 1998b.)

Evershed flow in this model continues radially outward along the magnetic canopy outside the sunspot, whereas observations tell us that only a small fraction of the flow does so. Also, the outward flow can attain unrealistically high speeds because the open outer boundary condition provides no impediment (adverse pressure gradient) to the flow.

One can imagine an improved thin-flux-tube model of the Evershed flow that combines the best features of the siphon-flow model (arched, returning flux tube; supersonic downflows decelerated at a tube shock) and the moving-tube model (time dependence and moving footpoints; heating of the flux tube at its base) into a very satisfactory model for the Evershed flow. In such a model, the pressure gradient that drives the flow could arise through a combination of heating (increasing the gas pressure) at the upstream footpoint and magnetic flux concentration (reducing the gas pressure) at the downstream footpoint.

5.5 Moving magnetic features in the moat

Intimately connected with the penumbra and the sunspot moat are the so-called *moving magnetic features (MMFs)*, small magnetic elements that move radially outward across the moat at speeds ranging from a few tenths to $3 \, \mathrm{km \, s^{-1}}$. These features were first detected by Sheeley (1969) as moving bright points in CN spectroheliograms. The bright points travel outward from the spot until they either disappear or reach the surrounding photospheric network and merge with it. Vrabec (1971, 1974) confirmed the magnetic nature of these features in sequences of Zeeman spectroheliograms and found what is, from a theoretical point of view, their most remarkable property: they come in both magnetic polarities around a single sunspot, with only a slight preference for the polarity of the spot. More extensive observations of these features were made by Harvey and Harvey (1973), who named them MMFs. The MMFs tend to spread and weaken as they move outward across the moat, and strong downdrafts have been observed in some MMFs (e.g. Nye, Thomas and Cram 1984). Individual MMFs move radially outward at nearly constant speed, but nearby MMFs may have quite different speeds (Brickhouse and LaBonte 1988).

Shine and Title (2001) have provided a useful classification of the MMFs into three types, according to the arrangement of their magnetic polarity. Type I MMFs are bipolar pairs of magnetic elements that move outward together at speeds of 0.5 to $1 \, \mathrm{km \, s^{-1}}$. Type II MMFs are single magnetic elements of the same polarity as the sunspot that move outward at speeds of 0.5 to $1 \, \mathrm{km \, s^{-1}}$. The latter features have generally been interpreted as flux tubes that have separated from the sunspot flux bundle and are being carried away by the moat flow: as such, they provide the primary mechanism for the decay of a sunspot. Type III MMFs are also single magnetic elements, but with polarity opposite to that of the sunspot. These features move outward more rapidly than the Type I or II MMFs, at speeds of 2 to 3 km s^{-1}.

The Type I bipolar pairs usually first appear just outside the penumbra and then move along a radial line extending outward from a dark penumbral filament. Often several bipolar pairs are seen to form and move outward sequentially along the same radial path. In the usual polarity arrangement, the element of the pair nearest the penumbra has the same magnetic polarity as the sunspot itself (Harvey and Harvey 1973). Sometimes pairs of magnetic elements with the opposite polarity arrangement are seen (Yurchyshyn, Wang and Goode 2001; Zhang, Solanki and Wang 2003), but some of these may be associated with newly emerging magnetic flux and hence not connected to the sunspot, as Type I MMFs are (V. Martínez Pillet, private communication). In general, it is difficult to assign MMFs to bipolar pairs in an unambiguous way because of the presence of many features and the possibility that features near the spot may be masked by the flux within the spot. Kubo, Shimizu and Tsuneta (2007) find that an MMF with polarity opposite that of the spot is often masked by ambient magnetic fields in the moat, and that one member of a pair (with polarity the same as the spot's) is often hidden within the outer penumbra itself. The bipolar MMFs are generally not visible at chromospheric levels, suggesting that they correspond to a magnetic loop less than 1500 km high (Nye, Thomas and Cram 1984; Penn and Kuhn 1995).

The bipolar Type I MMFs are associated with radial extensions of the more horizontal components of the penumbral magnetic field outward across the moat (Sainz Dalda and Martínez Pillet 2005; Kubo *et al.* 2007). The MMFs originate just inside the penumbra, cross the penumbral boundary, and move outward across the moat along the path of these filamentary extensions of the penumbral field. There is some evidence that the starting points of bipolar MMFs correspond to Evershed flow channels (Zirin and Wang 1991; Lee 1992), and that the launching of a bipolar MMF into the moat is associated with the arrival of a 'cloud' of more intense Evershed flow at the outer edge of the penumbra (Cabrera Solana *et al.* 2006). The Type II MMFs, single elements with the same polarity as the spot, originate at the outer edge of the radial spines of more vertical penumbral magnetic field, where they appear to be eroded away by granular convection at the edge of the penumbra (Kubo *et al.* 2007).

Hagenaar and Shine (2005) studied the behaviour of MMFs around eight different sunspots using sequences of high-resolution magnetograms from the Michelson Doppler Imager (MDI) on SOHO. They find an average MMF lifetime of 1 hr and average outflow speeds of 1.5–1.8 km s^{-1}, significantly faster than the moat flow speed of about 1 km s^{-1}. They also find that the moat flow itself is not purely radial; it also has a structured azimuthal component with radial lines of convergence and divergence, suggestive of convective rolls. The MMFs follow preferred paths across the moat which correspond to the radial lines of convergence of the moat flow.

Several different theoretical models have been proposed to explain the Type I bipolar MMFs: Figure 5.12 illustrates three of them. Harvey and Harvey (1973) proposed a 'sea-serpent' model in which the bipolar pairs correspond to small loops that form in a submerged horizontal flux tube, which remains attached to the main sunspot flux bundle near the surface. In this configuration, the element of the bipole nearest the spot has the same polarity as the sunspot. Wilson (1973) proposed instead that a flux tube detaches from the spot at some depth and is swept away by a subsurface flow, but remains attached to the spot above the surface, producing a submerged horizontal flux tube in the opposite direction to Harvey and

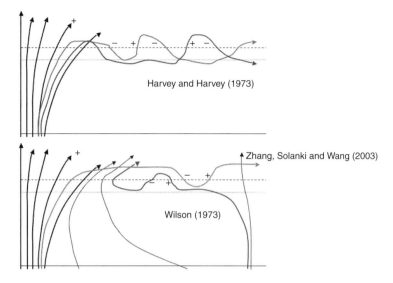

Fig. 5.12. Proposed models for bipolar MMFs: a flux tube detached near the surface (Harvey and Harvey 1973); a flux tube detached deeper down (Wilson 1973); and a depression in an elevated flux tube (Zhang, Solanki and Wang 2003).

Harvey's sea serpent. In this configuration the element nearest the spot has polarity opposite that of the spot. In a variant of this model, proposed by Spruit, Title and van Ballegooijen (1987), a submerged 'U-loop' rises to just below the surface where granular convection brings up smaller stitches of field and then reconnection leads to the formation of small loops. In this configuration the element nearest the spot has the same polarity as the spot. In a fourth alternative, Zhang, Solanki and Wang (2003) proposed an inverted sea serpent in which the bipolar pair corresponds to a depressed loop in an elevated flux tube in the canopy, the depression being caused by mass loading due to a flow. Here the element nearest the spot will have polarity opposite that of the spot. (The fact that the bipolar MMFs are not seen in the chromosphere argues against this picture.)

The sea-serpent models in Figure 5.12 raise an important question: what would keep the flux tube submerged in spite of its inherent magnetic buoyancy? The answer seems to be magnetic flux pumping by the compressible, turbulent granular convection (Thomas *et al.* 2002; Weiss *et al.* 2004). We discuss this process in detail in the next section.

5.6 Formation and maintenance of the penumbra

To a theoretician, the most puzzling features of the filamentary penumbra are its interlocking-comb magnetic structure and the presence of flux tubes that arch downwards and actually plunge below the solar surface in the outer penumbra. One might naively predict that an equilibrium configuration composed of adjacent magnetic sheets with substantially different inclinations should be unstable; the field reversals are even more surprising because one would expect flux tubes to rise and straighten out, owing to the combined effects of magnetic buoyancy and magnetic tension along the field lines. Apparently some force is tugging these fields down below the surface of the photosphere, and this process offers a key to understanding the complex magnetic structure of the penumbra.

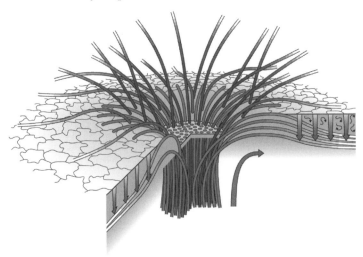

Fig. 5.13. Sketch of the interlocking-comb structure of the magnetic field in the penumbra. The flux tubes illustrate the orientations of the magnetic field in adjacent flux sheets. In bright filaments the field is less inclined to the vertical but the fields in darker filaments are much more steeply inclined. Some flux tubes emerge and hug the surface in a slightly elevated canopy but others dive below the surface either just inside or just outside the spot. Small-scale granular convection acts to pump this flux downwards, as indicated by vertical arrows, while the larger-scale moat flow carries magnetic features outwards. (From Weiss *et al.* 2004.)

Only one such process has so far been proposed: we believe that these magnetic fields are pumped downward by turbulent convection in the photospheric granulation layer that surrounds the sunspot, and stored in a less vigorously convecting layer that lies below (Thomas *et al.* 2002; Weiss *et al.* 2004; Thomas and Weiss 2004). This leads to the overall picture that is presented schematically in Figure 5.13.

Here we first outline the process of magnetic flux pumping and illustrate it with some idealized model calculations; then we explain how it may operate so as to maintain the interlocking sheets of magnetic field in a sunspot. Finally, we go on to consider the role of flux pumping in the formation of the sunspot penumbra.

5.6.1   *Magnetic flux pumping and returning flux tubes*

The origins of the concept of magnetic flux pumping trace back to earlier ideas of turbulent diamagnetism, magnetic flux expulsion, and topological pumping. In a turbulent flow field, magnetic flux tends to be transported from regions of stronger turbulence to regions of weaker turbulence, so that flux is pumped down a gradient in turbulent intensity ('turbulent diamagnetism'; Zeldovich 1956; Zeldovich, Ruzmaikin and Sokoloff 1983). Magnetic flux is expelled from the interiors of convective eddies through the combined actions of advection and diffusion, and concentrated between them, as was shown in simple kinematic calculations (Parker 1963; Clark 1965, 1966; Weiss 1966; Clark and Johnson 1967; see Appendix 2). In incompressible convection magnetic flux is transported preferentially in the direction of the flow within the connected downflow channels that surround the isolated upflow channels, an effect called topological pumping (Drobyshevski and

Yuferev 1974). However, in highly compressible, turbulent convection the up–down symmetry of Boussinesq convection is broken and the dominant effect that leads to downward flux pumping is the asymmetry of the flow pattern, which consists of strong, concentrated downdrafts and weaker, broad updrafts (e.g. Weiss *et al.* 2004). This pumping has been shown to be important in the lower part of the solar convection zone, where it can transport magnetic flux into the underlying stable layers (Tobias *et al.* 2001; Dorch and Nordlund 2001) as part of the solar dynamo process (see Section 11.1.2 below).

Idealized model calculations

The flux pumping process, in its simplest form, is best demonstrated by considering the effect of a vigorously convecting layer on an initially horizontal magnetic field. The configuration studied by Weiss *et al.* (2004) has two stratified layers, one above the other; the upper layer is strongly unstable, while the lower one is only very weakly unstable. (In the absence of convection, the polytropic indices m_1, m_2 of the two layers are related by a stability parameter $S = (m_2 - 3/2)/(3/2 - m_1)$, with $S = -0.01$.) The initial unidirectional field, introduced after the convection has reached a statistically steady state, is contained in a thin slab near the upper boundary, and Figure 5.14 illustrates its subsequent evolution. There is always a tangled field in the upper layer (see Fig. 3 of Thomas *et al.* 2002 or Fig. 12 of

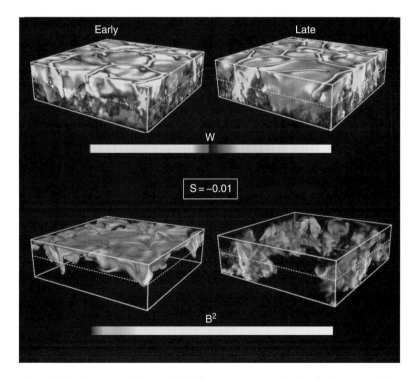

Fig. 5.14. Volume renderings of the instantaneous vertical velocity w and the magnetic energy density B^2 (left) near the beginning and (right) near the end of a run with the stiffness parameter $S = -0.01$. The velocity pattern shows broad upwellings enclosed by narrow sinking sheets and plumes. Magnetic energy is initially concentrated in the upper layer but later the surviving flux is concentrated in the lower region. (After Weiss *et al.* 2004.)

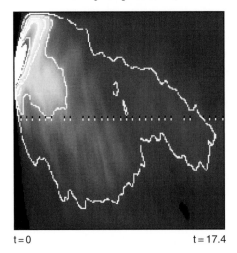

t = 0   t = 17.4

Fig. 5.15. Space-time diagram showing the redistribution of the horizontally averaged field for the calculation in Figure 5.14. Light shading indicates strong fields. Much of the flux escapes upwards but the remainder is pumped down into the lower layer. (From Weiss *et al.* 2004.)

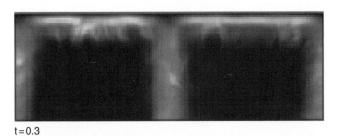

t = 0.3

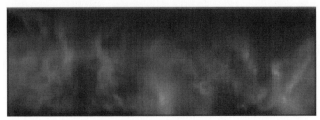

t = 4.4

Fig. 5.16. Flux pumping with an arched magnetic field. The initial field lies in the yz-plane. The grey-scale image in the upper panel shows the x-averaged distribution of $|\mathbf{B}|^2$ near the beginning of the calculation. The lower panel shows $|\mathbf{B}|^2$ later in the calculation, when the more horizontal fields have been pumped downwards, towards the bottom of the strongly unstable upper portion of the layer. (From Brummell *et al.* 2008.)

Thomas and Weiss 2004), but the horizontally averaged field is pumped downwards into the weakly stratified layer, where the field is more uniform. Since magnetic flux can escape from the top and bottom of the layer, this is a run-down calculation, as shown in Figure 5.15. At first, flux is expelled upwards and ejected; later, the remaining flux is pumped downwards into the weakly unstable layer, and gradually diffuses out.

While Figures 5.14 and 5.15 illustrate how magnetic flux can be held down below a layer of turbulent granular convection, they do not relate closely to the geometry of a sunspot. Figure 5.16 shows similar results for a slightly more realistic model, with an arched magnetic field that is pumped downwards in spite of both buoyancy and curvature forces. These various model calculations confirm that the pumping process is indeed robust.

Flux pumping at the solar photosphere

The calculations discussed above provide support for the sunspot model sketched in Figure 5.13, in which strongly inclined flux tubes near the edge of the penumbra are captured by turbulent granules and mesogranules and pulled down below the solar photosphere. The convection becomes less vigorous with increasing depth, and so a balance between pumping and magnetic buoyancy can be achieved. The latter may be aided by the larger-scale upflow associated with the radial outflow in the moat cell that surrounds a sunspot. As in the idealized models, not all the magnetic flux is pumped downwards; some is expelled upwards and accumulates in the magnetic canopy, which is supported by a layer of convective overshoot extending for several hundred km above the level where $\tau_{500} = 1$ (e.g. Rutten, de Wijn and Sütterlin 2004; Puschmann *et al.* 2005; Cheung, Schüssler and Moreno-Insertis 2007). The rest, however, is submerged and may only surface again at the periphery of the moat, which is often surrounded by plage with magnetic fields of the same sign as that of the spot itself.

Observations of Evershed flows in a δ-sunspot (i.e. one with a pair of umbrae with oppositely directed fields, and a shared penumbra) offer some additional evidence for downward pumping. Lites *et al.* (2002) found an example with radial outflows from each umbra that converged and plunged downwards below the surface, implying that the field lines to which they were attached were pulled downwards below the photosphere, instead of running horizontally from one umbra to the other. Further corroboration comes from the behaviour of moving magnetic features in the moat. If flux tubes emerging from the spot are pumped down by granular convection, and stored in a subsurface layer subject to the supergranular-scale moat flow, then Type I MMFs are easily explained as stitches of the submerged field that are brought up to the surface, either by an upwelling granular plume or through magnetic buoyancy. In that case, the inner element will have the same polarity as the spot, as in the sea-serpent model of Harvey and Harvey (1973). Type II MMFs are straightforward escapees. Type III MMFs, on the other hand, are submerged but rising upward and therefore travelling outwards. Since their flux tubes make an acute angle with the horizontal at the photosphere, the speed at which they travel is significantly enhanced. These different configurations are illustrated in Figure 5.17.

5.6.2 Formation of the penumbra

A sunspot forms through the coalescence of pores and smaller magnetic flux tubes into a single, growing pore (see Section 7.3 below). When the pore has grown to sufficient size (with a diameter of about 3.5 Mm) or, more likely, to sufficient total magnetic flux (about 1×10^{20} Mx), it forms a penumbra and becomes a fully fledged sunspot. The

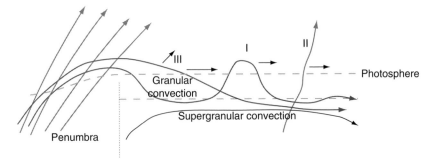

Fig. 5.17. Sketch showing how moving magnetic features of Types I, II and III can be related to flux pumping by granular convection at the photosphere and to the underlying moat flow. (From Weiss *et al.* 2004.)

penumbra often forms in sectors, usually beginning on the side of the umbra away from the opposite-polarity magnetic flux of the active region, but sometimes, because of the proximity of another sunspot of like polarity, beginning on the side toward the magnetic neutral line. The formation of a penumbral sector is a sudden event, generally occurring in less than 20 minutes, and both the interlocking-comb configuration of the magnetic field and the Evershed flow pattern of a mature sunspot are established within this same short time (Leka and Skumanich 1998; Yang *et al.* 2003). The formation of the penumbra is not associated with any abrupt increase in total magnetic flux (Zwaan 1992).

The sudden transition from pore to sunspot strongly suggests that the penumbra forms as a consequence of the onset of a fluting instability of the magnetic field configuration in the growing pore. Simple equilibrium models indicate that the inclination of the magnetic field at the edge of a pore increases as the total magnetic flux increases and suggest that the configuration becomes unstable when this inclination reaches a critical value, at which point the pore develops a penumbra and becomes a sunspot (Simon and Weiss 1970; Rucklidge, Schmidt and Weiss 1995; Hurlburt and Rucklidge 2000). Observations indicate that the critical inclination angle (to the local vertical) is about $35°$, which interestingly is the same as the inclination of the mean magnetic field at the umbra–penumbra boundary in a fully formed sunspot (Martínez Pillet 1997). There is apparently some hysteresis associated with the transition from a pore to a sunspot, because observations show that the largest pores are bigger than the smallest sunspots (Bray and Loughhead 1964; Rucklidge, Schmidt and Weiss 1995; Skumanich 1999).

We have suggested the following scenario[9] for the formation and maintenance of a sunspot penumbra (Thomas *et al.* 2002; Weiss *et al.* 2004; Thomas and Weiss 2004). The magnetic flux concentration in a growing pore eventually becomes convectively unstable to filamentary perturbations because of the increasing inclination of the field near the outer edge of the pore. The nonlinear development of this instability results in a moderate fluting of the

---

[9] Rosner (2000) has given a useful definition of the term 'scenario' in this context, as "a verbal description of a sequence of physical processes; typically, limited aspects of this sequence, as well as limited numbers of physical processes in this sequence, may be mathematically or computationally approachable calculations, but the entire sequence is usually completely inaccessible to realistic computations."

outer edge of the pore and the formation of a rudimentary penumbra, as observed in proto-spots (Leka and Skumanich 1998). The transition to a fully developed penumbra then occurs when the depressed, more nearly horizontal spokes of the mildly fluted magnetic field are grabbed by the sinking plumes in the surrounding layer of granular convection and dragged downward by magnetic flux pumping. Some of the magnetic field in the dark filaments is kept submerged in the moat by granular flux pumping as the sunspot evolves and, later, when it decays, thus maintaining the penumbra even when the total magnetic flux in the sunspot becomes somewhat less than that in the pore when the penumbra first formed. In this way, flux pumping provides the physical mechanism for the subcritical bifurcation proposed by Rucklidge, Schmidt and Weiss (1995; see also Tildesley and Weiss 2004) as the explanation for hysteresis in the pore–sunspot transition.

The convective filamentary instability that produces an embryonic penumbra in the above scenario has been investigated in the context of idealized models, but as yet these calculations have been inconclusive. What is needed is to set up a nonlinear equilibrium con-figuration with a magnetic field that is enclosed by moat-like convective motion and spreads out towards the upper surface, and then to probe its stability to filamentary modes. Tildesley (2003) considered a two-dimensional Boussinesq model in Cartesian geometry and demon-strated the presence of unstable three-dimensional modes whose growth rates are enhanced by the magnetic field. These modes eventually saturate to give a broad spoke-like pattern (Tildesley and Weiss 2004). The stability of an axisymmetric configuration in a compress-ible layer, with a central flux rope at its core (Hurlburt and Rucklidge 2000; Botha, Rucklidge and Hurlburt 2006) has been studied by Botha, Rucklidge and Hurlburt (2007): the non-axisymmetric instabilities that they find turn out, however, to be driven by external con-vection rather than by the magnetic field. In the nonlinear domain, they develop into a strik-ing spoke-like pattern (Hurlburt, Matthews and Rucklidge 2000; Hurlburt and Alexander 2002). In due course, it should become possible to extend these preliminary studies so as to develop a more realistic pore model and to confirm the onset and nonlinear development of a non-axisymmetric fluting instability of the magnetic field that is driven by convection.

6

Oscillations in sunspots

Various kinds of wave motions have been observed in sunspots. These include characteristic umbral oscillations with periods around 3 minutes, umbral oscillations with periods around 5 minutes (which differ in several respects from the 5-minute p-mode oscillations in the quiet photosphere), and large-scale propagating waves in the penumbra. These oscillatory phenomena are of considerable interest because they are the most readily observable examples of magnetohydrodynamic waves under astrophysical conditions. In addition, observations of oscillations in a sunspot and its nearby surroundings can be used to probe the structure of a sunspot below the solar surface ('sunspot seismology').

Interest in sunspot oscillations began in 1969 with the discovery of periodic umbral flashes in the Ca II H and K lines by Beckers and Tallant (1969). These flashes were soon attributed by Havnes (1970) to the compressive effects of magneto-acoustic waves. In 1972 three other types of sunspot oscillations were discovered: running penumbral waves in Hα (Giovanelli 1972; Zirin and Stein 1972); 3-minute velocity oscillations in the umbral photosphere and chromosphere (closely connected to the umbral flashes: Giovanelli 1972; Bhatnagar and Tanaka 1972); and 5-minute velocity oscillations in the umbral photosphere (Bhatnagar, Livingston and Harvey 1972). For some time these three types of oscillations were considered as distinct phenomena, but recent work suggests that they might actually be different manifestations of the same coherent oscillations of the entire sunspot (Bogdan 2000). Here we shall follow the historical development of the subject by discussing the three types of oscillations separately before attempting to present a unified picture. Progress in the subject can be traced through several useful review articles (Moore 1981; Thomas 1981, 1985; Moore and Rabin 1985; Lites 1992; Chitre 1992; Staude 1994, 1999; Bogdan 2000; Bogdan and Judge 2006).

6.1 Magneto-atmospheric waves

Before discussing observations and theoretical models of oscillatory phenomena in sunspots, it is helpful to consider briefly the general theoretical framework in which they can be understood. Oscillations in sunspots are manifestations of *magneto-atmospheric waves* (or *magneto-acoustic-gravity waves*), which occur in a compressible, gravitationally stratified, electrically conducting atmosphere permeated by a magnetic field (Thomas 1983). These waves are supported by some combination of three different restoring forces: a pressure force due to compression or expansion, a buoyancy force due to stratification under gravity, and a Lorentz force due to distortion of the magnetic field. The relative contributions of these restoring forces are different for different wave modes and vary with position for a given wave mode. These waves can be considered as magneto-acoustic waves modified by stratification and buoyancy, or as acoustic-gravity waves modified by the magnetic field. Most studies of magneto-atmospheric waves, beginning with the work of Ferraro and Plumpton (1958), have been limited to small-amplitude, linearized waves in a non-dissipative atmosphere. Even with these idealizations, the analysis is complicated because the atmosphere is both inhomogeneous and anisotropic, with gravity and the magnetic field each imposing a preferred direction.

Magneto-atmospheric waves occur over a wide range of length scales in the solar atmosphere and no doubt occur in the atmospheres of most magnetic stars. A sunspot, with its relatively well ordered magnetic field and reduced turbulence, makes an ideal 'laboratory' for studying these waves in detail. A broad spectrum of such waves is excited within a sunspot, either by interactions with the acoustic oscillations in the surroundings or by subsurface convective motions within the spot itself. Because of the strong magnetic field in a sunspot, the waves may best be thought of as magneto-acoustic waves modified by stratification and buoyancy. As with pure magneto-acoustic waves, there are fast and slow compressional modes and also a pure Alfvén mode with incompressible motions. (See Appendix 2 for a discussion of pure magneto-acoustic waves.) However, because of the temperature and density stratification of the sunspot atmosphere and the spreading magnetic field geometry, the sound speed c and Alfvén speed v_A vary with height, with (roughly speaking) $c > v_A$ below the umbral surface and $c < v_A$ above the umbral surface. Thus, the characterization of a particular wave mode as slow or fast has only local meaning at a particular height, and a wave of fixed frequency and wavenumber can change its character from slow to fast (or vice versa) as it propagates vertically over a range of heights. For example, a nearly longitudinal, acoustic-like wave propagating vertically upward in the umbra would be classified as a fast mode while it is below the surface but as a slow mode as it propagates through the umbral photosphere and chromosphere. The reader should keep in mind that the distinction between fast and slow modes is potentially confusing, and probably less useful, for waves in a highly stratified structure like a sunspot.

6.2   Umbral oscillations

The general characteristics of umbral oscillations are illustrated in Figure 6.1, which shows space-time plots of Doppler velocity along a spectrograph slit crossing a sunspot, in frequency bands centred on periods of 5 minutes and 3 minutes. Fairly regular oscillations are evident even in the simple, filtered velocity signals presented here, and the 3-minute oscillations, which are strongest in the chromosphere, nevertheless show up clearly in this photospheric signal. The 5-minute oscillations are generally coherent over a large fraction of the umbral area, while the 3-minute oscillations are more localized and more intermittent.

Typical temporal power spectra of umbral velocity oscillations in the photosphere and chromosphere are shown in Figure 6.2. The photospheric oscillations have a broad hump of power in the frequency range $2\,\text{mHz} \leq \nu \leq 5\,\text{mHz}$, peaking at about $3\,\text{mHz}$ (period near 5 minutes). The chromospheric oscillations show an even broader distribution of power, from 2 to $8\,\text{mHz}$, but with most of their power concentrated in the band $4.5\,\text{mHz} \leq \nu \leq 7\,\text{mHz}$, centred on a period of about 3 minutes. (Note the different scales for power in the two

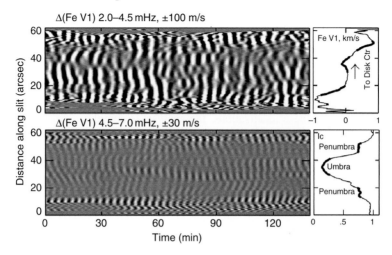

Fig. 6.1. Grey-scale plots of the space-time behaviour of fluctuations in Doppler velocity in Fe I 630.15 nm along the spectrograph slit. In the upper and lower panels the time series has been filtered to pass only oscillations in the 5-minute band ($2\,\mathrm{mHz} \leq \nu \leq 4.5\,\mathrm{mHz}$) and the 3-minute band ($4.5\,\mathrm{mHz} \leq \nu \leq 7\,\mathrm{mHz}$), respectively. The smaller plots to the right show mean values of the Doppler velocity and continuum intensity for the entire run, with thicker segments of the curves indicating the location of the umbra and penumbra along the slit. Note the difference in scaling of the velocity amplitudes in the two panels. (From Lites *et al.* 1998.)

power spectra in Fig. 6.2; the velocities are typically much higher in the chromosphere where the density is much lower.) Within these broad bands of power, the photospheric and chromospheric power spectra typically show several individual peaks of power whose significance will be discussed below.

In the rest of this section we discuss the 5-minute and 3-minute umbral velocity oscillations separately, then consider the magnetic field variations associated with these oscillations, and finally discuss theoretical models for the oscillations.

6.2.1   *Five-minute umbral oscillations*

Because a sunspot's magnetic flux bundle floats in the surrounding convection zone, it is to be expected that there will be oscillations in the umbral photosphere with periods around 5 minutes, in response to buffeting of the sunspot by the *p*-mode oscillations in the quiet Sun (Thomas 1981). Indeed, 5-minute oscillations do occur in the umbral photosphere, but with smaller amplitude than those in the quiet photosphere. Early detections of these oscillations were uncertain because of the possibility of contamination of the signal by the strong 5-minute oscillations in the surroundings, either by means of scattered light or through the use of the quiet-photosphere line profile as a wavelength reference. Beckers and Schultz (1972) were the first to detect 5-minute oscillations in sunspots, but they concluded that the oscillations were most likely caused artificially by oscillations in their wavelength reference, which was based on an average line profile at a position outside the sunspot. Subsequent observations employed techniques to overcome the problems associated with scattered light and the wavelength reference and firmly established the existence of 5-minute oscillations in the umbra. Bhatnagar, Livingston and Harvey (1972) reduced the scattered

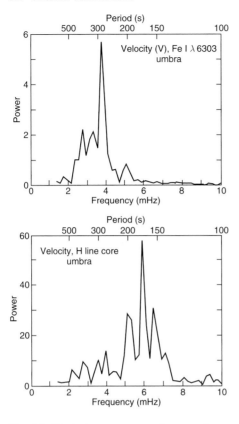

Fig. 6.2. Typical space-averaged temporal power spectra of umbral oscillations in Doppler velocity in the photosphere (in Fe I 630.3 nm) and in the chromosphere (in the Ca II H line core). The photospheric oscillations were measured using the Doppler shift of the Stokes *V* profile of the Fe line in order to reduce the influence of stray light. (From Thomas, Cram and Nye 1984.)

light by using molecular spectral lines that are present in the cool umbra but nearly absent in the hotter photosphere. Rice and Gaizauskas (1973) established a more stable wavelength reference by averaging line profiles over a very large area of quiet photosphere. Soltau, Schröter and Wöhl (1976) and Livingston and Mahaffey (1981) used umbral molecular lines referenced to nearby non-solar telluric lines. Thomas, Cram and Nye (1982, 1984) also used telluric lines as a wavelength reference and measured umbral velocities as Doppler shifts of the Stokes *V* profile of a magnetically sensitive line (Fe I 630.3 nm), thereby largely avoiding contamination by stray light from the surroundings where the magnetic field is weak or absent.

The 5-minute umbral oscillations are nearly coherent over most of the umbra and often extend into the surrounding penumbra, as can be seen in the upper panel of Figure 6.1. Compared to the 5-minute oscillations in the quiet Sun, their power extends over the same range of frequencies (roughly 2.0–4.5 mHz) but their power level is reduced by a factor of two or three. They have rms velocities in the range of 40–100 m s^{-1}. When these oscillations are measured with sufficient time resolution, one can detect rapid radial phase propagation outward from the centre of the umbra extending well into the penumbra, at speeds of

$50-100 \, \text{km s}^{-1}$. This rapid phase propagation produces a characteristic 'herringbone' pattern in space-time plots of velocity (Thomas, Cram and Nye 1984; Lites *et al.* 1998), such as that seen in Figure 6.1.

Because of the limited spatial extent of a sunspot, it is not possible to produce an accurate space-time ($k-\omega$) power spectrum for umbral oscillations alone. However, Abdelatif, Lites and Thomas (1986) presented crude $k-\omega$ diagrams for an umbra and for an equivalent small patch of quiet Sun, which indicated that the 5-minute oscillations in the surroundings are transmitted selectively into the umbra with an accompanying increase in horizontal wavelength. A simple theoretical model suggests that the shift of power to longer wavelengths is due to the faster propagation speed in the magnetized umbra, and that the selective transmission is due to variations of the transmission coefficient along a p-mode ridge because of resonances (Abdelatif and Thomas 1987).

6.2.2   Three-minute umbral oscillations and umbral flashes

The 3-minute umbral oscillations are primarily a chromospheric phenomenon, although they also appear in spectral lines formed in the upper photosphere (such as the Fe I 630.15 nm line). Compared to the 5-minute oscillations, they have greater rms velocities (largely as a consequence of the lower densities at greater heights) and they are coherent over smaller portions of the umbra. The rms velocities of the 3-minute oscillations can be as large as several kilometres per second, and the velocity time series has the sawtooth structure characteristic of nonlinear compressive waves forming shocks as they propagate into regions of lower density.

The nonlinear, highly compressive nature of the 3-minute umbral oscillations causes the related *umbral flashes*, which are sudden brightenings in the core of a chromospheric spectral line (such as the Ca II K line) followed by gradual dimming back to the unperturbed state (Beckers and Tallant 1969; Wittmann 1969). These flashes repeat quite regularly with periods in the range of 140–190 s for different sunspots. The flashes are best seen in a time-resolved spectrum (intensity on a grey scale plotted on a wavelength vs. time diagram), as shown in Figure 6.3 for the Ca II K line. Here we see that an umbral flash consists of a strong, rapid brightening in the blue side of the emission core, which decays rapidly (in about 50 s) leaving a weak, narrow emission peak that shifts uniformly to the red side of line centre in about 100 s. This is followed by a new brightening on the blue side, and the pattern repeats regularly, forming a series of Z-shaped signatures (Thomas, Cram and Nye 1984). This pattern can also be discerned in a time series of individual K-line profiles, as described by Beckers and Tallant (1969) and illustrated by Schultz (1974) and by Kneer, Mattig and von Uexküll (1981). The left panel of Figure 6.3 shows that the umbral flashes are considerably weaker in a light bridge, where instead there is a sporadic pattern of broader, more intense, and more symmetric brightenings of the K-line core, possibly associated with surges observed in light bridges in Hα filtergrams (Roy 1973).

The relation between umbral oscillations and flashes is well illustrated in the space-time plots of Ca II H intensity in Figure 6.4, from Rouppe van der Voort *et al.* (2003).[1] Within the umbra the 3-minute oscillations appear as stacks of nested bowl-shaped figures of enhanced intensity. The bowl shapes indicate that the horizontal propagation speed decreases outward

---

[1] The behaviour illustrated in Figure 6.4 is seen even more clearly in the movies that accompany the online version of this paper.

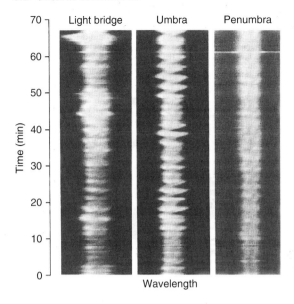

Fig. 6.3. Time-resolved Ca II K-line spectra for three different spatial points in a sunspot: in a prominent light bridge, in the umbra, and in the penumbra. Note the characteristic, repeating, Z-shaped signature of the umbral flashes in the umbral spectrum. The flashes are weaker in the light bridge and essentially absent in the penumbra. The light bridge displays a more sporadic pattern of strong brightenings and broadenings of the K-line emission core. (From Thomas, Cram and Nye 1984.)

from the centre of each bowl, starting at $20\,\mathrm{km\,s^{-1}}$ or more at the bottom of the bowl. Some of the bowls are seen to extend across the penumbra with propagation speeds as low as $5\,\mathrm{km\,s^{-1}}$. The umbral flashes are seen here to correspond to restricted regions of the bowls where the intensity variation is particularly great. In the case of one sunspot observed at high spatial resolution (Nagashima *et al.* 2007), the umbral flashes appear to be suppressed at a 'node' at the centre of the umbra.

Early observational studies of umbral oscillations generally produced temporal power spectra with discrete peaks, such as those in Figure 6.2, and these power peaks were interpreted as representing discrete resonant oscillation modes of the umbra. It gradually became apparent, however, that the oscillation periods of the power peaks differ in no systematic way in different sunspots and even shift significantly in the same sunspot during a long time sequence (Lites 1992). Assuming that resonant modes exist, it is now understood that limited spatial resolution prevents us from separating out individual eigenmodes of oscillation, and that the individual peaks in temporal power spectra actually represent a superposition of several modes. Alternatively, if there are no resonances, any data set of finite temporal extent will produce peaks and troughs superimposed on a continuous temporal power spectrum.

The extension of the 3-minute chromospheric oscillations upward into the transition region has been examined thoroughly in ultraviolet observations by the Oslo group (Brynildsen *et al.* 1999a,b, 2000, 2002; Maltby *et al.* 2001) and by Fludra (1999, 2001). Intensity oscillations with relative amplitudes up to 10% are seen in the O V 629.2 Å line and other transition-region lines formed in the temperature range $1.7\text{–}4.0 \times 10^5\,\mathrm{K}$ (Fludra 1999). These oscillations are most evident in the bright 'sunspot plumes' but are also seen in less

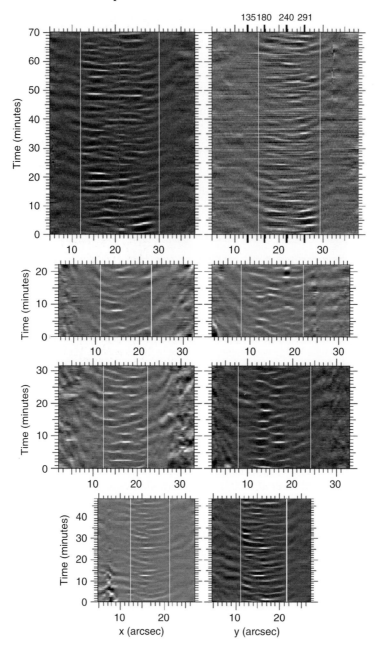

Fig. 6.4. Grey-scale plots of the space-time behaviour of Ca II H brightness variations associated with umbral oscillations and flashes, along image cuts across a sunspot in two perpendicular horizontal directions (left and right panels), for four different data sets (top to bottom). The vertical white lines mark the umbra–penumbra boundary. (From Rouppe van der Voort *et al.* 2003.)

bright areas of the transition region above a sunspot (Brynildsen *et al.* 2000). Phase relations between velocity and intensity in these transition-region oscillations in some cases indicate upward-propagating acoustic waves but in other cases show evidence of partial downward reflection of the waves (Brynildsen *et al.* 2000). Three-minute oscillations are also seen in bright transition-region features associated with magnetic structures outside of sunspots (Lin *et al.* 2005).

Umbral oscillations in the transition region have also been detected in microwave observations (Gelfreikh *et al.* 1999; Shibasaki 2001). For example, Nindos *et al.* (2002) used the Very Large Array (VLA) to detect 3-minute oscillations in microwave intensity (Stokes *I*) and circular polarization (Stokes *V*) above a sunspot umbra. These oscillations are intermittent and localized, with the strongest fluctuations recurring in the same locations. Nindos *et al.* interpreted the oscillations as being caused either by variations in magnetic field strength (of amplitude ∼40 G in the photosphere, which seems unlikely based on the results discussed in the next subsection) or by variations in the height of the base of the transition region (of ∼25 km).

Bogdan and Judge (2006) call attention to the striking disappearance of the 3-minute *intensity* oscillations as they progress across the umbral transition region into the low corona. They attribute this to the rapid increase in the local temperature scale height in going from the chromosphere to the corona. In the chromosphere the temperature scale height is substantially less than the vertical wavelength of the oscillations, and hence the motion and compression (or rarefaction) is essentially uniform over the height range in which a particular chromospheric spectral line is formed, producing strong Doppler and intensity variations. In the corona, however, the temperature scale height is much greater than the vertical wavelength of the oscillations and coronal emission lines are optically thin, so that several layers of alternating compression and rarefaction contribute to the line formation and cancel each other out, thus producing little or no variation in intensity. It is interesting to note, however, that umbral intensity oscillations can be seen to continue along coronal loops in EUV observations from the TRACE satellite (Schrijver *et al.* 1999).

6.2.3 Magnetic field variations

It is only natural to imagine that the velocity oscillations in the umbra might be accompanied by oscillations in the strength and inclination of the umbral magnetic field. However, detection of such magnetic oscillations has proved to be very difficult, and conflicting results have been obtained. From a theoretical viewpoint, this situation is understandable because any magnetic field perturbations would be expected to be quite small at the height of formation of magnetically sensitive photospheric spectral lines, where the plasma beta is small and the inertia of the moving plasma is insufficient to produce significant bending or compression of the magnetic field lines.

Some observers have reported positive detections of oscillations in umbral magnetic field strength, at various frequencies in the 3- and 5-minute bands (Mogilevskii, Obridko and Shel'ting 1973; Gurman and House 1981; Horn, Staude and Landgraf 1997; Rüedi *et al.* 1998; Balthasar 1999, 2003; Norton *et al.* 1999) or at higher frequencies (Efremov and Parfinenko 1996), while others have found no significant oscillations in field strength at the limit of their sensitivity (Schultz and White 1974; Thomas, Cram and Nye 1984; Landgraf 1997). The detection is confounded by possible spurious sources of variations in the measured magnetic field. One such source is the opacity variation due to the compressive nature of the velocity oscillations, which produces a variation in the effective magnetic response height of the spectral line (Lites *et al.* 1998). Even if the oscillations are due to pure acoustic waves with motions everywhere aligned with the undisturbed magnetic field, apparent oscillations in field strength and inclination will be detected simply because these quantities vary with the response height in the spreading magnetic field configuration. Rüedi and Cally (2003) have actually modelled this effect and find that it can account for nearly all of the magnetic variations measured with the Michelson Doppler Imager on SOHO. Another possible spurious source of magnetic field oscillations is stray light from the surrounding photosphere (Landgraf 1997).

The most reliable measurements of magnetic oscillations are those based on the full set of Stokes profiles. Lites *et al.* (1998), using the Advanced Stokes Polarimeter (ASP) at the Dunn Telescope, found very weak oscillations in field strength in the 5-minute band, with rms amplitude of about 4 G, which they attributed at least in part to instrumental and inversion cross-talk between the Doppler and magnetic signals. These authors also presented a theoretical model of the umbral oscillations that predicts field strength variations of at most 0.5 G in the spectral lines used in their observations (Fe I 630.15 and 630.25 nm). Settele, Sigwarth and Muglach (2002), also using the ASP, found only a marginal detection of oscillations of average amplitude 5.8 G in restricted parts of the umbra. Kupke, LaBonte and Mickey (2000), using the Mees CCD Spectrograph at Haleakala, detected oscillations in field strength concentrated only near the umbra–penumbra boundary (as found earlier by Balthasar 1999), with rms amplitude of 22 G. Staude (2002), using the Fabry–Perot interferometer at the VTT on Tenerife, also found the most significant oscillatory power in field strength to lie near the umbra–penumbra boundary, but he could not rule out the effects of opacity oscillations or cross-talk with the velocity oscillations. Bellot Rubio *et al.* (2000), using the Tenerife Infrared Polarimeter (TIP) at the German Vacuum Telescope, also presented a marginal detection of field strength oscillations, but they argued, based on the phase lag they found between the velocity and magnetic field fluctuations (105°), that the magnetic oscillations are mostly the result of opacity fluctuations.

Altogether, it seems fair to say that there have been at best only marginal detections of magnetic field oscillations in the umbral photosphere, at the level of only a few gauss, and that this result is fully consistent with theoretical considerations.

6.2.4   *Theoretical interpretations of umbral oscillations*

The most significant feature of the 5-minute umbral oscillations is their reduced power (by a factor of two or three) compared to the 5-minute oscillations in the quiet Sun. There are at least four different mechanisms that might contribute to this suppression of power (Hindman, Jain and Zweibel 1997). First, the reduced power might be due in part to weaker generation of *p*-modes by turbulent convection within the spot itself, where the convection is strongly inhibited by the magnetic field. (This effect has been investigated in recent numerical simulations by Parchevsky and Kosovichev 2007.) However, the *p*-modes are known to propagate over horizontal distances much larger than a sunspot, so the local effects of sunspots on the excitation will be mostly smeared out (but might produce a solar-cycle dependence of *p*-mode amplitudes). Ignoring internal excitation altogether, the 5-minute umbral oscillations may be interpreted as the passive response of the sunspot to forcing by the *p*-modes in the surrounding convection zone (Thomas 1981).

A second mechanism for the power suppression is that which produces the observed absorption of p-modes by the sunspot; this mechanism, which is well established by observations but still not completely understood, is discussed in some detail in Section 6.4.1 below. A third mechanism, which may in fact be closely related to the p-mode absorption, is selective filtering of incident p-mode acoustic waves by the sunspot, which transmits only selected modes into the interior (Abdelatif and Thomas 1987) and alters the eigenfunctions as the acoustic waves couple to magneto-acoustic waves in the umbra (Hindman, Jain and Zweibel 1997). Some of the transmitted modes are not vertically trapped; instead, they propagate wave energy downward along the sunspot's magnetic flux tube. These wave modes have been proposed as the mechanism for the absorption of p-modes by a sunspot. Finally, a fourth mechanism is simply due to the Wilson depression, which implies that velocities measured in a spectral line correspond to a greater geometric depth in the umbra (compared to the quiet photosphere), where the amplitude of the p-mode oscillations is naturally lower.

Turning now to the 3-minute umbral oscillations, we find that a large body of theoretical work has been devoted to interpreting them as resonant modes of oscillation of the sunspot itself, excited either by oscillatory convection within the umbra or by the high-frequency tail of the spectrum of incident p-modes. Possible resonant cavities for magneto-atmospheric waves exist in the stratified umbral atmosphere due to variations of the sound speed and Alfvén speed with height and the consequent reflection or refraction of the waves (see Fig. 6.5). Two such resonant cavities have been proposed: one for fast modes in the low umbral photosphere (Uchida and Sakurai 1975; Antia and Chitre 1979; Scheuer and Thomas 1981; Thomas and Scheuer 1982; Cally 1983; Abdelatif 1990; Hasan 1991), and one for slow modes in the umbral chromosphere (Zhugzhda, Locans and Staude 1983; Zhugzhda, Staude and Locans 1984; Gurman and Leibacher 1984). Waves can tunnel between the lower and upper cavities, and a unified theory of the 3-minute oscillations as coupled wave modes of the two cavities can be constructed (Thomas 1984a, 1985; Zhugzhda, Locans and Staude 1987).

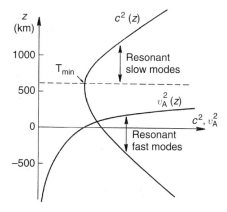

Fig. 6.5. Schematic plot of the variations with height of the square of the sound speed, c^2, and the square of the Alfvén speed, v_A^2, in a sunspot umbra. Also shown are the locations of the possible resonant cavities for 3-minute umbral oscillations. Wave modes in the two cavities may be coupled due to vertical tunnelling through the space between them. (From Thomas 1985.)

The theoretical work on resonant modes of umbral oscillation was motivated primarily by observational results showing temporal power spectra with individual peaks that might be associated with particular resonant modes (cf. Fig. 6.2). As discussed in Section 6.2.2 above, it is now understood that if a spectrum of resonant modes exists, then because of insufficient spatial resolution an individual peak in a temporal power spectrum of umbral oscillations actually corresponds to a superposition of a number of different modes with slightly different frequencies and spatial properties, and the frequency of this composite peak shifts about as the distribution of power among the different modes varies. Here the lack of spatial resolution is due to the limited spatial extent of the umbra, which makes it difficult to construct a space-time (k-ω) power spectrum of the oscillations and thus distinguish the various resonant modes (see, however, Abdelatif, Lites and Thomas 1986 and Penn and LaBonte 1993). Although individual resonant modes cannot be resolved in temporal power spectra, other evidence of the resonant nature of the oscillations exists. Small measured phase differences between the 3-minute oscillations in the low photosphere and the low chromosphere indicate a nearly standing wave in the vertical direction between these two levels (Lites and Thomas 1985). Simultaneous measurements in the umbral photosphere, chromosphere and transition region reveal a coherent 3-minute oscillation with an apparent node in the chromosphere (Thomas *et al.* 1987). Using cross-spectral analysis, O'Shea, Muglach and Fleck (2002) find both upward and downward propagating waves in the umbral chromosphere, consistent with the behaviour of waves in a resonant cavity.

What drives the 3-minute umbral oscillations? No doubt the mechanism lies below the solar surface, but then there are two possibilities: oscillatory convection in the subsurface layers of the sunspot itself (Moore 1973; Mullan and Yun 1973; Antia and Chitre 1979; Knobloch and Weiss 1984), or the high-frequency tail of the spectrum of p-mode oscillations in the surrounding convection zone (Moore and Rabin 1985). The question of which is the dominant mechanism is not fully settled, but evidence has accumulated in favour of the latter. For example, Rouppe van der Voort *et al.* (2003) find that there is no close connection between umbral flashes (due to the umbral oscillations) and umbral dots (due to the underlying magnetoconvection). There is little doubt that the 5-minute oscillations in the umbra are driven largely by the external p-modes (e.g. Penn and LaBonte 1993), and it is likely that this coupling extends to higher frequencies as well.

Bogdan and Judge (2006) argue that the umbral oscillations in the photosphere and chromosphere, in both the 5-minute and 3-minute bands, are manifestations of trains of slow magneto-acoustic-gravity waves with motions directed along the ambient, nearly vertical magnetic field. In the optically thin umbral atmosphere the Alfvén speed exceeds the sound speed and hence these slow waves are essentially pure acoustic waves with both plasma motions and propagation directed along the vertical magnetic field lines. The dominance of 3-minute oscillations over 5-minute oscillations in the chromosphere is then explained by the existence of the cutoff frequency for these acoustic-like waves, which in an isothermal atmosphere is given by $\omega_c = c/2H$ where c is the adiabatic sound speed and H is the density scale height. The umbral atmosphere is of course not isothermal, but roughly speaking we can think of it as having a local cutoff frequency $\omega_c = c/2H$ that varies with height because of the variations of c and H with temperature. This local cutoff frequency has a maximum value of about 5 mHz at the height of the temperature minimum in the low photosphere, a frequency that lies just between the 5-minute and 3-minute bands. Bogdan and Judge suggest that the 3-minute chromospheric oscillations correspond to oscillations in the high-frequency tail of the underlying photospheric oscillations that are allowed to propagate upward into the chromosphere, where they then dominate over the evanescent waves in the 5-minute band. Of course this interpretation is not incompatible with the existence of certain resonant modes of the slow waves, and it is in some respects similar to the 'tunnelling' mechanism associated with the resonant-mode models.

6.3   Penumbral waves

The most readily apparent waves in a sunspot are the so-called *running penumbral waves* (Giovanelli 1972; Zirin and Stein 1972). These waves are seen as dark wavefronts in Hα movies, in the form of concentric circular arcs that start from the outer edge of the umbra and propagate radially outward across the entire penumbra. The wavefronts often extend azimuthally around a substantial fraction of a full circle enclosing the umbra.

The radial propagation speed of a typical wavefront starts out at some 15–$20\,\mathrm{km\,s^{-1}}$ at the umbra–penumbra boundary but decreases monotonically across the penumbra, slowing to 5–$7\,\mathrm{km\,s^{-1}}$ at the outer edge of the penumbra. (A similar slowing trend is seen in the much faster radial propagation of waves across the umbra.) The oscillation period of the waves increases outward across the penumbra, from about $200\,\mathrm{s}$ to $300\,\mathrm{s}$ or longer. The wavefronts tend to become more ragged as they move radially outward across the penumbra, most likely due to different propagation speeds and directions in the light and dark filaments. The velocity amplitude of the penumbral waves is about $1\,\mathrm{km\,s^{-1}}$ in the chromosphere. These waves are less energetic than the 3-minute chromospheric oscillations in the umbra, and their oscillation periods are distinctly longer.

The running penumbral waves are primarily a chromospheric phenomenon: they are readily apparent over a range of heights in the chromosphere (in the spectral lines Na I D, Hα, and Ca II H and K), but they appear only weakly and intermittently at photospheric heights. Spectral lines formed in the penumbral photosphere generally show 5-minute oscillations moving through the spot but little evidence of the systematic penumbral waves. Running penumbral waves of a sort have been detected in the upper photosphere (in Fe I 557.61 nm; Musman, Nye and Thomas 1976) and even in the lower photosphere (in Fe II 722.45 nm; Marco *et al.* 1996), but they are more intermittent than the chromospheric waves and the relation between them is not clear.

Despite many attempts, the connection between umbral oscillations and running penumbral waves, if any, is still not well understood. Some studies have found no clear connection (e.g. Christopoulou, Georgakilas and Koutchmy 2000), while others have found evidence of penumbral waves originating from oscillating regions within the umbra (Alissandrakis, Georgakilas and Dialetis 1992; Tsiropoula *et al.* 1996; Tsiropoula, Alissandrakis and Mein 2000; Christopoulou, Georgakilas and Koutchmy 2001). On the basis of sequences of high-resolution Ca II H and K filtergrams, Rouppe van der Voort *et al.* (2003) found that umbral flashes and running penumbral waves are manifestations of a common oscillatory phenomenon with wavefronts spreading outward over the umbra and penumbra. They suggested that the wavefronts actually propagate vertically but spread out horizontally because of the diverging magnetic field.

Early theoretical models of running penumbral waves attributed them to vertically trapped fast magneto-atmospheric waves (Nye and Thomas 1974, 1976; Antia, Chitre and Gokhale 1978; Cally and Adam 1983) or to interfacial magneto-atmospheric waves running along the sunspot's magnetopause (Small and Roberts 1984; Roberts 1992). However, these models assumed a shallow penumbra with a nearly horizontal magnetic field overlying field-free gas. Since we now know that the penumbra is a deep structure with a strong vertical component of magnetic field, these models are no longer tenable. An alternative suggestion was that the penumbral waves are a manifestation of travelling-wave magnetoconvection in an inclined magnetic field (Galloway 1978). If this were true, the convective wave pattern should also be clearly visible at photospheric heights in the penumbra, which is not the case; instead, we see a more complicated convective pattern with bright features moving radially inward and outward (see Section 5.1).

The most effective theoretical approach to explaining penumbral waves and their relation to umbral oscillations would be to carry out numerical simulations of wave propagation in a more realistic representation of the sunspot magnetic field configuration. A promising start along these lines are the simulations of Bogdan *et al.* (2003) of waves in a two-dimensional stratified magneto-atmosphere with a spreading magnetic field and fairly realistic representations of the photospheric and chromospheric layers. These simulations reveal how fast and slow magneto-acoustic-gravity waves interact within such a configuration.

6.4   Sunspot seismology

Thomas, Cram and Nye (1982) were the first to suggest that sunspot oscillations could be used to probe the subsurface structure of a sunspot, introducing the concept of 'sunspot seismology' based on the interaction of the solar *p*-modes with the sunspot and the way in which different *p*-modes sample different depths below the solar surface. They interpreted a temporal power spectrum of 5-minute umbral oscillations in terms of a naive model of the interaction of these oscillations with the sunspot, producing a seismic measurement of the diameter of the sunspot's flux bundle at a depth of about 10 Mm that was consistent with sunspot models. This crude result was intended only to illustrate what might be achieved with better spatial and temporal resolution of the oscillations and better theoretical models. Alas, in spite of the development of a number of new observational techniques and theoretical ideas, reliable results from sunspot seismology have proved to be remarkably difficult to obtain.

An early approach was to compare space-time power spectra of oscillations inside and outside a sunspot, as in the work of Abdelatif, Lites and Thomas (1986) and Penn and LaBonte (1993) discussed in Section 6.2.1 above. This approach had some success in clarifying the interaction between the *p*-modes and a sunspot, but failed to produce any firm results on subsurface structure.

A different approach, based on observations of ingoing and outgoing waves in an annular region outside the sunspot, was developed by Braun, Duvall and LaBonte (1987, 1988). They made the remarkable discovery that a sunspot is a net absorber of the power of the incident *p*-modes, absorbing as much as half of the power at certain frequencies and horizontal wavenumbers. We discuss this phenomenon in some detail in Section 6.4.1 below. More recently, attention has focused on local helioseismology of sunspots, employing time–distance and holographic techniques; this work is discussed in Section 6.4.2.

6.4.1   *Absorption of p-modes by a sunspot*

When the *p*-mode oscillations in a circular annulus surrounding an isolated sunspot are decomposed into waves propagating radially inward and radially outward (in the form of a Fourier–Hankel decomposition), it is found that there is a relative deficit in power in the

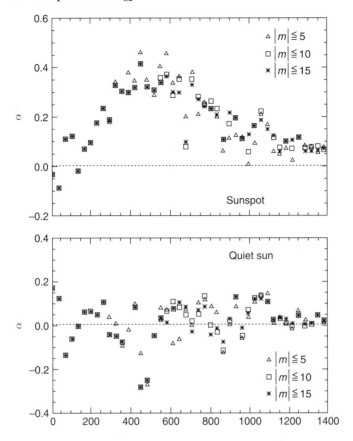

Fig. 6.6. *p*-mode absorption by a sunspot. Shown here is the absorption coefficient α integrated over frequency from $\nu = 1.5\,\mathrm{mHz}$ to $\nu = 5.5\,\mathrm{mHz}$ and summed over different sets of azimuthal order *m*. The upper panel is for an isolated sunspot centred within the circular annulus, while the lower panel is for a field of view containing only quiet Sun. (From Bogdan *et al.* 1993.)

outward-propagating waves (Braun, Duvall and LaBonte 1987, 1988; Bogdan *et al.* 1993; Chen *et al.* 1996, 1997). This power deficit is as high as 50% at some horizontal wavenumbers (see Fig. 6.6). No such power deficit is found for a circular annulus surrounding only quiet Sun.

Figure 6.7 shows the dependence on horizontal wavenumber of the absorption coefficient integrated over a range of frequencies and summed over a range of azimuthal orders, for two different sunspots (Braun, Duvall and LaBonte 1988; Bogdan *et al.* 1993). The agreement between the two data sets up to wavenumber $0.8\,\mathrm{Mm}^{-1}$ is striking, considering that different techniques of observing and data reduction were used. For higher wavenumbers, the higher-resolution results of Bogdan *et al.* show the absorption coefficient decreasing with increasing wavenumber *k*. The absorption coefficient can also be evaluated along individual *p*-mode ridges for radial orders up to $n = 5$: it is found that the absorption is greatest for $n = 1$ and decreases with increasing *n*, and that the absorption along each ridge peaks at an intermediate value of the spherical harmonic degree *l* in the range $200 \leq l \leq 400$ (Bogdan *et al.* 1993).

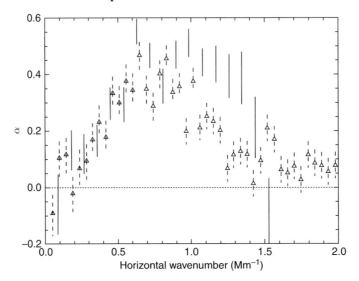

Fig. 6.7. The absorption coefficient α, integrated over frequency from $\nu = 1.5$ mHz to $\nu = 5.5$ mHz and summed over azimuthal orders $m = -5, -4, ..., +4, +5$, as a function of the horizontal wavenumber for an annular region surrounding a sunspot. The vertical line segments are the error bars for the original measurements of Braun *et al.* (1988) for a sunspot observed on 18 January 1983. The triangles (with dashed error bars) are for the later observations of Bogdan *et al.* (1993) of sunspot SPO 7983 on 19 March 1989. (From Bogdan *et al.* 1993.)

With sufficient temporal resolution (from data sets spanning longer times), one can determine not only the amplitudes of p-modes absorbed by a sunspot, but also their phase shifts. Braun *et al.* (1992a) presented the first measurements of these phase shifts using a 68-hour data set obtained at the South Pole. They found that the sunspot causes a phase shift δ that increases linearly with spherical harmonic degree l from $0°$ at $l = 125$ to about $150°$ at $l = 400$, while the absorption coefficient α increases nearly linearly from 0% to 40%. (Here α and δ are averaged over azimuthal orders $-5 \leq m \leq 5$.)

Following the discovery of acoustic absorption by sunspots, a number of different possible mechanisms for the absorption were soon put forward (see the review by Bogdan 1992). Hollweg (1988) proposed the resonant absorption of acoustic waves in a thin surface layer of the sunspot's magnetic flux tube. This suggestion was followed up in several subsequent papers (e.g. Lou 1990; Rosenthal 1990, 1992; Chitre and Davila 1991; Sakurai, Goossens and Hollweg 1991; Goossens and Poedts 1992; Keppens, Bogdan and Goossens 1994); these models are based on equilibrium configurations that do not include density stratification, and so they are not suitable for direct comparison with observations. Ryutova and colleagues (Ryutova, Kaisig and Tajima 1991: LaBonte and Ryutova 1993) proposed the enhanced dissipation due to inhomogeneities in a close-packed bundle of magnetic flux tubes. Another proposed mechanism involves mode mixing, in which incoming acoustic energy gets dispersed into a wide range of magnetic wave modes within the sunspot (e.g. D'Silva 1994).

Perhaps the most convincing model, and certainly the one that has been worked out in the most detail, is based on the suggestion by Spruit (1991) and Spruit and Bogdan (1992) that a purely acoustic oscillation impinging on a sunspot will be partially converted to a slow magneto-acoustic wave that propagates downward along the sunspot's magnetic flux tube and hence causes a leak of energy out of the near-surface acoustic cavity in which the resonant *p*-mode resides. With this mechanism, *p*-mode power is not actually absorbed within the surface layers of the sunspot, but instead is carried downward into the convection zone where presumably it is diffused and becomes part of the overall convective energy. The partial conversion of the incoming acoustic wave to the slow MHD wave takes place a little below the surface, at a depth where the sound speed and the Alfvén speed are equal (or equivalently, where the plasma beta is near unity).

A detailed normal-mode description of this mechanism has been developed in a series of papers by Cally and collaborators, first assuming a uniform vertical magnetic field (Cally and Bogdan 1993; Cally, Bogdan and Zweibel 1994; see also Rosenthal and Julien 2000) and then allowing the uniform field to be tilted (Crouch and Cally 2003; Crouch *et al.* 2005; Schunker and Cally 2006), which substantially enhances the process of mode conversion for higher-order modes. Numerical simulations of the process generally confirm the results of the analytical models (Cally and Bogdan 1997; Cally 2000). These results are in fairly good agreement with the measured variations of the absorption coefficient and phase shift with spherical harmonic degree (Cally, Crouch and Braun 2004). The mode conversion of incident *p*-modes into magneto-acoustic-gravity waves in a sunspot will produce not only downward-propagating waves, which are largely responsible for the absorption of *p*-mode power, but also upward-propagating waves that might be associated with umbral oscillations and running penumbral waves.

6.4.2 *Time–distance and holographic seismology of sunspots*

Time–distance helioseismology, introduced by Duvall *et al.* (1993), provides an effective technique for detecting sound speed variations and flow patterns beneath the solar surface. The technique in its simplest form is based on ray theory, but it can be formulated more generally (Gizon and Birch 2002). It was first applied to sunspots by Duvall *et al.* (1996), who found downflows beneath the spots extending down to a depth of about 2 Mm, with speeds of about $2 \, \mathrm{km \, s^{-1}}$. Since then, the technique has been applied to active regions and sunspots with several interesting results.

Using the time–distance technique, Gizon, Duvall and Larsen (2000) detected the sunspot moat flow using just the surface gravity mode (the *f*-mode), which is sensitive to the flow velocity over the first 2 Mm beneath the solar surface. They found a radially directed horizontal outflow with speeds up to $1 \, \mathrm{km \, s^{-1}}$ in an annular region extending out to 30 Mm from the centre of the sunspot. Outside the moat they detected an annular counter-flow, implying a downflow at the moat boundary. Using the time–distance technique with a range of acoustic waves, Zhao, Kosovichev and Duvall (2001; see also Kosovichev 2006) found a horizontal inflow in a sunspot moat at depths of 0–3 Mm, in disagreement with the outflow found by Gizon *et al.* (and with direct Doppler measurements), as already mentioned in Section 3.6. This conflict is likely to be resolved as the spatial resolution of the time–distance technique improves. Zhao *et al.* also found downflows beneath the sunspot to depths of about 6 Mm and horizontal outflows away from the sunspot at depths of about 5–10 Mm, a pattern suggestive of the 'collar flow' proposed as a mechanism for stabilizing the sunspot flux tube (see Section 3.5).

Most of the work on time–distance helioseismology of sunspots has ignored the effects of the sunspot's magnetic field on the acoustic waves passing through it. These acoustic waves will be partially converted to magneto-acoustic waves, especially at depths where the local sound speed is comparable to the local Alfvén speed, and this mode conversion will depend on frequency and also on the angle of incidence (Cally 2005). More detailed models of this interaction are clearly needed in order to interpret the results of time–distance helioseismology of sunspots and active regions.

Another approach to local helioseismology of sunspots is that known as helioseismic holography (Lindsey and Braun 1990, 1997; Braun and Lindsey 2000), which involves a computational reconstruction of the acoustic wave field in the solar interior based on the disturbances it creates at the surface. This technique has revealed that a sunspot is typically surrounded by an 'acoustic moat', a region with a deficit of acoustic power (of order 10–30%) extending radially outward from the spot over a distance of 30–60 Mm (Braun *et al.* 1998). The acoustic moat is roughly contiguous with the traditional sunspot moat defined by the surface velocity pattern (see Section 3.6).

6.4.3   Acoustic halos

The effect of near-surface magnetic fields, both inside and outside sunspots, on the *p*-modes is an important consideration in helioseismology. It has been known since their discovery (Leighton, Noyes and Simon 1962) that the amplitude of the 5-minute oscillations is reduced in regions of strong magnetic field (see e.g. Howard, Tanenbaum and Wilcox 1968; Woods and Cram 1981). More recently, however, it was discovered that the amplitude of higher-frequency acoustic waves, with frequencies in the range 5.5–7.7 mHz (just above the acoustic cutoff frequency in the low photosphere), is actually increased in areas surrounding regions of strong magnetic field. In the chromosphere so-called 'halos' of excess high-frequency acoustic power (in Ca II K intensity) surround active regions and often extend well into the surrounding quiet Sun (Braun *et al.* 1992b; Toner and LaBonte 1993), while in the photosphere more compact and fragmented halos of excess acoustic power (in Doppler velocity) surround small patches of strong magnetic field (Brown *et al.* 1992; Hindman and Brown 1998; Jain and Haber 2002). Both the suppression of *p*-mode power within a developing pore and the surrounding halo of enhanced higher-frequency acoustic power can be detected even before the pore begins to appear as a dark patch in the photosphere (Thomas and Stanchfield 2000).

The origin of the acoustic halos and the connection between the chromospheric and photospheric halos are still not fully understood. The large chromospheric halos might be caused by a general enhancement of acoustic emission from the convection zone in regions adjacent to a strong magnetic field. The more localized photospheric halos might be associated with some sort of magnetic flux-tube wave concentrated around the surface of the flux tube, such as the 'jacket modes' found by Bogdan and Cally (1995) or a surface Alfvén wave excited by resonant absorption of incident acoustic waves. Hindman and Brown (1998) found that the photospheric halos show up in Doppler velocity but not in continuum intensity, suggesting either nearly incompressible motions or compressible acoustic motions aligned vertically by the magnetic field, increasing the line-of-sight velocity without increasing the intensity variations. Muglach (2003) found no enhancement of 3-minute chromospheric power around active regions in UV intensity variations measured with broad-band filters on the TRACE satellite, suggesting that the halos in Ca II K intensity observed with narrow-band filters may be more a product of Doppler shifts of the line than of intensity variations within the line.

7

Sunspots and active regions

Sunspots are the most conspicuous but not the only product of solar magnetic activity. In this chapter we relate sunspots to other manifestations of solar activity, such as the emergence of magnetic flux at the solar surface, the formation of active regions, and the organization of magnetic flux into small flux tubes, pores and sunspots. We also consider the evolution of an individual sunspot, from its formation by the coalescence of small magnetic flux tubes and pores to its decay through the loss of magnetic flux at the periphery. Most of this chapter is devoted to features seen in and immediately above the photosphere, but we also attempt to relate these features to processes occurring beneath the solar surface.

7.1   Description of active regions

Magnetic activity on the Sun is not uniformly distributed over the solar surface, but instead is concentrated into *active regions* where sunspots, pores, faculae, plages and filaments are gathered. The underlying cause of all these features of solar activity is the Sun's magnetic field; an active region is basically a portion of the solar surface through which a significant amount of magnetic flux has emerged from the interior. Magnetic fields are also found everywhere on the solar surface outside of active regions, in weaker, more diffuse form or as small flux concentrations in the intergranular lanes, but those fields are not organized into the structures that so clearly define an active region. Active regions can be easily identified in full-disc magnetograms, such as the one shown in Figure 7.1.

Active regions may be as large as 100 000 km across and may live for several months, but they have a broad range of sizes and lifetimes, extending down to the small *ephemeral active regions* only 10 000 km across and lasting only several hours. Figure 7.2 shows a white-light photograph of three large active regions, on an unusually active day on the Sun. Throughout most of this chapter we shall be concerned with the properties of such large active regions, which typically contain sunspots and take part in the Sun's cyclic activity, with magnetic fields that follow Hale's polarity laws. We shall, however, return to ephemeral active regions, and yet smaller scale flux emergence, in Section 7.6.

7.1.1   Pores and their relation to sunspots

Pores are essentially the smallest sunspots, umbrae without penumbrae. The smallest pores are about the size of an individual granule, with diameters of about 1000 km, while the largest have diameters of 7000 km or more, larger than the smallest sunspots with penumbrae. Pores have continuum intensities ranging from 80% down to 20% of the normal photospheric intensity, and maximum (central) magnetic field strengths of 1500 to 2000 G.

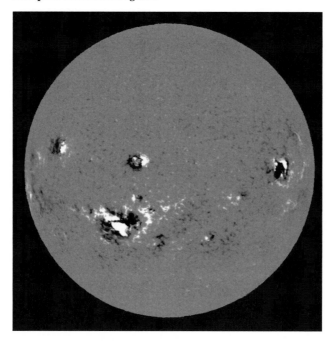

Fig. 7.1. Full-disc SOLIS magnetogram from 27 October 2003 showing several active regions. (Courtesy of NSO.)

(The detailed magnetic structure of a pore is described in Section 3.4.2.) They are generally darker than the intergranular lanes but somewhat brighter than the umbrae of larger sunspots. Large pores display a pattern of umbral dots that are essentially the same as those in sunspot umbrae, but perhaps a little brighter and longer lived. Several examples of pores can be seen in the white-light images in Figures 1.2 and 7.2.

Pores represent an intermediate state of magnetic flux concentration between the smaller intense magnetic flux tubes produced by granular convection (described in Section 7.2.3 below) and proper sunspots with penumbrae. Pores begin to form shortly after the emergence of the active-region magnetic flux. The local environment just before the formation of a pore consists of a coherent concentration of nearly vertical magnetic field, a few arcseconds across, that as yet shows no signature in continuum intensity (Keppens and Martínez Pillet 1996). The radiative disturbance that signals the formation of a pore is accompanied by a redshift (Leka and Skumanich 1998), suggesting that a convective collapse mechanism is at work (see Section 7.2.3). A newly formed pore can grow through the coalescence of nearby intense magnetic flux tubes, driven by an annular convective flow on a somewhat larger scale than the granulation pattern (Wang and Zirin 1992). The flow pattern consists of a radial inflow (at a speed of about $0.5\,\mathrm{km\,s^{-1}}$) across an annulus about $2''$ wide around the pore and a downflow at the edge of the pore (Keil *et al.* 1999; Sobotka *et al.* 1999; Tritschler, Schmidt and Rimmele 2002). This 'collar' cell confining the pore has been modelled in the numerical simulations of Hurlburt and Rucklidge (2000).

The lifetimes of pores in active regions range from about ten minutes (comparable to the lifetime of a granule) up to a day or more, with a typical value being about an hour.

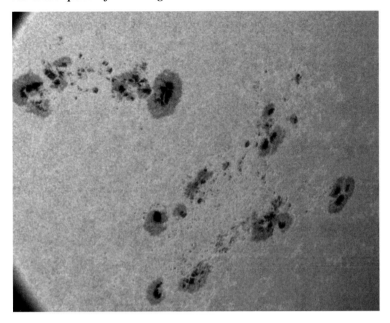

Fig. 7.2. Three closely spaced, large active regions on the Sun (NOAA ARs 10039, 10044 and 10050), photographed in white light on 31 July 2002 by the TRACE satellite. (Courtesy of Stanford-Lockheed Institute for Space Research.)

(Pores also form in the quiet photosphere, but their lifetimes are typically only 10–15 minutes; McIntosh 1981.) Many small pores simply disappear, but often two or more pores (of the same magnetic polarity) move together and coalesce to form a larger pore (Vrabec 1971, 1974; McIntosh 1981; Zwaan 1985, 1992). Occasionally a growing pore, accumulating magnetic flux through coalescence with intense magnetic flux tubes and other pores, reaches sufficient size to form a penumbra at its periphery and become a fully fledged sunspot (as described in Section 5.6.2). We will return to the subject of the formation of a sunspot in Section 7.3 below.

7.1.2 Faculae and plages

Whenever a sunspot forms, it is accompanied by the formation of nearby, irregularly shaped bright patches called *faculae* (from Latin, meaning 'little torches'). Apart from sunspots, faculae are the most apparent features on the solar disc observed in white light, where they are best seen near the limb. Several faculae are visible in the active region shown in Figure 7.2. At high resolution ($<0.5''$) the faculae are seen to consist of numerous small, close-packed bright elements, or *facular points*, in the intergranular lanes (see Fig. 7.3). The facular points correspond to magnetic flux concentrations with diameters of order 200 km (Keller and von der Luhe 1992).

Although about 90% of all faculae are closely associated with active regions, small facular points also exist outside of active regions, distributed more sparsely over the entire solar surface in the photospheric network corresponding to the boundaries between supergranules. Facular bright points tend to cluster into crinkly rows within the intergranular lanes, forming bright structures known as *filigree* (Dunn and Zirker 1973).

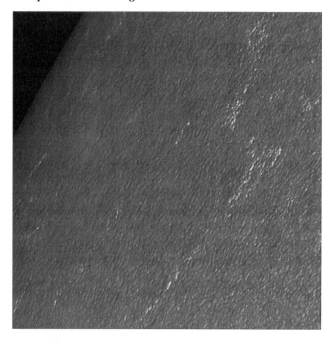

Fig. 7.3. Speckle-reconstructed image of faculae near the limb in the continuum at wavelength 487.5 nm, taken with the Swedish Solar Telescope. The field of view is approximately $80'' \times 80''$. (From Hirzberger and Wiehr 2005.)

In white light, faculae are barely detectable at disc centre but they are easily seen near the limb, indicating that their contrast depends on the viewing angle.[1] In line radiation, however, faculae are quite visible even at disc centre, especially in strong chromospheric lines such as the Ca II H and K lines where the corresponding facular brightenings, known as *plages*, have much higher contrast. There is a continuous transition (in height) between the photospheric faculae and the plages, and indeed they almost certainly represent the same magnetic structures seen at different heights. At disc centre the photospheric faculae are clearly visible as bright points in the G-band (CH absorption band), serving as a useful indicator of strong magnetic fields.

The temperature of faculae is about 100 K higher than their surroundings at the photosphere. The number of faculae varies in phase with the sunspot cycle, and the enhanced radiation from the large area of hot faculae at solar maximum more than compensates for the reduced radiation from cool sunspots, thereby producing a slightly greater total solar irradiance at sunspot maximum (see Section 12.1). This fact alone makes it important to understand the origin of faculae.

The granulation pattern in plage regions is distorted by the prevalent magnetic fields. As expected, the individual cells are smaller and strong fields fill the lanes between them. At the centre of the disc, flux elements appear only as magnetic bright points, without any overall brightening, but bright faculae appear at the limb (Berger, Rouppe van der Voort

---

[1] As a result of this, and the non-uniform distribution of faculae with latitude, it is very difficult to determine the true shape of the solar limb, and hence the oblateness of the Sun.

and Löfdahl 2007). Observations at the highest available spatial resolution (0.12″) of granules and faculae near the limb begin to reveal their three-dimensional structure (Lites *et al.* 2004; Hirzberger and Wiehr 2005). The $\tau_{500} = 1$ surface is elevated over bright granules and depressed in intergranular lanes (Lites *et al.* 2004), as shown in Figure 7.4. A typical facular brightening appears as an extended feature lying on the inclined, Sun-centre side face of an individual elevated granule, just limbward of the concentrated facular magnetic field in the intergranular lane. Adjacent to this bright face is a dark, narrow lane separating the facular brightening from the next granule toward Sun centre. This configuration has been well explained in realistic numerical simulations of photospheric magnetoconvection (Carlsson *et al.* 2004; Keller *et al.* 2004; Steiner 2005). The bright face appears because of the greater transparency in the partially evacuated magnetic flux concentration, allowing a clear view of

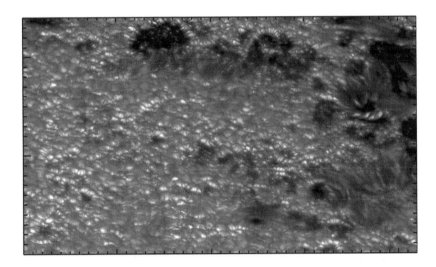

Fig. 7.4. Upper panel: the three-dimensional structure of facular granules in an active region near the solar limb. Perspective view obtained with the SST in an 8 nm bandpass centred at 487.7 nm; tick marks have a 1″ spacing. (Courtesy of B. W. Lites.) Lower panel: corresponding image of the emergent intensity, as viewed from a 60° inclination to the vertical, from a realistic numerical simulation of granular magnetoconvection. (Courtesy of Å. Nordlund.)

the adjacent hot granule on the limbward side, as first explained in the 'bright wall' model of Spruit (1976). The dark lane corresponds to cool, dense, downflowing gas between the flux concentration and the adjacent granule on the centreward side.

Within a facular region individual facular elements form and dissolve on a time scale of a few hours, but the facular regions themselves live longer than sunspots. Facular regions often appear before their associated spots, by as much as several days, and can outlast them by a factor of two or three. On the other hand, sunspots are almost never seen without nearby faculae. Faculae and plages are associated with enhanced magnetic field strengths and they are broadly distributed across an active region, so that their extent gives a good measure of the overall size of an active region. Initially, faculae are compact, bright, and irregularly distributed, but as they evolve they fragment and lose contrast as they are carried into the active-region magnetic network by the supergranules and then eventually disperse into the quiet-Sun network.

The main zones of faculae correspond to the two sunspot zones, although they tend to be about 15° broader and extend a bit farther toward the poles. These facular zones move equatorward along with the sunspot zones over the course of the solar cycle. There is also a separate class of faculae that occur at higher latitudes (above about 60°), poleward of the activity belts. These polar faculae are most numerous near the minimum of the solar cycle (Waldmeier 1955).

7.2   Birth and evolution of active regions

The spatial scale and systematic properties of a solar active region indicate that it is produced by a fairly coherent bundle of toroidal magnetic flux that rises buoyantly from deep in the convection zone, in the form of an Ω-shaped loop, and breaks through the photosphere in fragmented form (see Fig. 7.5). The relatively quick emergence of this magnetic flux, the

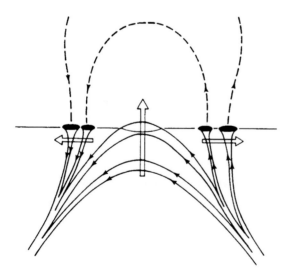

Fig. 7.5. A schematic model for the formation of an active region. An emerging flux region is formed as an Ω-shaped magnetic loop rises through the solar surface. Broad open arrows indicate local displacements of the flux tubes. (From Zwaan 1992, courtesy of Springer Science and Business Media.)

rapid formation of sunspots, and the immediate onset of the slow decay process together suggest that the fragmented flux that first appears is well organized into a tighter, more coherent bundle somewhere below the surface (Zwaan 1978, 1992). As it approaches the surface, the flux bundle is shredded into many separate strands which, upon emergence, are quickly concentrated into small, intense (kilogauss strength) magnetic flux bundles (or elements) by the vigorous thermal convection occurring in the thin superadiabatic layer at the top of the convection zone. These small flux elements then accumulate at the boundaries between granules or mesogranules, and some of them coalesce to form small pores (Keppens and Martínez Pillet 1996; Leka and Skumanich 1998). Some pores and flux elements in turn coalesce to form sunspots (as we shall describe in more detail in Section 7.3).

7.2.1 Magnetic flux emergence in active regions

The process of *magnetic flux emergence*, in which magnetic flux that has risen through the convection zone under magnetic buoyancy reaches the solar surface and penetrates into the solar atmosphere, is a fundamental aspect of solar activity. Flux emergence is a highly dynamical process, in which the configuration of the magnetic field changes rapidly as it emerges from the dense subsurface region into the rarefied atmosphere.

A large active region is formed over a period of a few days by a succession of emerging and expanding magnetic bipoles known as *emerging flux regions* (Zirin 1972, 1974). The emerging magnetic field produces characteristic signatures in the chromosphere and corona. The first evidence in the chromosphere of an emerging flux region is a compact, bright bipolar plage (Fox 1908; Waldmeier 1937). Soon thereafter an *arch filament system* is seen to form in Hα, appearing on the disc as a number of parallel dark fibrils connecting regions of opposite polarity. Each fibril is actually a rising magnetic loop, carrying plasma that subsequently flows down along the field lines under gravity. The top of the loop ascends at speeds up to $10 \, \mathrm{km \, s^{-1}}$ while matter flows downward from the top along both legs of the arch at speeds up to $50 \, \mathrm{km \, s^{-1}}$ in the chromosphere, as measured by Doppler shifts in Hα (Bruzek 1969), and speeds of about $2 \, \mathrm{km \, s^{-1}}$ in the photosphere, where of course the density is much greater (Zwaan, Brants and Cram 1985).

An emerging flux loop has bright faculae of opposite polarity marking its two footpoints. These footpoints move apart as the loop rises, initially at speeds of $2–3 \, \mathrm{km \, s^{-1}}$ but slowing down to less than $1 \, \mathrm{km \, s^{-1}}$ over the next few hours and decelerating further over the next several days (Harvey and Martin 1973; Strous *et al.* 1996; Shimizu *et al.* 2002). In longitudinal magnetograms the footpoints first appear in the photosphere as a pair of small magnetic patches of opposite polarity, and in white light the granulation pattern between these patches has abnormally dark intergranular lanes which presumably mark the locations of other magnetic loops just reaching the surface.

The behaviour of an emerging flux region has been beautifully clarified by Kubo, Shimizu and Lites (2003), who followed the evolution of the full vector magnetic field in the region using Stokes polarimetry. The field that first appeared was nearly horizontal, as expected, and weak ($B \sim 500 \, \mathrm{G}$), with a high filling factor ($>80\%$) and an upward motion of less than $1 \, \mathrm{km \, s^{-1}}$. Over the next several hours, as the loop rose and the bipole expanded, the magnetic field strength at the two footpoints increased to about $1500 \, \mathrm{G}$ and the filling factor dropped to 40% as the magnetic field became more vertical there and, presumably, the convective collapse mechanism kicked in. The magnetic field inclination was not symmetric about the centre of the emerging flux region, indicating that the rising magnetic loop was affected by pre-existing magnetic fields in the region.

Magnetic flux emergence is also responsible for generating many of the bright structures seen in the corona, as evidenced by many observations from Yohkoh, SOHO and TRACE. The average X-ray brightness of the corona begins to increase above an emerging flux region almost immediately after its first appearance and remains high for several days (Kawai *et al.* 1992; Yashiro, Shibata and Shimojo 1998). Transient phenomena such as flares and jets are often triggered by the emerging flux, perhaps through reconnection with pre-existing magnetic fields at the site (Yokoyama and Shibata 1995, 1996). In established active regions, bright coronal loops seen in X-rays or the EUV connect the opposite magnetic polarities seen at the surface.

New magnetic bipoles emerge within old ones and their footpoints also move outward away from the neutral polarity line as the corresponding magnetic loop rises. An arch filament system in Hα exists as long as new magnetic loops are emerging, which can be up to several days, but individual fibrils in the system live only about a half hour, fading away as material is drained from the corresponding loop and then replaced by fibrils associated with newly emerging loops. The several emerging flux regions that form an active region initially have a somewhat random orientation, but they gradually align to form a large bipole oriented nearly E–W (Weart 1970, 1972; Frazier 1972).

K. Harvey carried out a careful study of the emergence of active regions based on Kitt Peak magnetograms (Harvey-Angle 1993). She defined the *emergence time* of an active region as the time between the first appearance of a bipolar magnetic field on the magnetograms and the time of maximum development, when flux emergence apparently ceases. From the magnetograms, the emergence time can be determined to within about half a day. The 978 active regions in Harvey's sample all had emergence times of five days or less. Both the emergence time and the total lifetime of an active region tend to increase with increasing total area (and flux) at maximum development, although there is a wide range of these time scales for any given maximum area. In all cases, the emergence time is short compared to the total lifetime of the active region.

Although most active regions end their lives by dispersing and weakening to the extent that they no longer show a measurable bipolar field, a significant fraction of them are disrupted by the emergence of new flux leading to the formation of a new active region at the same site. This fits into a more general pattern of *activity nests*, in which a new active region emerges in the same location as a previous active region even though the site had been inactive for a period of two or three solar rotations (Brouwer and Zwaan 1990). More than one-third of all active regions appear within such nests, and at solar maximum sunspot groups tend to appear within nests in rapid succession (van Driel-Gesztelyi, van der Zalm and Zwaan 1992). Activity nests are manifestations of the more general phenomenon of *active longitudes* in which new active regions tend to develop at the same Carrington longitude as old regions over periods of up to several years.

Most active regions exhibit some degree of twist in their magnetic fields. A useful measure of the amount of such twist is the *magnetic helicity density* $\mathbf{A} \cdot \mathbf{B}$, where $\mathbf{A}$ is the vector potential for $\mathbf{B}$ (i.e. $\mathbf{B} = \nabla \times \mathbf{A}$), or the total magnetic helicity $H_{\mathrm{m}} = \int_V \mathbf{A} \cdot \mathbf{B} \, dV$ in a volume V. The flux of magnetic helicity into the solar atmosphere has two components, one due to the vertical advection of an already-twisted magnetic field into the photosphere, and the other due to local creation of helicity through the 'braiding' of the field by differential rotation or other large-scale shearing motions in the photosphere (Berger and Field 1984). As discussed in the next subsection, at least a small amount of twist is needed in a rising flux tube in order for it to resist fragmentation and preserve its identity as it rises through the convection zone. Once the magnetic flux rises into the photosphere, however, additional helicity is created by the braiding process. The contribution due to braiding is known to be much greater than might be expected from surface differential rotation alone, based on direct photospheric observations (DeVore 2000; Chae 2001; Moon *et al.* 2002; Nindos, Zhang and Zhang 2003) and also on the observed loss of helicity through coronal mass ejections (Démoulin *et al.* 2002; Green *et al.* 2002). The additional vortical photospheric motions responsible for producing helicity are most likely driven by magnetic forces due to twist in the subsurface magnetic field and can be thought of as torsional Alfvén waves transporting helicity along the flux tube (Longcope and Welsch 2000; Pevtsov, Maleev and Longcope 2003).

7.2.2   Theoretical models of emerging magnetic flux

The magnetic fields in surface active regions on the Sun are thought to originate from a strong global toroidal field generated by a dynamo mechanism near the base of the convection zone. Loops of magnetic flux must occasionally break away from this region of strong toroidal field and rise through the convection zone to the surface as reasonably coherent structures in order to produce the observed patterns of surface activity, as characterized by empirical relations such as Hale's polarity law and Joy's law. The individual loops might be produced directly by turbulent dynamo action, or they might be created by an instability of a larger-scale, continuous toroidal field produced by a dynamo. The buoyant rise of an individual magnetic loop through the convection zone has been modelled intensively using the thin flux tube approximation, in which the magnetic field behaves essentially as a one-dimensional, flexible, buoyant 'string'. However, this approximation is inappropriate for modelling the emergence into the atmosphere, where the magnetic field must expand rapidly as the gas pressure drops rapidly with height (the pressure scale height being only about 150 km in the photosphere). Instead, flux emergence at the surface has of necessity been studied through direct numerical solutions of the full MHD equations.

Thin flux tubes rising through the convection zone

In the standard picture of the solar dynamo, the strong toroidal flux is produced by differential rotation in the tachocline, coincident with the subadiabatic region just below the base of the convection zone. When this toroidal field reaches a sufficiently high strength, it becomes buoyantly unstable and a loop of strong, buoyant magnetic flux rises through the convection zone to form an active region at the surface. Beginning with the seminal work of Moreno-Insertis (1983, 1986) and Choudhuri and Gilman (1987), there have been a number of calculations of the rise of these buoyant magnetic flux tubes through the convection zone (see the reviews by Moreno-Insertis 1992, Fisher *et al.* 2000, and Fan 2004[2]); these calculations are mostly based on the thin flux tube approximation (explained in Appendix 2). In almost all such models, the convection zone is modelled as a static, adiabatically stratified gas, ignoring the effects of the turbulent convection. The inclusion of solar rotation in the

---

[2] This last review is part of an ever-expanding series of *Living Reviews of Solar Physics*, accessible on-line at the website www.livingreviews.org.

model is crucial, as the Coriolis force has a significant effect on the shape and path of the rising flux tube. Also, it is generally found that the rising flux tube must have some degree of twist in order for magnetic tension forces to prevent fragmentation of the tube during its rise to the surface.

A key result of these calculations is that in order to reproduce the observed properties of sunspot groups (latitude of emergence, tilt angle, asymmetry of leader and follower spots, proper motions) the field strength of the flux tube at the base of the convection zone must be of order 10^5 G.

Numerical simulations of flux emergence

To study flux emergence at the surface, we must abandon the thin flux tube approximation and solve the full MHD equations numerically. The early numerical models of flux emergence were two-dimensional (Shibata *et al.* 1989, 1990). For example, Shibata *et al.* (1990) showed how an initial magnetic slab in the convection zone rises into a stable, unmagnetized atmosphere under the combined effects of convection and magnetic buoyancy and forms an expanding magnetic loop. Their results essentially illustrate the *Parker instability* (Parker 1966, 1979a), in which plasma slides down the two legs of the rising loop, thereby evacuating the top of the loop and increasing its buoyancy, causing it to rise farther. Other studies have shown that an initially weak field $B \sim 500$ G in the upper convection zone is intensified locally in the photosphere to $B \sim 1000$ G by the convective collapse triggered by the downflows (Nozawa *et al.* 1992), and that Rayleigh–Taylor instabilities occur in the flattened horizontal part of the emerging flux tube (Magara 2001). Yokoyama and Shibata (1995, 1996) carried out two-dimensional simulations of magnetic flux emerging into a pre-existing coronal field, which produced magnetic reconnection and coronal jets suggestive of observed X-ray jets.

However, the complex magnetic structures formed by flux emergence are inherently three-dimensional, and three-dimensional numerical simulations, beginning with those of Matsumoto *et al.* (1993), have indeed revealed more complicated and realistic behaviour. A number of these studies have examined the buoyant rise of magnetic flux sitting initially just below the surface and rising into a non-magnetized atmosphere (e.g. Magara and Longcope 2001, 2003; Manchester *et al.* 2004; Cheung, Schüssler and Moreno-Insertis 2007). Several independent studies of the rise of thin flux tubes through the convection zone have shown that the tubes must have a significant amount of twist in order to maintain their integrity and not fragment in the face of hydrodynamic forces, and indeed observations show that magnetic flux usually emerges at the surface already in a significantly twisted state (e.g. Lites *et al.* 1995). This suggests that studies of flux emergence should begin with a twisted magnetic flux tube somewhere below the solar surface, as indeed has now been done in the analytical model of Longcope and Welsch (2000) and in several numerical simulations (e.g. Fan 2001; Abbett and Fisher 2003; Archontis *et al.* 2004).

The numerical simulations by Archontis *et al.* (2004) and Galsgaard *et al.* (2005) treat the rise of magnetic flux into an atmosphere with a pre-existing coronal magnetic field, taken to be uniform and horizontal. The results show the formation of a concentrated, arched current sheet and magnetic reconnection accompanied by strong local heating, which drives strong, collimated jets along the ambient magnetic field lines. Such jets are indeed seen in Yohkoh and TRACE observations. Recently, Archontis, Hood and Brady (2007) simulated the rise of two twisted flux tubes arriving and emerging successively at the solar surface and producing current sheets, high-speed reconnection jets, and arcades of magnetic loops with flare-like brightenings.

7.2.3   Intense magnetic elements

Beginning in the 1970s, high-resolution observations revealed that most of the magnetic flux seen at the solar surface outside sunspots is in the form of intense field concentrations with diameters less than about 300 km and field strengths of order 1500 G (see the review by Spruit and Roberts 1983). The standard theoretical picture of an intense magnetic element is an isolated, vertical magnetic flux tube that is highly evacuated and in total (gas plus magnetic) pressure balance with its surroundings.

During the emergence of an active region, magnetic flux appears in the form of many separate, small flux bundles. A typical flux bundle emerges with a field strength of order 500 G and radius of order 200 km (for a total magnetic flux of order 10^{18} Mx), but is quickly compressed to a field strength of order 1500 G and radius of order 100 km. The magnetic fields in these separate flux bundles expand with height and merge somewhere in the chromosphere. The flux bundles exhibit a strong tendency to cluster into pores and sunspots, in spite of the opposing magnetic stresses that this clustering must generate in the chromosphere where their fields merge. The clustering must be driven by hydrodynamic forces due to a converging flow near the surface (Meyer *et al.* 1974; Meyer, Schmidt and Weiss 1977), established by a strong downdraft immediately adjacent to the flux bundle. The flow creates a vortex ring around each flux bundle, and the clustering can be understood as being the result of the mutual attraction of interacting vortices (Parker 1992).

The surprisingly high field strengths of these magnetic elements approach the limiting magnetic field strength $B_p = (8\pi p_e)^{1/2}$ of a totally evacuated flux tube in static pressure balance with the external pressure p_e. This *vacuum field strength* has a value of about 1700 G at the solar surface (continuum optical depth $\tau_{500} = 1$). Field strengths greater than this value have been reported, which might be explained either by the addition of dynamical pressure due to an exterior flow converging on the flux tube or as a consequence of the enhanced transparency of the flux tube, which lowers the geometric level of surfaces of constant optical depth. In any case, the high field strengths of the magnetic elements reflect convective processes occurring very near the solar surface, rather than the deeper processes of dynamo generation of the field and its buoyant rise to the surface.

Convective collapse of a thin flux tube

Parker (1978) and Zwaan (1978) first proposed that the intense, kilogauss magnetic flux concentrations observed in the photosphere are formed by a convective collapse mechanism occurring in a magnetic flux tube within the superadiabatic layer just below the solar surface. This mechanism was first modelled as an instability-driven process in an isolated, vertical, thin magnetic flux tube (Webb and Roberts 1978; Spruit 1979; Spruit and Zweibel 1979; Unno and Ando 1979). Although we now know that the actual mechanism of flux concentration is more complicated than this, involving fully nonlinear magnetoconvection, it is nevertheless instructive to examine this linear instability as an illustration of the strong dynamical effect of the superadiabaticity of the near-surface layer of the Sun. The basic mechanism for the instability is illustrated in Figure 7.6. The thin flux tube is initially in thermal equilibrium with the surrounding convection zone, so the temperature

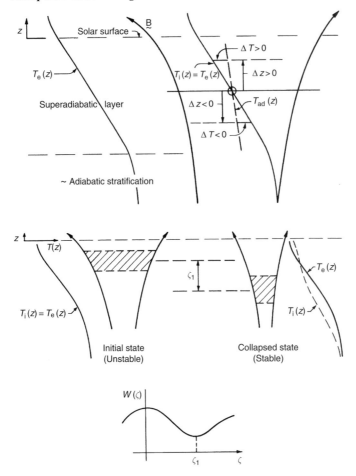

Fig. 7.6. Schematic diagram of the convective collapse mechanism for a magnetic flux tube. Upper panel: the temperature profile $T_i(z)$ inside the flux tube (solid curve) is initially the same as the profile $T_e(z)$ outside the tube, but the temperature of a vertically displaced fluid parcel will follow an adiabatic profile (dashed line). For a downward displacement, the parcel will be cooler than the surroundings and the tube will collapse and concentrate. (From Thomas 1990.) Lower panel: in the initial unstable state, the flux tube is in thermal equilibrium with its surroundings, but in the collapsed state it is cooler than the surroundings over most of the superadiabatic layer. The cross-hatched regions correspond to the same mass of plasma within the flux tube. The schematic plot of total energy $W(\zeta)$ of the system as a function of the vertical displacement ζ illustrates the stability properties of the two equilibrium states. (After Spruit 1979.)

profile (solid curve) is superadiabatic both inside and outside the tube near the solar surface. Because of the inhibiting effect of the magnetic field on small-scale convection, the temperature of a parcel of fluid displaced vertically within the tube will closely follow an adiabat (dashed line). After a downward displacement, the parcel will be cooler than its surroundings, and if the magnetic pressure is not too high the parcel will also be denser than its surroundings. The resulting negative buoyancy force on the parcel will then accelerate the displacement, thereby evacuating the upper part of the tube and causing it to collapse and intensify its magnetic field. If the initial magnetic field strength is sufficiently high, however, lateral pressure balance can be maintained by an increase in magnetic pressure, and both the gas pressure and the density of the parcel will be less than in the surroundings, giving a positive (restoring) buoyancy force and producing an oscillation rather than an instability. This argument suggests that there is a critical magnetic field strength above which the tube is stable and below which it is unstable to convective collapse. This is indeed confirmed by a calculation for a model atmosphere (Spruit and Zweibel 1979), which gives a critical field strength of around 1350 G, typical of the observed field strengths in intense magnetic elements. (Interestingly, the linear instability accelerates an upward displacement too, leading to an expansion of the tube and a weakening of the magnetic field, suggesting that both collapse and expansion of magnetic flux tubes occur at the solar surface.)

After the initial collapse, the nonlinear development of the instability leads to a new, stable, collapsed equilibrium state with higher magnetic field strength (Spruit 1979). If radiative heat exchange and viscosity are included, however, the collapsed state is found to be overstable, with oscillation periods of order 1000 s or more (Venkatakrishnan 1985; Hasan 1985; Massaglia, Bodo and Rossi 1989). Other influences occurring on shorter time scales, such as the formation and decay of adjacent granules, make these oscillations somewhat irrelevant. The detailed behaviour of the collapse instability depends rather sensitively on the boundary conditions imposed at the top and bottom of the model (Schüssler 1990).

Formation of intense flux elements in nonlinear magnetoconvection

In reality, the processes of magnetic flux concentration are occurring continually near the solar surface, and they should be looked upon as different facets of magnetoconvection (Hughes and Proctor 1988; Schüssler and Knölker 2001). Models of the formation of intense flux elements in a turbulent convecting layer based on mixing-length theory have been presented by Schüssler (1990, 1991) and his collaborators (Deinzer *et al.* 1984; Knölker, Schüssler and Weisshaar 1988; Knölker and Schüssler 1988). The interactions between intense flux sheets and turbulent convection were subsequently modelled in two dimensions (Steiner *et al.* 1998; Grossmann-Doerth, Schüssler and Steiner 1998). Strong localized magnetic fields also appear both in direct numerical simulations of granular magnetoconvection, whether anelastic (Nordlund 1983, 1986) or fully compressible (Carlsson *et al.* 2004; Vögler *et al.* 2005; Stein and Nordlund 2006), and in idealized numerical experiments on three-dimensional compressible magnetoconvection (Weiss *et al.* 1996; Weiss, Proctor and Brownjohn 2002; Bushby and Houghton 2005).

There are two relevant processes that lead to the appearance of fields that are locally intense: the first is the expulsion of magnetic flux from a convective eddy (Parker 1963; Clark 1965, 1966; Weiss 1966), while the second – which is probably more relevant here – is the transport of flux tubes or sheets along the intergranular network. Recent calculations reveal the asymmetry between rising and falling plumes, owing to the effects of stratification and buoyancy braking. Strong, local magnetic flux concentrations (or flux sheets or tubes) are formed in the narrow, intense downdrafts in the cool intergranular lanes. The strong magnetic field inhibits heat transport along and into the flux tube, and radiative losses at the surface may cause a tube to collapse even more. Once the magnetic pressure becomes significant, the tube must be partially evacuated in order to achieve an overall lateral pressure balance. Such flux tubes can reach field strengths exceeding the equipartition value $B_e = (4\pi\rho v^2)^{1/2}$ and approaching – or even exceeding – the limiting vacuum field strength $B_p = (8\pi p_e)^{1/2}$

(where p_e is the external gas pressure). The tube is then contained by a balance between the combined magnetic and (reduced) gas pressure within, and the sum of the gas pressure p_e and the dynamic pressure $\frac{1}{2} \rho |\mathbf{u}|^2$ without, giving a value of B greater than B_p (Bushby 2007; Bushby *et al.* 2008).

7.3   Formation, growth and decay of sunspots

Sunspots form through the coalescence of pores and smaller magnetic flux tubes into a single, growing pore (Vrabec 1974; Zwaan 1978, 1992; McIntosh 1981). If a growing pore reaches a sufficient size (a diameter of about 3500 km but sometimes as much as 7000 km), or perhaps more significantly, sufficient total magnetic flux (of order 1×10^{20} Mx; Leka and Skumanich 1998), it forms a penumbra at its periphery and becomes a fully fledged sunspot. The formation of a penumbra is a rapid event, occurring in less than 20 minutes, and the characteristic sunspot magnetic field configuration and Evershed flow are both established within this same short time period (Leka and Skumanich 1998; Yang *et al.* 2003). (The process of the formation of a penumbra has already been described in some detail in Section 5.6 above.)

A substantial fraction of the magnetic flux in an active region ends up in sunspots, especially the flux of leading polarity of which up to 60% may be concentrated into one or more sunspots at the maximum stage of development of the region (Zwaan 1992). The flux of leading polarity usually forms a dominant spot near the leading (western) edge of the region. A large spot forms through the coalescence not only of pores but also of small sunspots already possessing penumbrae. Figure 7.7 shows an example of the formation of a large sunspot. Several other good illustrations of sunspot formation are presented by McIntosh (1981). The coalescence of small spots into large spots often continues for several days, and it can take a week or more for the dominant leader spot to form. However, even when a large spot is still experiencing net growth, it may already be losing magnetic flux from part of its periphery.

After individual pores have coagulated to form a sunspot, the boundaries between them sometimes persist, as in a ball of putty formed from individual strands. These visible seams preserve some memory of the spot's formation and provide a recognizable pattern in the umbra that may persist over the lifetime of the spot. (In H. U. Schmidt's words, sunspots have recognizable faces.) Eventually, light bridges form along these same seams as the spot begins to break up into roughly the same segments from which it formed (Garcia de la Rosa 1987a,b). On the other hand, some large sunspots develop very dark umbral cores virtually free of fine structure or any visible seams between the fragments that formed them (Zwaan 1992).

7.3.1   *Growth rates of sunspots*

The growth phase of a sunspot is generally much shorter than its decay phase. A sunspot forms in only a few days but will typically live for a few weeks and occasionally for several months (several solar rotation periods). A sunspot begins to decay almost immediately after it is fully formed, but the decay process is usually slow enough that it is reasonable to think of a steady-state 'middle age' for the spot, during which its structure is as we have described in Chapters 3–5.

Early studies of the growth rates of sunspots based on the Greenwich photoheliographic records found that the growth was fastest at the first appearance of a spot (Greenwich 1925),

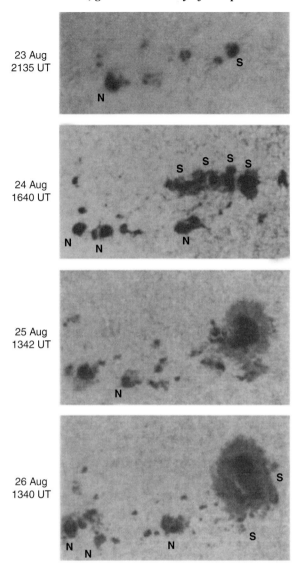

Fig. 7.7. A sequence of photographs showing the formation of a large sunspot through the coalescence of smaller spots and pores (in August 1966). Magnetic N and S polarities of the individual spots are indicated. This large, symmetric leader sunspot went on to have a very long lifetime of 137 days. (From McIntosh 1981.)

but Dobbie (1939) later found that the growth rate typically increased during the early stages of spot growth, especially for the largest spots.

An important step forward in the study of sunspot growth and decay was the digitization of the Mount Wilson white-light solar images (Howard, Gilman and Gilman 1984). This data set includes the areas and positions of every sunspot visible on the disc in the daily photographs taken from 1917 to 1985. Howard (1992) examined the daily changes in area of over 36 000 sunspot groups in this data set and found that the median percentage daily area change for all these groups was -25%, while the mean values for growing and decaying groups separately were $+502\%$ and -45%, respectively, showing the strong asymmetry between the rapid growth and slow decay of sunspots.

7.3.2   Lifetimes of sunspots

Sunspots have a broad range of lifetimes. Some spots form and disperse in a matter of hours, while a few during each sunspot cycle live for several months and thus are visible over several disc passages. Leader and follower spots generally have very different lifetimes. Follower spots usually live only a few days at most and seldom form a stable, long-lived configuration. Leader spots, on the other hand, tend to achieve a round, stable, slowly decaying configuration and can live up to several months (Bumba 1963; Bray and Loughhead 1964). As summarized by Zwaan (1992), a long-lived sunspot has the following properties: (i) it is a leader spot; (ii) it has one or more dark umbral cores; (iii) it is very nearly circular in shape; and (iv) it is surrounded by a moat.

Larger spots generally live longer, with lifetime T being roughly proportional to the maximum area A_{max} of the spot (Gnevyshev 1938; Waldmeier 1955),

$$A_{\mathrm{max}} = D_{\mathrm{GW}} T, \tag{7.1}$$

with $D_{\mathrm{GW}} \simeq 10 \times 10^{-6}\, A_{\odot/2}\, \mathrm{day}^{-1}$ (where $A_{\odot/2}$ is the surface area of a solar hemisphere). Petrovay and van Driel-Gesztelyi (1997) confirm this relationship and give the more precise value $D_{\mathrm{GW}} = (10.89 \pm 0.18) \times 10^{-6} A_{\odot/2}\, \mathrm{day}^{-1}$.

7.3.3   The decay of a sunspot

Cowling (1946) was the first to show that sunspots do not decay solely by Ohmic dissipation, which would occur on much too long a time scale (a few hundred years). Instead, the decay must be due in large part to fluid motions that carry magnetic flux away from the spot.

In discussing the decay mechanisms of sunspots, it is useful to distinguish between the rapid *fragmentation* of the short-lived (usually follower) sunspots and the *gradual decay* of longer-lived (usually leader) sunspots (Zwaan 1992). Follower sunspots are typically irregular in shape, with only partial penumbrae, and usually fragment within a few days after formation, as a result of being torn apart by convection. An example of this is shown in Figure 7.8. Here a stable, fairly symmetric leader spot formed and maintained its identity, while the pores and flux elements in the following-polarity region formed a rudimentary spot that lasted only several hours. The fragmentation normally starts with the appearance of very bright dots in the umbra, after which the umbra breaks up in less than a day (Zwaan 1968, 1987). Only a small percentage of the follower spots achieve a longer-lived, stable configuration. Other factors seem to affect the decay rate: spots with larger proper motion (Howard 1992) and spots at higher latitudes (Lustig and Wöhl 1995) generally decay faster.

During the gradual decay process, a long-lived spot slowly shrinks in size while roughly maintaining the same relative size, brightness, and magnetic field strength of its umbra and penumbra. The process occurs through the gradual loss of magnetic flux to the surroundings at the periphery of the spot and a redistribution of magnetic flux from the umbra to the penumbra within the spot. The spot is surrounded by an annular moat across which small, moving magnetic features (MMFs) stream away from the spot in the radial direction; some

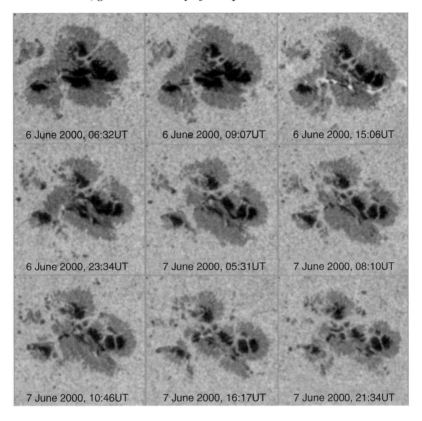

Fig. 7.8. Fragmentation of a large sunspot into a decaying cluster of small spots and pores. This sequence of white-light images, taken by the TRACE satellite, shows the central spot in NOAA AR 1926, which produced three X-class flares on 6–7 June 2000. (Courtesy of the Stanford-Lockheed Institute for Space Research.)

of these MMFs carry magnetic flux away from the spot. The decay process continues until the spot is reduced either to a pore or to a small spot that fragments.

Decay laws for sunspot area and magnetic flux

A number of observational studies have examined the rate at which the total area A of a long-lived sunspot, as seen in white light, decreases with time during its gradual decay phase. Measurements of this *photometric decay* are important because they offer clues to the nature of the decay mechanism, since different mechanisms produce different area decay laws (Martínez Pillet 2002). Most of the early studies were statistical analyses of large data sets, such as the long series of white-light photographs of sunspots taken at the Royal Greenwich Observatory beginning in 1874. These studies generally show that A decreases linearly with time, with a decay rate that is different for different spots (Greenwich 1925; Cowling 1946; Bumba 1963; Bray and Loughhead 1964; Gokhale and Zwaan 1972). For example, Bumba (1963) found a linear decay law $dA/dt = D$ for recurrent spots in the Greenwich plates, with a mean value of D equal to $-4.2 \times 10^{-6} A_{\odot/2}\,\text{day}^{-1}$ (where $1 \times 10^{-6} A_{\odot/2}\,\text{day}^{-1} = 3.321\,\text{Mm}^2\,\text{day}^{-1} = 3.843 \times 10^{11}\,\text{cm}^2\,\text{s}^{-1}$). Martínez Pillet,

Moreno-Insertis and Vázquez (1993) found a mean value of D equal to $-12.1 \times 10^{-6} A_{\odot/2}$ day^{-1} for recurring spots in the Greenwich data, and they also found that the values of D for individual spots have a lognormal distribution. They also found evidence that the decay law is slightly nonlinear, with the decay rate decreasing with time. More recently, Petrovay and van Driel-Gesztelyi (1997) studied the Debrecen photoheliograph data and found a parabolic area decay law, $dA/dt \propto \sqrt{A}$.

It is often assumed that the total magnetic flux Φ in a large sunspot is simply proportional to total area A, which then implies that the magnetic decay law $d\Phi/dt$ follows the area decay law dA/dt. Φ will be proportional to A if the distribution of magnetic flux over the spot remains self-similar during the decay, and there is evidence that this is nearly the case. Direct measurements of the total magnetic flux Φ of a spot as a function of time are desirable but difficult; they require vector magnetograms so that the flux can be measured accurately at different viewing angles as the spot moves across the solar disc. Martínez Pillet (2002) reported measurements for two slowly decaying leader spots in the group studied by Martínez Pillet, Lites and Skumanich (1997), and found in each case a very good fit to a linear decay law for both area and total magnetic flux over a 6-day time span. The $A(t)$ and $\Phi(t)$ data and linear fits for one of the sunspots are shown in Figure 7.9. The area decay rates for the two sunspots were -5.26 and $-3.87 \times 10^{-6} A_{\odot/2}$ day^{-1}, in good

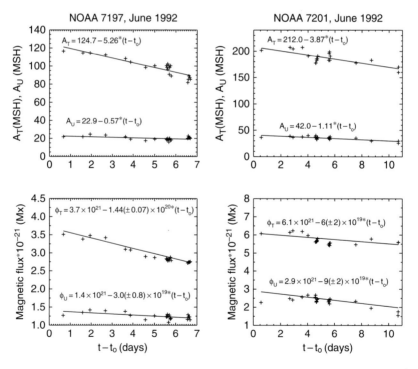

Fig. 7.9. Upper panels: total sunspot area and umbral area in units of MSH $= 1 \times 10^{-6} A_{\odot/2}$ as a function of time for two slowly decaying leader sunspots (in NOAA 7197 and 7201). Lower panels: total and umbral magnetic fluxes as functions of time for the same sunspots. (From Martínez Pillet 2002.)

agreement with Bumba's value of $-4.2 \times 10^{-6} A_{\odot/2} \, \mathrm{day}^{-1}$, and the magnetic flux decay rates were -1.44×10^{20} and $-0.6 \times 10^{20} \, \mathrm{Mx \, day}^{-1}$.

Diffusion models of sunspot decay

A linear decay law for sunspot area A is reproduced by a theoretical model involving a constant turbulent diffusion across the spot (Meyer *et al.* 1974; see also Krause and Rüdiger 1975). Petrovay and Moreno-Insertis (1997) modelled the decay with a turbulent diffusivity $\eta_{\mathrm{T}}(B)$ dependent on magnetic field strength in order to incorporate the quenching of the turbulent diffusion by a strong magnetic field ('η quenching'). Their model produces a parabolic decay law, as found by Petrovay and van Driel-Gesztelyi (1997). Rüdiger and Kitchatinov (2000) also present a model with η quenching, leading to a nonlinear decay law with decreasing decay rate.

7.4   Sunspot groups

Although individual, isolated sunspots are not uncommon, most sunspots occur in groups, which are often very large and complex. Studies of the properties and motions of spot groups reveal information about the dynamo-generated magnetic field in the solar interior. Such studies are quite demanding, however, requiring long-term synoptic observations and careful data analysis. Important contributions include the summary by McIntosh (1981) of his 25 years of sunspot observations.

Sunspot groups display a wide variety of sizes and configurations, with significant differences between groups and between different evolutionary stages of the same group. As with most complex phenomena, it has proved helpful to have a classification scheme for the various configurations. The standard scheme is the venerable Zurich classification system (Waldmeier 1947, 1955), which is essentially an evolutionary sequence for the longest-lived groups. Class A consists of a single small spot or group of spots without penumbrae and Class B consists of bipolar pairs of small spots without penumbrae; in the terminology we have used, these spots would all be called pores. Classes C and D are similar to A and B except that at least one of the spots in the group has a penumbra. Classes E and F are large bipolar groups with large spots with penumbrae and several pores, extending over at least $10°$ (Class E) or $15°$ (Class F) of longitude. Classes G, H and J consist of groups in successive stages of decay, with Class J consisting of single, round, long-lived spots. Half of all groups never progress beyond Class A or B. McIntosh (1981) developed a more detailed classification scheme, based on the Zurich system but with additional information on sunspot size, complexity and stability to aid in forecasting solar flares.

7.4.1   *The magnetic configuration of active regions and sunspot groups*

A large, fully developed active region has a bipolar magnetic configuration (see Fig. 7.10). The effective axis of the bipole is slightly inclined with respect to the solar equator, with the leading polarity being closer to the equator. The inclination angle α varies with latitude, ranging from at most a few degrees for regions nearest the equator up to $15°$ for regions at latitude $35°$ (*Joy's law*: Hale *et al.* 1919; Brunner 1930; Hale and Nicholson 1938). The smallness of the inclination angle suggests that active regions originate from a strong, buried toroidal magnetic field, with the slight inclination being produced by the Coriolis effect as the flux rises to the surface. The polarities of bipolar active regions follow *Hale's polarity law*: the east–west polarity orientation is opposite in the northern and southern

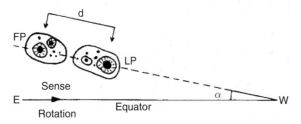

Fig. 7.10. Sketch of a large, fully developed, bipolar active region, showing areas of leading polarity (LP) and following polarity (FP). The inclination angle α varies with the latitude of the active region, ranging from $0°$ near the equator to about $15°$ at latitude $35°$. The separation distance d between the centroids of LP and FP is typically 100–150 Mm. (From Zwaan 1992, courtesy of Springer Science and Business Media.)

hemispheres, and the orientation reverses in the next 11-year sunspot cycle. Joy's law and Hale's polarity law play an important role in models of the solar dynamo (see Chapter 11).

There are systematic differences between leader and follower spots. While leader spots tend to be stable and long-lived, follower spots seldom achieve long-term stability. Of all sunspot groups that last long enough to return for a second disc passage, some 95% return with only a leader spot. Follower spots tend to be more irregular in shape than leader spots, often having only a partial penumbra, and they often break up within a few days after emergence. We have already seen an example of this rapid fragmentation in Figure 7.8; over most of its life, this active region consisted of a leader spot plus some plage of the same polarity and trailing plages of the opposite polarity.

7.4.2 Sunspots and solar rotation

Both the Sun's overall rotation and its surface differential rotation were discovered by observing the passage of sunspots across the solar disc. Here we discuss some subtler aspects of the relation between sunspots and solar rotation.

The rotation rate of sunspots around the Sun's rotation axis is observed to be slightly faster than that of the surrounding plasma (Snodgrass 1984; Stix 2002; Thompson *et al.* 2003). This difference is consistent with the idea that a sunspot's magnetic flux tube is anchored to denser material at some depth below the surface, where the rotation rate is slightly higher (in the latitude bands where sunspots live). The sunspots' rotation rate varies systematically through the solar cycle, with a distinct maximum rotation rate occurring near the time of minimum activity and a smaller maximum at the time of peak activity (Gilman and Howard 1984).

There is also evidence that the rotation rate of sunspots decreases with their age (Tuominen 1962; Nesme-Ribes, Ferreira and Mein 1993; Pulkkinen and Tuominen 1998; Hiremath 2002), perhaps due to a systematic change in the anchoring depth of the spots, or to a relaxation in the shape of the loop of emerging toroidal flux that creates the active region.

As mentioned above, in a bipolar sunspot pair the leader spot generally lies at a lower latitude than the follower spot. The solar surface differential rotation then causes the lower-latitude leader spot to rotate more rapidly than the higher-latitude follower spot, causing a slow longitudinal drift between the positions of the two spots, at a rate of about $0.1°$ per day

(Bray and Loughhead 1964). This systematic motion, due to a small difference in rotation rates between leader and follower spots, is superimposed on solar-cycle variations in rotation rate which the leader and follower spots share (Gilman and Howard 1985).

In addition to the systematic longitudinal drift due to the Sun's differential rotation, many spot groups show *proper motions* in longitude (relative to the solar rotation) and latitude (Waldmeier 1955), presumably caused by the dynamical behaviour of the corresponding magnetic flux tubes beneath the surface. Also, some sunspots are seen (in white light) to rotate slowly around their central axes. The total rotation can be significant, as much as $200°$ over a period of 3–5 days (Brown *et al.* 2003). This rotation causes the coronal loops emanating from the spot to twist and, in some cases, to erupt as a flare. The rotation of the sunspot is most likely due to the relaxation of a twist in the spot's magnetic flux bundle beneath the solar surface; in this case, one would expect a systematic preference for the sense of rotation in each hemisphere. There seems to be insufficient evidence on this point, although the results of Brown *et al.* (2003) show anticlockwise rotation in five of six northern-hemisphere spots.

Possible meridional (north–south) motions of sunspots and spot groups have been looked for in statistical analyses of the large Greenwich and Mount Wilson data sets, with results that are not all in accord (see Pulkkinen and Tuominen 1998 and Wöhl and Brajša 2001 and references to earlier papers therein). In these studies, the scatter in the data points is very large and the statistical significance of the results is not strong. We mention only two interesting results here. Howard and Gilman (1986), using the digitized Mount Wilson data, obtained the surprising result of a northward motion at intermediate latitudes (roughly from 10 to $25°$) in *both* hemispheres. Using the Greenwich data, Wöhl and Brajša (2001) found that in each hemisphere sunspot groups tend to move away from the latitude of the centroid of all spots in that hemisphere. Results from helioseismology, however, show a shallow flow toward the zones of maximum activity (Beck, Gizon and Duvall 2002), although this flow may reverse direction deeper down (Zhao and Kosovichev 2004).

7.5 Dissolution of active regions

By the time that all of the available magnetic flux has emerged, an active region has already begun to show signs of decay (Zwaan 1992). Most of the sunspots are already diminishing in size and total magnetic flux, and only one or two dominant spots continue to grow for a few more days. Then most of the pores disappear, followed by most of the sunspots, leaving perhaps just one stable leading spot, which decays away slowly over a period of a few weeks or occasionally up to a few months. During this whole time, the active region has been spreading over a greater part of the solar surface while the filling factor of its magnetic field has been decreasing. Holes appear in the plage regions as magnetic flux is swept to the boundaries of supergranules, forming an enhanced bright network which eventually fades until it is indistinguishable from the quiet-Sun network. The lifetime of an active region is roughly proportional to the maximum value of its total magnetic flux, with a constant of proportionality of about 1×10^{20} Mx day^{-1} (Golub 1980). Exceptions to this decay scenario are the active regions appearing in succession within an activity nest, where a new region emerges before the previous region has fully decayed.

What happens to the magnetic flux of an active region as it disappears? Part of the flux leaves the region by gradual dispersal over a wider area, the dispersed flux then forming part of the quiet-Sun magnetic field. However, this dilution cannot be the only process involved;

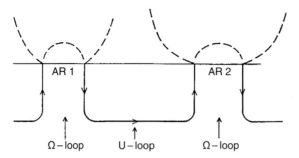

Fig. 7.11. The formation of a U-loop between two rising Ω-loops along a single toroidal flux tube. (From Zwaan 1992, courtesy of Springer Science and Business Media.)

surface magnetic flux is often seen to disappear without visible transport away from its location (e.g. Wallenhorst and Howard 1982; Wallenhorst and Topka 1982; Wilson and Simon 1983; Simon and Wilson 1985; Topka, Tarbell and Title 1986). Photospheric magnetic flux can seemingly disappear *in situ* through two mechanisms: the emergence of a U-loop or the re-submergence of an Ω-loop (Zwaan 1992). As shown in Figure 7.11, U-loops form naturally in a rising toroidal flux tube, between the Ω-loops that form active regions. While an Ω-loop can rise rapidly into the atmosphere by draining plasma down its two legs, a U-loop cannot rise so rapidly because the plasma is trapped within it, and some mechanism of moving mass across field lines is required for complete escape of the loop (Parker 1984). As Spruit, Title and van Ballegooijen (1987) have pointed out, a buoyant U-loop will rise, expand, and fragment near the surface, where subsurface convection will then dominate the weakened field and arrange it into the typical intranetwork pattern. Also, smaller-scale Ω-loops forming along a large U-loop might rise, expand laterally, and reconnect where they meet to form closed O-loops that decay away through Ohmic dissipation. Convincing observational evidence for the emergence of a U-loop between two active regions has been presented by van Driel-Gesztelyi, Malherbe and Démoulin (2000).

Re-submergence of an Ω-loop is no doubt an important process in the case of a long-lived activity nest, which survives through successive emergences and retractions of an Ω-loop whose coherence is maintained by the strong underlying toroidal flux tube.

7.6   Flux emergence in the quiet Sun

So far, we have only discussed large active regions, which contain sunspots and behave as part of the Sun's activity cycle (see Section 10.1 below). However, active regions come in a broad range of sizes. There is a hierarchy, running from large-scale activity in regions containing magnetic fluxes Φ that are greater than 3×10^{20} Mx, through bipolar *ephemeral active regions* with fluxes between 3×10^{20} and 3×10^{18} Mx, to small-scale fields that nestle between granules and have fluxes as low as 10^{15} Mx. In this section we consider both ephemeral regions and intergranular magnetic fields.

The intermediate-scale fields emerge across the entire solar disc, from pole to pole, but they are most apparent in the quiet Sun, away from large active regions. Here the emergence and reconnection of magnetic fields give rise to soft X-ray emission in the form of X-ray bright points (XBP). These were first discovered in rocket flights (Krieger, Vaiana

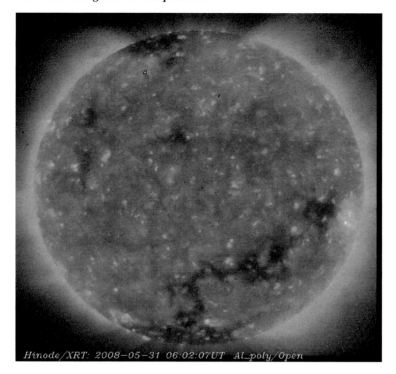

Fig. 7.12. Soft X-ray image from the X-Ray Telescope on Hinode, showing X-ray bright points distributed across the entire solar disc, at a time of very low large-scale activity. Note the dark coronal holes. (Courtesy of NAOJ/SAO/JAXA/NASA.)

and Van Speybroeck 1971) and then observed from space, with ever-increasing precision, by the Skylab, Yohkoh and Hinode satellites. Figure 7.12 shows the distribution of XBPs across the solar disc at sunspot minimum, and indicates the prevalence of ephemeral active regions.

These fields give rise to a 'magnetic carpet' that covers the surface of the Sun, as shown in Figure 7.13 (Title and Schrijver 1998; Title 2000). This carpet is composed of individual magnetic elements with fluxes around 10^{18} Mx and intense kilogauss magnetic fields (Stenflo 1973). The ephemeral regions appear all over the solar surface, and are largely uncorrelated with sunspots and large-scale activity; they seem to owe their origin to small-scale dynamo action (Durney, De Young and Roxburgh 1993; Cattaneo 1999; Hagenaar, Schrijver and Title 2003) at relatively shallow depths below the photosphere, as discussed in Section 11.5.

7.6.1 Ephemeral active regions

Ephemeral active regions emerge as small magnetic bipoles that last for a relatively short time (Harvey and Martin 1973; Harvey-Angle 1993). They generally have areas less than 2.5 square degrees and lifetimes that are less than a day and typically only a few hours. While at most only a few large active regions emerge each day, hundreds of ephemeral regions do so and together contribute as much surface magnetic flux as a single large active

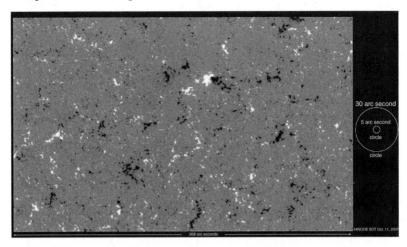

Fig. 7.13. The magnetic network and the mixed-polarity magnetic carpet on the quiet Sun. In this magnetogram, obtained with the Solar Optical Telescope on Hinode, regions with line-of-sight field strengths less than 5 G are shown in grey, and light (dark) regions indicate parallel (antiparallel) magnetic fields. The region shown has a width of about 200 Mm. (Courtesy of A. M. Title.)

region. The rate of emergence corresponds to an unsigned flux of 5×10^{23} Mx day^{-1} over the entire solar surface (Hagenaar 2001). The orientation of the emerging bipoles is scarcely related to that of the active regions that are involved in the solar cycle: Harvey-Angle (1993) and Hagenaar (2001) found that 60% of ephemeral regions have a net latitudinal orientation corresponding to that of the sunspot cycle, but Hagenaar, Schrijver and Title (2003) see no significant correlation. On the other hand, the rate of emergence does seem to vary, by $\pm 20\%$, in *antiphase* with the solar cycle (Harvey-Angle 1993; Hagenaar 2001; Hagenaar, Schrijver and Title 2003).

The fate of the flux that emerges in ephemeral regions is closely related to the pattern of supergranular convection at the solar surface, which was mentioned in Section 2.5. The motion in a supergranule corresponds to a radial outflow, carrying magnetic fields and granules with it, towards the magnetic network that forms the supergranule's boundary (Simon and Leighton 1964; Simon *et al.* 1988). Thus any small-scale fields are bound to accumulate in the network. Observations with the Michelson Doppler Imager (MDI) on SOHO show that ephemeral regions emerge in the interiors of supergranules and preferentially near their centres. Bipolar features with fluxes between 10^{18} and 10^{19} Mx appear first as horizontal fields, whose footpoints move rapidly apart, attaining a separation of about 7000 km in the next 30 minutes. Meanwhile, these flux concentrations are shredded by turbulent granular convection into fragments containing only about 10^{17} Mx but with locally intense fields of around 1500 G. These fragments drift apart and are transported within a few hours to different parts of the network; once there, they migrate along the supergranule boundaries until they meet flux elements of opposite polarity, reconnect and disappear (Title 2000). In a steady state, the total unsigned flux in the network amounts to 2×10^{23} Mx, implying that the lifetime of emerging flux is less than a day. Kinematic modelling allows the whole process of eruption, fragmentation, transport and cancellation to be represented numerically (Simon, Title and Weiss 2001; Parnell 2001).

7.6.2   *Intergranular magnetic fields and smaller-scale flux emergence*

The presence of small-scale, mixed-polarity fields within the supergranular net-work has been known for a long time (Livingston and Harvey 1971). These intranetwork[3] fields give rise to a 'pepper and salt' appearance in magnetograms like that in Figure 7.13. As the spatial resolution of magnetograms has improved, they have revealed the emergence of mixed-polarity magnetic fields on smaller and smaller scales. Recent progress can be fol-lowed both in ground-based observations, from Big Bear (Wang *et al.* 1995) and Sacramento Peak (Lin and Rimmele 1999; Lites 2002; Rimmele 2004; Socas-Navarro, Martínez Pillet and Lites 2004), as well as the SVST (Berger and Title 1996; Topka, Tarbell and Title 1997) and SST (Berger *et al.* 2004; Sánchez Almeida, Márquez and Bonet 2004; Rouppe van der Voort *et al.* 2005; Berger, Rouppe van der Voort and Löfdahl 2007) on La Palma, and in those from space, on Spacelab 2 (Simon *et al.* 1988), with MDI on SOHO (Schrijver *et al.* 1998; Title and Schrijver 1998; Ortiz, Solanki and Domingo 2002) and with the Solar Optical Telescope (SOT) on Hinode (Centeno *et al.* 2007; Lites *et al.* 2007).

The pattern of intranetwork fields is organized on two scales, by granules (with diameters around 1 Mm) and by mesogranules (with diameters of 5–8 Mm). The granules correspond to the vigorous energy-carrying scale of convection at the photosphere, but the mesoscale pattern, though driven by convection, is selected as a result of nonlinear interactions between the granules; mesocells are a characteristic feature of both Boussinesq and compressible convection (Stein and Nordlund 1998; Cattaneo, Lenz and Weiss 2001; Rincon, Lignières and Rieutord 2005). At the solar photosphere, mesogranules were discovered by November *et al.* (1981; see also Shine, Simon and Hurlburt 2000), and Spacelab observations showed that magnetic flux around an active region was carried into a mesogranular network (Simon *et al.* 1988). Higher resolution observations have confirmed that the intranetwork field has a mesogranular structure in the quiet Sun as well (Domínguez Cerdeña 2003; Lites *et al.* 2007).

The most striking feature of these small-scale fields is their relationship to granules. Intense vertical fields accumulate in the dark intergranular lanes, where flux sheets can form, and especially at corners, where isolated flux tubes can appear. These patterns are constantly changing: the fields move as a 'magnetic fluid' without any evidence of quasipermanent flux tubes (Berger and Title 1996; Berger *et al.* 2004; Rouppe van der Voort *et al.* 2005; Berger, Rouppe van der Voort and Löfdahl 2007). If sufficient magnetic flux is present (as in the network, or near sunspots), the fields have a ribbon-like structure at the photo-sphere, but with weaker overall flux ephemeral point-like features are more prevalent, in keeping with the numerical results described in Section 7.2.3 above. These magnetic ele-ments are associated with localized 'bright points' in the continuum (Dunn and Zirker 1973; Mehltretter 1974; Muller 1983) and especially in CH G-band emission (Berger *et al.* 1995, 1998), which acts as a reliable proxy for magnetic fields. Intergranular G-band emission is naturally most prominent in plage regions, and Figure 7.14 shows two very clear examples.

Berger, Rouppe van der Voort and Löfdahl (2007) emphasize the importance of the Wilson depression within magnetic flux elements. If they are sufficiently slender, the $\tau_{500} = 1$ level within them will be heated laterally by radiation from the hot walls of adjacent granules, making them appear bright at disc centre; only if they are wide enough to be optically thick

---

[3] Both *internetwork* and *intranetwork* appear in the literature. We prefer using the latter to describe the areas that are enclosed by the network.

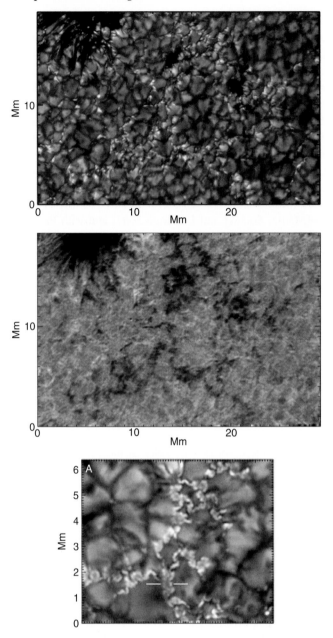

Fig. 7.14. Fine structure of intergranular magnetic fields. The top panel shows a G-band image of filamentary and point-like structures, clustered around mesogranules in the neighbourhood of a sunspot. The middle panel displays the corresponding magnetic fields. The lower panel shows a filigree-like pattern of G-band emission in greater detail. The ribbon indicated has a width of about 500 km. (From Berger *et al.* 2004, 2007.)

will they cool down and appear dark, as micropores. At the limb, the bright sides of the granules are visible through the transparent flux elements, giving rise to faculae as proxy evidence of magnetic fields.

The predominantly vertical fields move through intergranular lanes, as in the numerical models with an imposed magnetic field that were discussed in Section 7.2.3. Intragranular flux elements can also emerge as bipoles (Lites *et al.* 1996) or, after reconnection, disappear. Centeno *et al.* (2007) present an example of bipolar flux emergence on a granular scale, obtained using the SOT with a spectropolarimeter aboard Hinode. These measurements of the vector field **B** reveal the initial appearance of a horizontal field centred on a bright granule, which gradually spreads until two oppositely directed vertical fields appear at either end of a granule diameter. Thereafter, the horizontal field fades away (presumably rising upwards out of the region where the absorption lines are formed) while the vertical fields remain. Numerical simulations of flux loops rising into the photosphere reveal similar behaviour (Cheung, Schüssler and Moreno-Insertis 2007).

At the centre of the solar disc, the magnetic field in flux elements is predominantly longitudinal but flux conservation demands that oppositely directed longitudinal fields in the photosphere must be connected by transverse fields higher up. Harvey *et al.* (2007) reported measurements of "seething" horizontal fields on scales greater than $3''$. Lites *et al.* (1996) measured transverse, predominantly horizontal fields with typical lifetimes of 5 minutes associated with flux emerging on a $1''$ scale. Hinode observations of the vector field in an unusually quiet Sun have attained much higher resolution: Lites *et al.* (2007) and Orozco Suárez *et al.* (2007) find relatively strong horizontal fields – typically of order $100\,\mathrm{G}$ and predominantly at the edges of granules – while vertical fields are confined to intergranular lanes. This overall pattern is consistent with the observations of an individual event described above. The likely source of this disordered emerging flux is probably a turbulent small-scale dynamo operating on a granular (and mesogranular) scale and largely independent of the dynamo processes that give rise to cyclic activity and ephemeral active regions. This process has been demonstrated numerically both in Boussinesq calculations (Cattaneo 1999; Cattaneo, Lenz and Weiss 2001) and in more realistic models (Vögler and Schüssler 2007).

8

Magnetic activity in stars

If we were to observe the Sun from the distance of α Centauri (4.3 light years), we would not be able to resolve its spots directly or to detect luminosity variations as they came and went; nor could we measure its magnetic field. Although we naturally expect that there should be analogues of solar magnetic activity on lower main-sequence stars that are similar to the Sun (Tayler 1997; Rosner 2000), we can only detect this activity through indirect measurements of other effects that are known to be associated with active regions on the Sun itself. These effects include X-ray emission from stellar coronae, optical and radio emission from flares and enhanced chromospheric emission, notably in the H and K lines of singly ionized Ca II (Wilson 1994; Schrijver and Zwaan 2000). As we shall see, there are indeed also some stars that are much more active than the Sun, whose magnetic fields can be measured directly; such stars also exhibit substantial variations in luminosity that can be ascribed to the presence of starspots on their surfaces.

8.1 Stellar Ca II emission

The most widely used indicators of stellar magnetic activity are the emission cores of the Ca II H and K absorption lines. On the Sun, Ca II emission forms a chromospheric network, first observed by Hale (Bray and Loughhead 1974), which corresponds to the magnetic network that outlines supergranules in the photosphere. This emission could be detected if the Sun were viewed from a nearby star, and the cyclic variation of solar activity would also be apparent, as can be seen from Figure 8.1. The strong correlation between Ca II H and K emission and magnetic fields on the Sun (Skumanich, Smythe and Frazier 1975; Schrijver and Zwaan 2000) provides a firm basis for also using Ca II emission as a proxy measure of stellar magnetic fields. Early observations were reported by Eberhard and Schwarzschild (1913) but it was not until 1966 that a systematic programme of observations was initiated, by Olin Wilson on the 100-inch telescope at Mount Wilson in California. This programme was continued for almost 40 years, providing measurements of activity on about 2500 nearby stars as well as monitoring variations in activity (to be discussed in Chapter 10 below) on up to 400 of them.

Surveys of nearby main-sequence stars in both northern and southern skies reveal a wide range of magnetic activity (Vaughan and Preston 1980; Soderblom 1985; Baliunas *et al.* 1995; Henry *et al.* 1996; Wright *et al.* 2004). Figure 8.2 shows a measure, R'_{HK}, of normalized chromospheric Ca II emission as a function of spectral type (i.e. of effective surface temperature) for solar-type stars. The position of the Sun is indicated, together

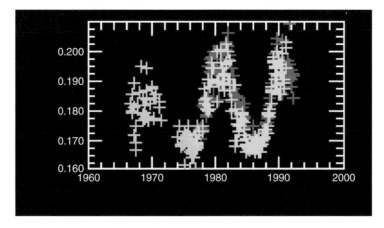

Fig. 8.1. The Sun as a star. Variation of Ca II H and K emission with the solar cycle. (Courtesy of the Mount Wilson H-K Project.)

with its range of variation with the solar cycle. It is apparent that there is a spread of an order of magnitude in $R'_{\rm HK}$ for stars of the same spectral type. Moreover, there appears to be an underpopulated gap, around $\log R'_{\rm HK} = -4.75$, between groups of more active and less active stars. Henry *et al.* conclude that this so-called 'Vaughan–Preston gap' is a real effect, with about 30% of the stars lying above it. They go further to distinguish a small subgroup of very active stars with $\log R'_{\rm HK} > -4.20$ and one of inactive stars with $\log R'_{\rm HK} < -5.1$; the former contains less than 3% of the stars, while the latter (which may include stars undergoing a grand minimum) contains 5–10% of the total sample. The hyperactive group includes rapidly rotating binaries, with components that have already evolved off the main sequence.

It follows from Figure 8.2 that a star of given mass, for instance a G2 star like the Sun with $B - V = 0.66$, may exhibit a wide range of magnetic activity. Further information can be gained by studying stellar clusters. All members of a cluster have the same age, and this age can be determined from lithium abundances or by noting where massive stars have evolved off the main sequence. It is found that stars in a given cluster – for instance the Hyades, with an age of about 600 Myr – lie on a roughly horizontal strip in diagrams like Figure 8.2 (Soderblom 1985) and that younger clusters show stronger magnetic activity, as measured by Ca II emission, than do older clusters. It follows that an individual star will trace a downward trajectory in such a diagram as it evolves and, in particular, that the Sun has moved downwards during its lifetime of 4.6 Gyr on the main sequence; thus its magnetic activity has (on average) decreased with age until it reached its present state.

8.2   Variation of activity with rotation rate and age

The crucial physical parameter that determines the strength of magnetic activity in a star is its angular velocity (Hartmann and Noyes 1987). For a rapidly rotating star the angular velocity can be estimated from the Doppler broadening of spectral lines, which is proportional to $v \sin i$, the projected component of its rotational velocity, where v is its equatorial velocity and i is the angle between the rotation axis and the line of sight. The angle i is generally not known. If there is some observable feature that rotates with the star, its rotation period P can be directly measured. Thus the rotation rates of stars with local sources

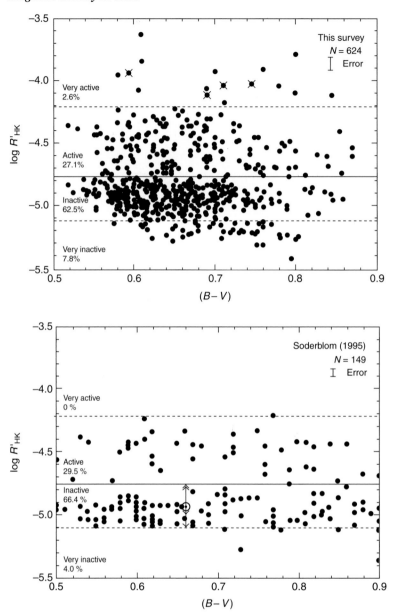

Fig. 8.2. Ca II H and K emission in solar-type (F, G and K) stars. A normalized measure of Ca II emission, R'_{HK}, for stars of different spectral types, measured by $B - V$, visible in the southern (upper panel) and northern (lower panel) hemispheres. The lower panel also shows the range of the Sun's variation. (From Henry *et al.* 1996.)

of Ca II emission can be determined and differential rotation can even be detected. This has been achieved for many of the stars in the Mount Wilson data set (Noyes *et al.* 1984; Baliunas *et al.* 1995; Saar and Brandenburg 1999). When R'_{HK} is compared with the angular velocity $\Omega = 2\pi/P$ it becomes apparent that, for stars of a given spectral type, Ca II emission

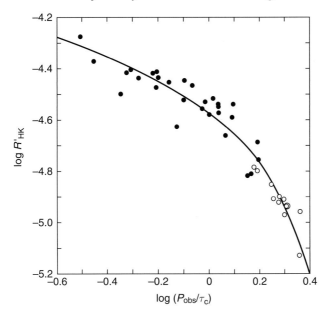

Fig. 8.3. Dependence of stellar Ca II emission on rotation rate. The measure $\log R'_{HK}$ of H and K emission is plotted against $\log Ro$ for a selection of nearby stars. The Rossby number Ro is inversely proportional to Ω, and the line is a cubic fit to the data. (From Noyes *et al.* 1984.)

increases with increasing angular velocity. In order to take some account of variations in stellar structure, Noyes *et al.* (1984; Hartmann and Noyes 1987; see also Montesinos *et al.* 2001) introduced the Rossby number $Ro = P/\tau_c = 2\pi/(\Omega\tau_c)$ as a parameter; here τ_c is a crude estimate (derived, for instance, from stellar models that rely on mixing-length theory) of the convective turnover time at the base of the star's convection zone. This *ansatz* helps to reduce the spread between stars of different masses and therefore of different spectral types. Figure 8.3 shows the relationship between R'_{HK} and Ro for 40 lower main-sequence stars, including the Sun. It is clear that this measure of magnetic activity decreases monotonically as Ro increases – or increases monotonically as the rotation period P decreases – though R'_{HK} tends to saturate for very short rotation periods.

The rotational history of a star such as the Sun can be established by measuring the rotation rates of younger G stars as a function of their ages. In general, it is found that pre-main-sequence T Tauri stars have a spread of angular velocities, depending in part on whether they possess discs and massive planets. As they approach the main sequence they collapse, conserving their angular momentum and therefore spinning up (Soderblom *et al.* 1993b). On arriving at the zero-age main sequence (ZAMS), after about 20 Myr, G stars have a range of angular velocities up to a maximum of $\Omega \approx 100\,\Omega_\odot$ (corresponding to a rotation period of only 6 hr), which is about twice the critical rotation rate for centrifugal break-up. Their rotational evolution can be followed by using observations of stars in clusters of different ages, starting with those that are very young. In the youngest clusters, such as IC2602 and IC2391 which are only 30 Myr old (Stauffer *et al.* 1997; Barnes *et al.* 1999), or α Persei which is 50 Myr old (Stauffer *et al.* 1985; Stauffer, Hartmann and Jones 1989; Prosser 1992),

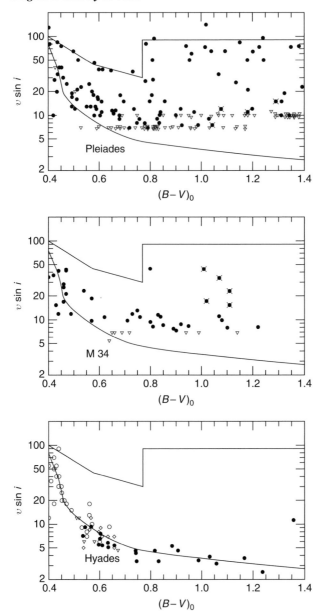

Fig. 8.4. $v \sin i$ vs. $B - V$ for the Pleiades, M34 and the Hyades, with ages 100 Myr, 140 Myr and 600 Myr, respectively. (From Soderblom, Jones and Fischer 2001.)

late-type stars display a wide range of rotational velocities, with $v \sin i$ ranging from 10 to 200 km s^{-1}. The Pleiades have an age of 100 Myr and the peak values of $v \sin i$ range from 50 km s^{-1} for G stars to 100 km s^{-1} for K and M stars, as shown in Figure 8.4 (Stauffer and Hartmann 1987; Soderblom *et al.* 1993a; Soderblom, Jones and Fischer 2001). These peak values continue to decline in clusters like NGC 2516 (Terndrup *et al.* 2002), NGC 6475

(James and Jeffries 1997) and M34 (Soderblom, Jones and Fischer 2001), with ages of 140, 220 and 250 Myr, respectively. By an age of 500 Myr, as in the U Ma cluster (Soderblom and Mayor 1993), the spread has almost disappeared, and in the Hyades, with an age of 600 Myr, there is very little scatter about a mean relationship between rotation rate and spectral type (Radick *et al.* 1987; Soderblom *et al.* 1993a; Soderblom, Jones and Fischer 2001), with an average value of $v \sin i$ for G stars of only $6 \, \text{km s}^{-1}$.

Figure 8.4 illustrates the rapid spin-down of fast rotators, which is followed by a more gradual decay in their rotation rates until, for the Sun at an age of 4.6 Gyr, $v \approx 2 \, \text{km s}^{-1}$. A significant early result was that of Skumanich (1972), who claimed that, for a broad sample of stars in the Pleiades, Hyades and Ursa Major, both chromospheric activity (as measured by Ca II emission) and stellar rotation rate decrease with age as $t^{-1/2}$. Since Ca II emission is roughly proportional to the surface magnetic field strength in a star, this would indicate that magnetic field strength is proportional to rotation rate and that both decay together as the inverse square root of time. With the huge increase in the number of measured rotational velocities over the last 30 years, it has become clear that Skumanich's simple law can only be applied, as a rough approximation, to stars that are younger than those in the Hyades.

The mechanism responsible for stellar spin-down relies on the star's magnetic field, which is generated by dynamo action in the star's interior (see Chapter 11) and depends, as we have seen, on the rotation rate itself. The magnetic field is responsible for heating a corona to temperatures of several million degrees (thus emitting X-rays as thermal radiation) and so for driving a stellar wind. The rate at which angular momentum is lost depends on the flow of plasma across the Alfvénic surface, where the radial velocity equals the Alfvén speed and the open field lines therefore act as a magnetic lever arm (Mestel 1999). Hence magnetic braking is a strongly increasing function of the field strength and so of the rotation rate itself. The star's internal magnetic field is probably strong enough to ensure that its convection zone remains coupled to the radiative interior and that the entire star is therefore almost uniformly rotating as it spins down (Mestel 1999). As the field grows weaker, so magnetic braking becomes less effective; apparently the angular velocity never sinks so low during the star's main-sequence lifetime that its field actually disappears.

8.3 Vigorous activity in late-type stars

The most active stars are those that are most rapidly rotating, and the consequent Doppler broadening of spectral lines (along with other effects) makes it difficult to detect the Zeeman signature of their magnetic fields. Until recently, most measurements have been based on the method first implemented by Robinson, Worden and Harvey (1980) in which the Zeeman effect is extracted through a comparison of line profiles of a magnetically sensitive line and a magnetically insensitive line. Measurements become easier in the infrared, where the Zeeman splitting is relatively stronger (Valenti, Marcy and Basri 1995). By using a number of lines with different Landé factors it is possible to obtain estimates of the filling factor, f, and the magnetic field strength, $|B|$ (Saar 1996). Typical values show increases in activity towards later spectral types, ranging from $|B| \approx 2 \, \text{kG}$ and $2\% \leq f \leq 20\%$ for main-sequence G and K dwarfs, through to $|B| \approx 4 \, \text{kG}$ and $f \approx 50\%$ for dwarf M emission-line (dMe) stars; for comparison, the Sun has $|B| \approx 1.5 \, \text{kG}$ and $f \approx 1\%$. These field strengths are all such that the magnetic pressure is able to balance the gas pressure at the photosphere. Note, however, that these values refer to network fields rather than to dark spots: Zeeman–Doppler imaging of starspots (Donati *et al.* 1997) will be discussed in Section 9.1.4.

It is empirically clear that any rotating star with a turbulent outer convection zone generates magnetic fields that lead to the presence of a hot corona which produces X-ray emission. Observations with the Einstein, ROSAT and XMM-Newton satellites have amply confirmed that all late-type stars are X-ray sources (Schmitt and Liefke 2004). Saar (1996; see also Shi and Zhao 2003) finds that the surface flux density of X-ray emission, F_X, varies approximately as the normalized magnetic flux density $f|B|$. As would be expected, both F_X and the X-ray luminosity L_X increase as the rotation rate Ω increases, or decrease as Ro increases (Hempelmann *et al.* 1995).

Flaring activity is prevalent in dMe stars, which have only a tenth of the Sun's surface area, and also in the active RS Canum Venaticorum (RS CVn) stars (Tayler 1997). The latter are rapidly rotating members of close binary systems, with their spin and orbital angular velocities tidally coupled; they have already evolved off the main sequence and are among the most vigorously active stars. These flares are visible in the optical range, and may correspond to a temporary increase of one or more magnitudes in luminosity. The optical emission is also accompanied by flaring at both radio and X-ray frequencies. Solar flares are insignificant by comparison with such displays.

8.4   Other magnetic stars

Since magnetic fields are omnipresent in galaxies and the interstellar medium, we should expect most stars to harbour some magnetic flux, whether dynamo-generated or a fossil relic of some earlier stage of evolution (Rosner 2000; Mestel and Landstreet 2005). Unfortunately, these fields are not always easy to detect. The strongest fields are found in neutron stars, which are compact remnants of exploding supernovae. Typical pulsars have fields of order 10^{12} G, while the subclass of magnetars exhibit magnetic fields with strengths of up to 10^{15} G – the strongest fields that are known (Lyne 2000; Rüdiger and Hollerbach 2004). The fields detected in white dwarfs are much weaker, ranging from 10^7 G to over 10^8 G (Mestel and Landstreet 2005). Among normal massive upper main-sequence stars, magnetic fields of around 1.5 kG have recently been measured in a couple of O stars, along with weaker fields in some B stars (Donati *et al.* 2001, 2006a; Wade *et al.* 2006).

The best-known magnetic stars are, however, those stars of spectral classes A and B that exhibit spectroscopic peculiarities corresponding to anomalous abundances, the Ap and Bp stars (Mestel 1999; Mestel and Landstreet 2005). The mean line-of-sight components of their magnetic fields, which can be determined by measuring circularly polarized components of magnetically sensitive spectral lines, range from a lower observational limit of a few gauss up to about 20 kG, with typical rms values of around 300 G. The corresponding values of the mean field strength, which can be measured directly from the Zeeman splitting for sufficiently strong fields, are significantly larger (Mathys *et al.* 1997; Bagnulo *et al.* 2003, 2004), ranging up to 34 kG for Babcock's star, HD 215441 (Babcock 1960). The periodic variation of these fields is ascribed to the rotation of the stars; compared to typical stars of the same spectral types, they turn out to be relatively slow rotators, with periods ranging from half-a-day up to half-a-century (but typically of several days). Their spectra show anomalous abundances of certain elements, including rare earths, that vary with the same rotational period. The magnetic field of such a star is adequately represented by a rotating dipole whose axis is inclined to the axis of rotation. Such an oblique rotator, with a long-lived fossil field, must be contrasted with the oscillatory fields, maintained by dynamo action, that will be discussed in Chapter 11.

9

Starspots

Starspots are analogues of sunspots that appear as dark patterns in a stellar atmosphere and modulate the radiative output over the visible hemisphere of the star. Early detections of starspots, beginning with the work of Kron (1947), are discussed in Section 2.8. Intensity patterns of various types have been detected on many stars, not all of which are analogous to sunspots; for instance, optical aperture synthesis has revealed convection cells on Betelgeuse. Here we shall restrict attention to those patterns that are most like sunspots, although in almost all cases the dark areas are substantially larger than spots on the Sun, for otherwise they would not have been detectable. A good example of such a large starspot has already been shown in Figure 1.8. We infer that starspots share with sunspots a magnetic origin.

The motivation to study starspots comes from many areas of investigation, including the study of magnetoconvection and the study of stellar activity and patterns of emergence of magnetic flux. Starspots and other surface intensity patterns provide the most accurate means of determining stellar rotation periods and also allow the detection of surface differential rotation, a key ingredient in understanding stellar dynamos. The possible effects of starspots often have to be considered in interpreting data related to stellar pulsations or searches for extrasolar planets.

Methods for detecting and mapping starspots on stellar surfaces have advanced rapidly over the past two decades. Techniques have been developed for determining starspot temperatures and areas and their location on the stellar surface, and for using starspots to determine surface differential rotation on stars much as we do for the Sun with sunspots. Even the direct detection of magnetic fields in starspots has been proposed. The study of starspots is moving fast, and a future breakthrough, such as direct interferometric imaging of stellar surfaces, may render some of the specific results we present here obsolete, but our principal aim in this chapter is to give a thorough introduction to the fundamentals of research on starspots that might remain useful for some time to come.

9.1 Observing techniques

9.1.1 Photometry

Starspots, or more general surface brightness patterns, can be detected by searching for photometric light-curve variations whose period is the same as the rotation period of the star, provided the surface inhomogeneities have lifetimes extending over several rotations. Multi-colour photometry and careful modelling of limb darkening allow one to estimate the temperatures and surface distribution of cool starspots. Although photometric studies of starspots generally provide less information than the spectroscopic methods to be described below, they can take advantage of the existence of long-term sequences of single-colour or multi-colour photometric data for many active stars, allowing studies of the long-term evolution of the spots and the detection of spot cycles. Photometric monitoring of starspots can be done from smaller telescopes, and a number of dedicated programmes are under way. The precision of photometry from ground-based telescopes can be 1 milli-magnitude or higher in several colour bands.

In order to deduce starspot properties from photometric data, both forward and inverse methods have been used. In the forward methods (sometimes called 'spot modelling' or 'light-curve modelling'), one attempts to reproduce the light curve of a spotted star by trial and error using trial distributions of spotted areas of certain simple shapes and sizes at assumed temperatures. It is often assumed that the spots are circular, with either uniform intensity (Budding 1977; Vogt 1981b; Poe and Eaton 1985) or with distinct umbral and penumbral intensities (Dorren 1987). Alternatively, the spots can be taken to be bounded by lines of latitude and longitude (Bopp and Evans 1973; Eaton and Hall 1979). In many cases a single spot is unable to fit the observed light curve, so two or more spots must be assumed. For example, Figure 9.1 shows an early two-spot model of the light curve of II Pegasi (HD 224085), a rapidly rotating (period 6.7 days) RS CVn binary of spectral class K2 IV. In general, it is not possible to arrive at unique values of the positions, sizes, shapes, and temperatures of the starspots by this method alone; additional information is required,

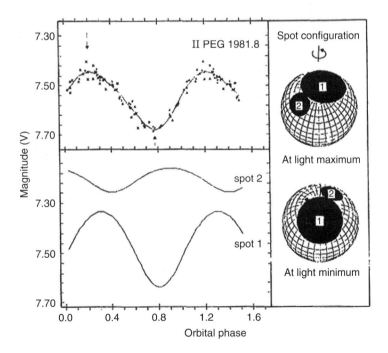

Fig. 9.1. Two-spot fit to the light curve of II Peg. The spot configuration is shown at two rotational phases, corresponding to light maximum and light minimum (indicated by arrows on the light curve on the left). The spots have an assumed surface temperature of 3300 K, whereas the unspotted surface is at temperature 4500 K. (From Rodonò *et al.* 1986.)

such as that provided by multi-colour photometry or independent measurements of spot temperatures using molecular lines or line-depth ratios.

Good constraints on starspot properties can be obtained using multi-colour photometry and modelling simultaneously the variations in total light and in colour. Separating geometric and temperature effects requires photometric light curves in at least two colours and careful account of the wavelength dependence of limb darkening. Vogt (1981a,b) developed a forward method for determining both temperature and effective area of starspots from standardized V and R light curves obtained over a star's rotation period. The effective spot temperature is obtained from a calibrated dependence of T_{eff} on the colour difference $V - R$. Light curves assuming a single, circular spot are computed for many different combinations of spot size and latitude and different inclinations of the rotation axis and are compared automatically with the actual light curves to produce a best-fit model. Spot modelling has been automated with user-friendly computer codes (e.g. Ribárik, Oláh and Strassmeier 2002) in order to handle the large amount of photometric data collected by dedicated robotic telescopes.

Various inverse methods for deducing starspot properties from light curves have been developed. For example, the method of matrix light-curve inversion (Harmon and Crews 2000), originally developed for imaging the surface of Pluto or an asteroid (Wild 1989, 1991), divides the stellar surface into a large number of patches bounded by circles of latitude and meridians of longitude. Each patch is assumed to radiate uniformly, and the method then seeks the set of patch intensities that produces the best fit to the observed light curves.

Messina *et al.* (2006) applied two-band photometry (V and B) to three active K dwarf stars (AB Doradus, LQ Hydrae and DX Leonis), producing synthetic V and $B - V$ curves based on Dorren's (1987) two-component spot model with a range of spot sizes, latitudes and temperatures. While this method does not lead to a unique solution, it does yield a narrow range of spot temperatures for each star at each epoch, and a mean temperature that varies from epoch to epoch as the starspots evolve. They also found that the best fit to the data is obtained with a two-component (two-temperature) spot model consisting of either dark umbrae and less dark penumbrae, or dark spots and bright faculae.

9.1.2   *Spectroscopy*

With photometric data alone one cannot completely distinguish between the effects of spot area and spot temperature on the light curves, but with the addition of spectroscopic measurements to determine spot temperature this ambiguity can be resolved.

Line-depth ratios

Gray and collaborators showed that line-depth ratios are an effective diagnostic of temperatures in stellar atmospheres, capable of determining absolute temperatures to within a few tens of degrees and detecting temperature differences of less than $10\,K$ (Gray 1994). Line-depth ratios also provide a useful measure of temperatures in starspots (Gray 1996; Catalano *et al.* 2002a). The passage of dark spots across the visible hemisphere of a star produces a variation of the depth of absorption line profiles, and the magnitude of this variation is different in lines of different temperature sensitivity. Thus the ratio of the depths of a temperature-sensitive line and a temperature-insensitive line is a good diagnostic of temperature variations across the stellar surface. This method has been used to detect temperature variations associated with activity cycles (Gray *et al.* 1996) and rotational modulation by large surface features (Toner and Gray 1988).

Catalano *et al.* (2002b) and Frasca *et al.* (2005) have shown how photometric light curves and spectroscopic line-depth ratios can be combined to yield unique values for spot areas and temperatures, even on slowly rotating stars where Doppler imaging does not work. They applied their method to data from three slowly rotating active binaries of RS CVn type and obtained spot temperatures in the range 3800–4000 K (comparable to sunspot umbral temperatures) and area coverages of 36–45% of the stellar disc. Recently, however, O'Neal (2006) has cautioned that the use of atomic line-depth ratios by Catalano *et al.* (2002a,b) tends to overestimate spot temperatures because their atomic lines blend with TiO molecular lines in cooler spots.

Molecular lines

For stars of high enough effective temperature, molecular absorption lines are absent in the normal photospheric spectrum and can form only in the cooler atmospheres of starspots. Hence these lines serve as a useful diagnostic of spots on these stars. Molecular lines have long been used as a diagnostic tool for sunspots (especially their umbrae), where they have the distinct advantages of high temperature sensitivity and greatly reduced scattered light from the quiet photosphere.

The first detection of molecular bands in a starspot was made by Vogt (1979), who found TiO and VO bands in the spectrum of HD 224085, a star of spectral type K2 which in an unspotted state would be too hot to show molecular bands. From the relative strengths of these molecular lines, Vogt was able to determine a 'spectral type' of M6 for the starspot spectrum. Soon thereafter Ramsey and Nations (1980) observed the TiO band system at 886.0 nm in the active binary system HR 1099 (V711 Tau) and found that these lines strengthened greatly when this photometrically variable system was near its minimum intensity. The spectral classes of the two stars in this system, G5 IV and K1 IV, are incompatible with the formation of the TiO bands in their normal photospheres; instead, these bands can be understood to form on the K1 star in spots that are at least 1000 K cooler than the normal photosphere. Further confirmation of the association of molecular lines with starspots was provided by Huenemoerder, Ramsey and Buzasi (1989), who found a phase-dependent variation in the strength of the TiO band in II Pegasi with the highest strength occurring in phase with the minimum in overall photometric intensity.

A technique for determining the temperatures and area filling factors of starspots using two or more molecular bands was first suggested by Huenemoerder and Ramsey (1987) and has since been considerably refined. For example, the absorption bands of TiO near 705.5 nm and the band at 886.0 nm can be used to measure the temperatures and filling factors of starspots on late-type stars (Neff, O'Neal and Saar 1995; O'Neal, Saar and Neff 1996; O'Neal, Neff and Saar 1998; O'Neal *et al.* 2004a). The spectrum of an inactive M star is used to model the spotted regions of the star, and the spectrum of the appropriate inactive G or K star is used to model the unspotted regions. These proxy spectra are weighted by their relative continuum fluxes and by surface-area filling factors in producing a net spectrum of the spotted star. The strengths of the two TiO band systems both increase with decreasing temperature but at different rates, and hence their relative strength is a measure of temperature. Their absolute strengths, however, are functions of the fractional projected area f_s of spots on the visible hemisphere, weighted by limb darkening. Hence, f_s is a flux-weighted filling factor for the starspots.

Applying this technique to the very active star II Peg, Neff, O'Neal and Saar (1995) derived a spot temperature T_s of about 3500 K and a filling factor f_s varying from 54% to 64% as the star rotated. In subsequent observations of II Peg (O'Neal, Saar and Neff 1996), they obtained similar results ($T_s = 3500 \pm 100$ K and f_s ranging from 43% to 56%). Later observations of II Peg using Doppler imaging (Berdyugina *et al.* 1998, 1999) produced much smaller filling factors f_s, in the range of 10% to 15%, suggesting that perhaps this star has a large area of permanent, uniformly distributed spots (with f_s of 30% to 40%) that contribute to the molecular bands but not to the Doppler images, together with a varying spot coverage (with f_s of 10% to 15%) that contributes to both the molecular bands and the Doppler-imaging signal. O'Neal, Saar and Neff (1996) also reported results for four other spectroscopic binary stars: EI Eri ($T_s = 3700$ K and f_s ranging from 17% to 38%); V1762 Cygni ($T_s = 3450$ K and $f_s = 24\%$ for one observation); V1794 Cygni ($T_s = 3800$ K and f_s ranging from 16% to 37%); and σ Gem ($T_s = 3850$ K and f_s ranging from 14% to 33%). The latter two stars were too hot to produce a measurable depth in the 886.0 nm band, so T_s had to be estimated from photometry.

The use of molecular lines for determining spot temperatures can be extended to other molecular bands, in particular to bands in the near infrared where spots are relatively brighter than the unspotted photosphere and hence contribute more strongly to the star's overall spectrum (O'Neal and Neff 1997). Using the absorption lines of the OH molecule near 1.563 μm, O'Neal *et al.* (2001) detected starspots on several active stars of the RS CVn and BY Dra classes. The OH lines are formed at higher temperatures (up to 5000 K) than the TiO lines and hence can only be used to detect starspots on hotter stars.

O'Neal *et al.* (2004b) compared three methods of determining starspot temperatures for the same data sets: (1) fitting TiO-band spectra using spectra of proxy stars; (2) fitting TiO-band spectra using model atmospheres; and (3) fitting line-depth ratios of different bands.

It has also been suggested that molecular lines might be used as diagnostics of the magnetic fields in starspots (Berdyugina and Solanki 2002; Berdyugina, Solanki and Frutiger 2003). The TiO lines at 7055 Å are magnetically sensitive, with effective Landé factors approaching 1, but their separation is small making magnetic splitting difficult to determine. The possibility of using Stokes polarimetry in the TiO lines to measure the magnetic field in starspots has been discussed by Berdyugina (2002). The expected Stokes V signal is only at a level of about 0.3%, so very high sensitivity is required, but it may soon be possible to detect the magnetic field of a single, large, unipolar spot on a stellar disc.

9.1.3   *Doppler imaging*

Although modelling of a star's photometric variations can yield good estimates of a spot's size and temperature, the spot models are usually highly idealized and the solutions are never unique. Fortunately there is a newer and more powerful technique, Doppler imaging, that can better determine a spot's size, shape, and position on the disc and follow its rotational and migratory motion.

The concept of Doppler imaging traces back to the methods of Deutsch (1958, 1970), Falk and Wehlau (1974) and Goncharskii *et al.* (1977) for producing surface maps of magnetic field strength and element abundance anomalies in Ap stars, based on the fact that the combination of surface inhomogeneities and Doppler shifts due to the star's rotation produces line profiles that vary in shape with the phase of the rotation. These early methods assumed uniform surface brightness on the star. In the early 1980s, Vogt and Penrod introduced the technique of Doppler imaging of surface brightness and produced the first images of starspots (Vogt 1981c; Vogt and Penrod 1983).[1] Their technique was soon improved by employing maximum-entropy methods for solving the inverse problem of constructing images of the spots from line profiles (Vogt, Penrod and Hatzes 1987). Several other inversion techniques for Doppler imaging have been developed since then.

The powerful technique of Doppler imaging revolutionized the study of starspots. It allows one to determine the longitude and latitude of an individual starspot, and in cases where a long time series of observations is available, to detect surface differential rotation, active longitudes and cycle periods.

Technique

Doppler imaging is based on the way in which dark starspots at different locations on the surface of a rapidly rotating star produce distinctive deviations from an ideal rotationally broadened absorption line profile (see the review by Rice 2002). For a rapidly rotating star, there is a one-to-one correspondence between wavelength position across a rotationally broadened absorption line profile and spatial position across the stellar disc. Lines of constant line-of-sight radial velocity are chords across the stellar disc parallel to the rotation axis, and there is a one-dimensional mapping between the position of the chord across the disc (perpendicular to the rotation axis) and Doppler-shifted wavelength across the spectral line. A dark spot on the stellar surface causes an associated bump in the absorption line profile at the wavelength position corresponding to the spot's position on the stellar surface. This effect has been clearly explained by Vogt and Penrod (1983) using the simple model illustrated in Figure 9.2. In this model the stellar disc is divided into five sections by chords parallel to the rotation axis, and the radial velocity is assumed to be constant in each section. For the unspotted star (left-hand column), ignoring limb-darkening effects, all five sections contribute equally (per unit area) to the total line profile. For simplicity, the intensity profile in each section is assumed to drop to zero at line centre and the continuum intensities are normalized so as to add up to unity when summed over the five sections. Thus the individual line profiles for the five sections are all identical except for their Doppler shifts, and their sum produces the Doppler-broadened line profile shown at the bottom left. For the spotted star (right-hand column), there is a decreased continuum contribution, and hence a decreased absorption of continuum photons, at the wavelength corresponding to the spot's velocity section. Here the spot has been placed in section III and is assumed to occupy half the area of that section. The individual line profile in this section is unchanged but the continuum level is halved because of the presence of the spot. When the individual line profiles are summed, they produce a Doppler-broadened line profile with an *apparent* emission bump at the wavelength position corresponding to its section. This bump does not actually represent enhanced emission, but rather a lack of absorption at that wavelength position. The bump will move across the line profile from blue to red as the spot is carried across the visible hemisphere by the star's rotation, allowing us to determine the spot's longitude at any time.

---

[1] It is fitting that the first paper (Vogt 1981c) to present the Doppler imaging technique (called "spectral imaging" in that paper) was actually presented at a conference devoted to sunspots (Cram and Thomas 1981).

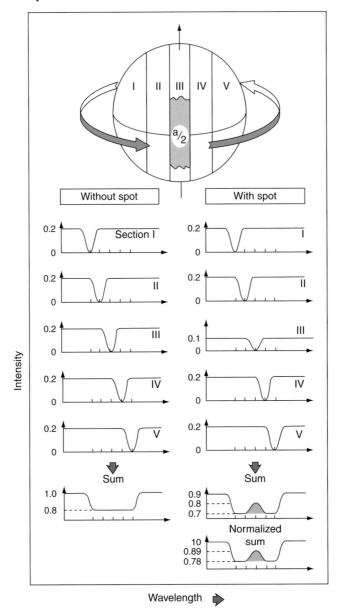

Fig. 9.2. Schematic illustration of the formation of a bump in a spectral absorption line by a dark spot on the surface of a rapidly rotating star. (From Vogt and Penrod 1983.)

With real, observed line profiles the situation is of course more complicated than the simple model of Figure 9.2. The continuum flux and line profile from the spot itself cannot be ignored, and the centre-to-limb variation of the continuum flux (limb darkening) and the line profile must be accounted for. Obviously, the technique requires spectra of high resolution and high signal-to-noise ratio, but the factors that most often limit the effective resolution of the technique are the rotation velocity of the star and the intrinsic width of the spectral line.

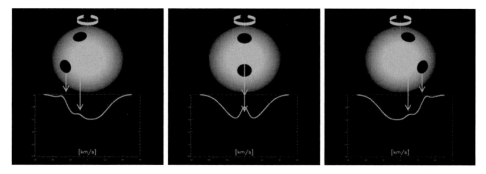

Fig. 9.3. Illustration of the motion of the bumps across a rotationally broadened spectral line profile for starspots at different latitudes. (From Rice 2002.)

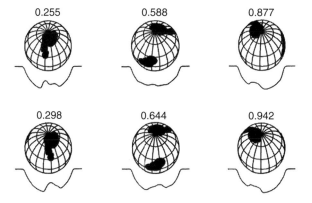

Fig. 9.4. Doppler images of starspots on HR 1099 at different phases of its rotation. Beneath each image is the corresponding theoretical spectral line profile (for Fe I 643.09 nm), which provides a best fit to the observed line profile. (From Vogt and Penrod 1983.)

Doppler imaging also allows us to determine the latitude of a spot from the range of wavelengths over which the 'emission' bump moves during each disc passage, the range being wider for spots nearer the equator, as shown in Figure 9.3. For spots closer to the poles, the bump enters and leaves the line profile closer to line centre, and a spot centred on the visible rotation pole will produce a bump that sits at line centre throughout the rotation period. If several spots appear at different latitudes on a star, and we can trace their motions across the disc separately, we then have a means of determining the rotation rate of the star as a function of latitude, i.e. we can measure the star's surface differential rotation.

The first spotted star to be studied by Doppler imaging was the primary star in HR 1099 (V711 Tau), one of the brighter members of the RS CVn class (Vogt and Penrod 1983). Starspots had already been postulated for this star to explain its photometric variability (Eaton and Hall 1979), and their existence had been confirmed by the presence of molecular TiO bands in the spectrum (Ramsey and Nations 1980). In autumn 1981, Doppler imaging revealed two large spots on HR 1099, separated by about 120° in longitude (see Fig. 9.4): a large, nearly circular polar spot with a narrow appendage extending down to about 30°

latitude, and a large spot near the equator (latitude $\sim12°$). Each of these spots covered nearly 10% of the star's surface.

The technique of Doppler imaging is only possible when the rotational Doppler shift of light from the approaching and receding hemispheres of the star exceeds the wavelength shifts due to other effects, that is, when Doppler broadening due to rotation is the dominant mechanism of line broadening. This technique is thus limited to rapidly rotating stars, which are generally much more active than the Sun. Many of these rapidly rotating stars are in close binaries, in which the rotation rates remain high during the stars' evolution because of tidal synchronization with the orbital motion.

Doppler imaging poses the basic inversion problem of determining the intensity pattern on the stellar surface that produces the best fit, in some statistical sense, to a time sequence of observed line-profile distortions. Such a procedure is liable to various systematic errors. First of all, the angle i between the star's rotation axis and the line of sight is unknown and has to be estimated; values of i assigned by different observers may differ by as much as $20°$ (Strassmeier 2002). Then, latitudinal resolution is poor near the equator, where $v \sin i$ is extremal, and features therefore appear to be smeared out in latitude. Conversely, the rotational velocity drops to zero at the pole and high-latitude features are poorly determined. Any errors in the inversion procedure could therefore accumulate in polar regions, where they are least constrained by observations. Many stars do in fact exhibit polar spots, like that in Figure 9.4. Although the existence of such spots was initially controversial, it is now accepted that they are real. Fortunately, there are examples of active stars without polar spots (e.g. HD 31993) and others (e.g. AB Doradus) where the polar spot is not always present.

As of this writing, more than 70 stars have been successfully investigated using Doppler imaging. Strassmeier (2002) presents a useful table of the 65 late-type stars that had been Doppler imaged by 2002, of which 29 are single stars and 36 are in close binary systems. The sample includes stars at almost all evolutionary stages, from pre-main-sequence T Tauri stars to evolved giants. The closest solar analogue among these stars is the G1.5 dwarf EK Draconis (HD 12933), which rotates ten times faster than the Sun, illustrating the restriction of Doppler imaging to fast rotators.

Doppler imaging can only detect fairly large starspots. On the Sun, however, there is a wide range of spot sizes with a lognormal distribution. Extrapolating this distribution to higher activity levels suggests that even on the most heavily spotted stars there are many smaller spots that are unresolved by Doppler imaging (Solanki and Unruh 2004). This suggestion is supported by the fact that, in cases of stars where both techniques have been applied, photometric light-curve modelling often indicates a greater total spot coverage than the Doppler-imaging technique (e.g. Unruh, Collier Cameron and Cutispoto 1995).

In very rapidly rotating stars, such as G, K and M stars in young clusters, the rotational broadening of photospheric spectral lines far exceeds their intrinsic line width. In principle, this permits very high resolution Doppler imaging, but the signal-to-noise ratio of measurements in a single spectral line is quite low because the lines are very broad and shallow. In order to overcome this problem, Donati *et al.* (1997) developed a least-squares deconvolution technique using the profiles of many different lines (thousands) to produce a single Doppler imaging signal with high signal-to-noise ratio and surface angular resolution of a few degrees.

Barnes *et al.* (1998) used both Doppler imaging and photometry to observe spots on two rapidly rotating G dwarfs at two epochs separated by a month. Since these stars are too faint for conventional Doppler imaging in a single spectral line, they used many photospheric metal lines and deconvolved them into a single line profile to obtain a high signal-to-noise ratio. On both of the stars they found both a dark polar cap and isolated spots at lower latitudes. For one of the stars (He 699, rotation period 0.49 days) the polar cap was essentially unchanged after a month, whereas the pattern of spots at lower latitudes was quite different.

9.1.4 Zeeman–Doppler imaging

The possibility of applying the methods of Doppler imaging to polarimetric measurements of spectral lines in order to detect the surface magnetic field distribution on stars was first discussed by Semel (1989). The principle behind the method can be simply stated. Suppose there are two equivalent magnetic starspots of opposite polarity on a stellar disc. In the absence of any rotation of the star, their circular polarization signals will cancel. If the star rotates, however, and the spots are at different longitudes, then there will be a relative line-of-sight velocity between the spots and the Doppler effect will separate in wavelength their contributions to the circular polarization signal. In principle, the line-of-sight magnetic field component of each spot can then be determined. In practice, several different spectral lines must be used to achieve a sufficient signal-to-noise ratio.

Brown *et al.* (1991) developed the first inversion code for Zeeman–Doppler imaging, based on the maximum entropy principle. Their code assumed a purely radial magnetic field and used only the rotationally modulated Stokes *V* (circularly polarized) and *I* profiles (Rice 2002). It is clear, of course, that small bipolar features (compact starspot groups) will not be resolved with this technique. Subsequently, Donati and colleagues implemented a least-squares deconvolution method that combined Stokes *V* and *I* profiles from about 1500 different lines across the visible spectrum; from these data they were then able to calculate distributions of all three components of the magnetic field across the surface of the star (Donati and Brown 1997; Donati and Collier Cameron 1997). It should be noted that the maps obtained with this technique are incomplete. Although regions with azimuthal fields are adequately detected and distinguished from those where the field is radial or meridional, the latter components cannot be unambiguously separated, especially at low latitudes (Donati 1999; Donati *et al.* 2003). Consequently, the magnetic field may appear not to be solenoidal. Attempts are now being made to include measurements of the Stokes *Q* and *U* parameters, which measure linear polarization, in order to gain a better estimate of the three-dimensional vector magnetic field (Rice 2002).

Zeeman–Doppler imaging has been applied intensively to three stars: the young dwarfs AB Dor and LQ Hya, and the RS CVn star HR 1099 (Donati *et al.* 2003 and references therein; Petit *et al.* 2004a). Figure 9.5 shows both Doppler and Zeeman–Doppler images of HR 1099. A common feature of the three stars is that the strongest magnetic fields do not coincide with the darkest areas in intensity. The radial magnetic field component is concentrated in patches of both polarities at mid-latitudes, while the azimuthal component of the field is concentrated in nearly axisymmetric rings around the star, one at high latitude and the other, of opposite polarity, at a lower latitude. Donati *et al.* (2003) interpret this pattern as reflecting the poloidal and toroidal components of the large-scale mean field generated by a dynamo distributed throughout the convection zone. Solanki (2002) suggested that instead the azimuthal field may reside in large-scale penumbral regions. Another reason why the strongest magnetic fields are not seen in the darkest regions is that the contrast at optical

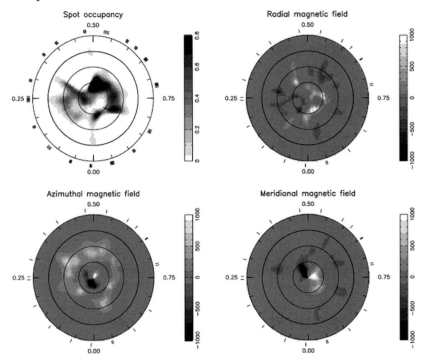

Fig. 9.5. Starspot and magnetic field distributions on HR 1099 in December 1998, shown in a flattened polar projection. The concentric circles correspond to latitudes $-30°$ (outer circle), $0°$, $+30°$, $+60°$. The upper-left panel shows a Doppler image of brightness; the other three panels show the radial (r), azimuthal (ϕ) and meridional (θ) components of the magnetic field, referred to spherical polar co-ordinates, as constructed by Zeeman–Doppler imaging. (From Petit *et al.* 2004a.)

wavelengths between these regions and the bright photosphere is so strong that the Zeeman signal is suppressed in the dark spots (Donati and Collier Cameron 1997).

Zeeman–Doppler imaging is not restricted to stars like the Sun, whose radiative interiors are enclosed by outer hydrogen convection zones. Donati *et al.* (2006a) recently produced a map of the magnetic field of a rapidly rotating M4 dwarf (V374 Peg), which is fully convective, using tomographic imaging from a time series of spectropolarimetric measurements. Donati *et al.* (2006b) have also reported measurements of 0.5 kG fields on the much more massive early type B0.2 V star τ Sco, which has a convective core but no significant convective envelope.

9.2   Case studies of starspots

In this section we discuss some observations of starspots on specific stars or types of stars. As indicated above, these are all rapid rotators and many of them belong to close binary systems, so that their rotation periods are tidally synchronized with their orbital periods.

Table 9.1 presents some basic properties of the spotted stars discussed in this section: spectral and luminosity class, type, effective temperature T_{eff}, rotational period P_{rot}, and line-of-sight equatorial velocity $v \sin i$ (where i is the angle of inclination of the star's rotation

Table 9.1 *Properties of spotted stars discussed in this chapter*

Star	Type	Spectral class	T_{eff} (K)	P_{rot} (days)	$v \sin i$ (km s^{-1})
Sun	Single	G2 V	5780	25.4	2.0
EK Dra	Single	G1.5 V	5870	2.60	17.3
HD 171488	Single	G0 V	5800	1.337	38
AB Dor	Single	K0 V	5250	0.5148	91
HR 1099	RS CVn	K2 V	4800	2.84	41
HD 12545	RS CVn	K0 III	4750	24.0	21
II Peg	RS CVn	K2 IV	4600	6.72	23
FK Com	FK Com	G4 III	5080	2.40	155
HD 199178	FK Com	G5 III–IV	5450	3.32	71.5
BY Dra	BY Dra	K7 IV–V	4100	3.8	7.4
AG Dor	BY Dra	K0 V	4900	2.56	18
V410 Tau	T Tau	K4 IV	4400	1.872	77
HDE 283572	T Tau	G8 IV–V	5500	1.55	80

axis to the line of sight). Most of the data in Table 9.1 are taken from a more extensive table, compiled by Strassmeier (2002), of properties of all stars that have been Doppler imaged.

9.2.1 EK Draconis and other solar analogues

EK Draconis (HD 129333) is a young, active, nearby, effectively single[2] star of class G1.5 V. It is perhaps the closest analogue of the early Sun among the stars that have been studied extensively for their activity and spots. Its rotation rate (period 2.6 days) is about ten times faster than that of the current Sun. Considerable attention has been given to EK Dra ever since its strong Ca II H and K emission was discovered in the survey of solar-neighbourhood stars by Vaughan and Preston (1980). It displays the highest level of surface activity of any known early G star that is not in a close binary system (Soderblom 1985), and its X-ray luminosity is about 300 times that of the Sun. From its spatial motion, as measured by the Hipparcos satellite, we know that EK Dra is almost certainly a member of the Pleiades moving group, implying that its age is 70–100 Myr (Soderblom and Clements 1987; Soderblom, Jones and Fischer 2001). It is thus a young star, very near the ZAMS, and it gives us a glimpse of what our Sun might have been like at the same early stage of its evolution.

Photometric variations of EK Dra associated with its rotation were discovered independently by Chugainov, Lovkaya and Zajtseva (1991) and Dorren and Guinan (1994a). These variations have a range of about 0.05 magnitude in the visible, a relatively large range for a star that is not a member of a close binary system. Additionally, there has been a slow decrease in brightness by about 0.005 magnitudes per year since about 1975 (Messina and Guinan 2002; Fröhlich *et al.* 2002).

---

[2] Although EK Dra has generally been considered to be a single star, it seems that it actually belongs to a very long period binary with a low-mass companion (Duquennoy and Mayor 1991; König *et al.* 2005).

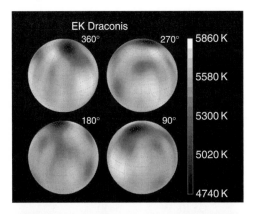

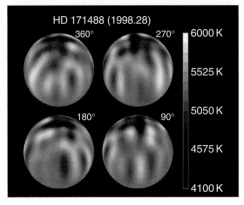

Fig. 9.6. Doppler images of starspots on the solar analogues EK Draconis and HD 171488 at different phases of their rotation. Temperatures are indicated according to the grey scale beside each image. (Courtesy of K. Strassmeier.)

Strassmeier and Rice (1998a) have produced Doppler images of spots on EK Dra, using spectra obtained in 1995 (see Fig. 9.6); more recent images have been obtained by Järvinen *et al.* (2007). Strassmeier and Rice detected several spots at low and middle latitudes, but the dominant feature was a large spot at latitude 70–80°, which might have been an appendage of a spot at the pole (which could not be seen because of the small value of $v \sin i$). They found temperature deficits for the spots in the range $\Delta T = 400 - 1200$ K.

Long-term photometry of EK Dra has shown that it has a long-lived, non-axisymmetric distribution of spots appearing preferentially at two active longitudes separated by about 180° (Järvinen, Berdyugina and Strassmeier 2005). The activity switches between these two longitudes at intervals of 2–2.25 years, producing a flip-flop cycle with period of about 4–4.5 years. There also appears to be a longer-term periodic variation of the total area of spot coverage, with period 10.5 years, during which the active longitudes migrate around the star.

EK Dra is just one of several G stars that are being studied as proxies for the Sun at various stages of its evolution. Another close solar analogue is the star HD 171488, a single, rapidly rotating (period 1.337 days) G0 V star. Doppler images of its surface (Strassmeier *et al.* 2003) are shown in Figure 9.6. The intensity pattern shows a high-latitude spot and alternating, elongated (in latitude) bright and dark bands suggestive of the 'banana cells' seen in numerical simulations of convection in a rotating spherical container. Biazzo *et al.* (2007), using a combination of photometric and spectroscopic measurements, modelled photospheric and chromospheric inhomogeneities and found a close association between spots and plage in three young solar analogues.

There are specific observational programmes devoted to solar analogues. For example, the *Sun in Time* programme (Dorren and Guinan 1994b) involves multi-wavelength observations of single G0–G5 V stars with ages ranging from about 70 Myr (near the ZAMS) to 9 Gyr (near the end of their main-sequence lifetimes).

9.2.2 RS CVn binaries and FK Com stars

Close binary systems in which at least one of the components has an outer convection zone often display intense magnetic activity, presumably driven by a strong dynamo sustained by the fast rotation of the stars that is maintained by tidally enforced spin-orbit synchronization. Among these systems, the detached RS Canum Venaticorum binaries have been studied most extensively. In these systems the more massive primary star is a G or K giant or subgiant (which has evolved off the main sequence) and the secondary star is a subgiant or dwarf of spectral class M, G or K. In many cases the secondary has much lower luminosity than the primary and the system shows a single-line spectrum, which simplifies the analysis. The large photometric variability of these non-eclipsing binaries indicates the presence of very large starspots (such as that shown in Figure 1.8 for the red giant star HD 12545) which may cover as much as half of the visible hemisphere. The intense activity, high luminosity and rapid rotation of the RS CVn stars have made them ideal subjects for photometric, spectral, and Doppler-imaging studies of starspots, and much of our knowledge of starspots indeed comes from these stars.

The RS CVn star HR 1099 (V711 Tau) was the first star to be Doppler imaged, by Vogt and Penrod (1983). Their images (for late 1981), already shown in Figure 9.4, reveal two large spots, one at the pole and the other near the equator. A more recent Doppler image of the same star, by Strassmeier, displays a single, large, high-latitude spot.

One of the most active and best studied RS CVn variables is the K2 IV star II Peg, whose photometric variation has been discussed already in Section 9.1.1. During the period 1992–8, the constantly evolving surface pattern was dominated by two high-latitude spots but showed no polar cap (Berdyugina *et al.* 1998, 1999). Figure 9.7 shows a later Doppler image of this star, with several spots at different latitudes, accompanied by a striking Zeeman–Doppler image of the radial magnetic field at the stellar surface.

Closely related to the RS CVn stars are the FK Com stars (Bopp and Stencel 1981), which are single, but rapidly rotating, G–K giants exhibiting strong and variable Ca II H and K emission and visual magnitude variations with periods of a few days. As single giant stars, their rapid rotation is puzzling; they might, for example, be recently coalesced binaries. The prototype star, FK Comae Berenices itself, is a G4 III star with a rotation period of 2.40 days. Small rotationally modulated variations in its visual magnitude were first discovered by Chugainov (1966) and later interpreted by Bopp and Rucinski (1981) as being caused by starspots. Dorren, Guinan and McCook (1984) confirmed this interpretation and found the spots to be some 600–800 K cooler than the unspotted surface. Subsequently, Doppler imaging has shown that the spots on FK Com can occur at both high and low latitudes

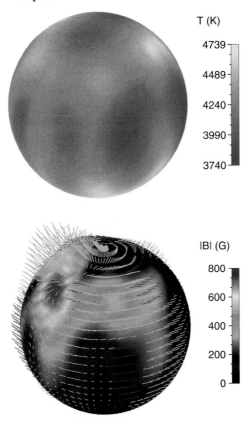

T (K)

4739
4489
4240
3990
3740

|B| (G)

800
600
400
200
0

Fig. 9.7. Upper panel: Doppler image of starspots on II Peg. Lower panel: the corresponding Zeeman–Doppler image, showing the radial field; the vectors represent the orientation and magnitude of a potential field extrapolation. (From Carroll *et al.* 2007; courtesy of K. G. Strassmeier.)

(Piskunov, Huenemoerder and Saar 1994; Korhonen *et al.* 2000). FK Com was the first of several stars discovered to have spots occurring at preferred longitudes about 180° apart and to have the relative strength of spots at the two longitudes switch back and forth on long time scales (the so-called 'flip-flop' phenomenon; see Section 9.3.2).

Another FK Com star known to be spotted is HD 199178, a G5 III–IV star with a rotation period of 3.32 days. Figure 9.8 shows Zeeman–Doppler images of HD 199178 made in July 2003. (See also the back cover of this book.) This star has a very large polar spot along with several smaller spots near its equator.

9.2.3   *BY Draconis*

The dMe flare star BY Draconis shows nearly sinusoidal variations in brightness which early on were interpreted as due to rotational modulation of surface features loosely called starspots. Changes in the amplitude of these variations were attributed to changes in the area, temperature, and location of these spots (Chugainov 1966; Krzeminski 1969; Bopp and Evans 1973; Torres and Ferraz Mello 1973). BY Draconis is now considered the prototype of a class of rapidly rotating G, K or M dwarfs that show intense magnetic activity, flares and starspots. They show chromospheric emission lines, and many belong to binary

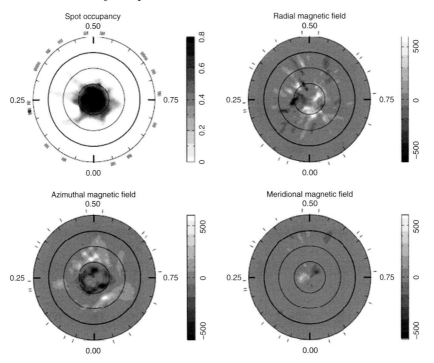

Fig. 9.8. Zeeman–Doppler images of starspots on HD 199178 in July 2003. (From Petit *et al.* 2004b.)

systems. It follows that BY Dra stars are similar in many respects to RS CVn systems, though they are still on the main sequence.

Early multi-colour photometry on BY Dra itself suggested the presence of spots (e.g. Poe and Eaton 1985). The relatively large amplitude of the photometric variations indicated that the spots were almost certainly on the brighter of the two companion stars, the K7 IV primary. Doppler imaging of BY Dra is fairly unreliable because of its low inclination angle i. A better candidate for Doppler imaging is the BY Dra star AG Doradus. Figure 9.9 shows Doppler images of AG Dor obtained by Washuettl and Strassmeier in 2001; it has a large, dark, near-polar spot and a less-contrasting pattern of bright and dark regions at lower latitudes.

RS CVn and BY Dra binary systems contain some of the most active of all stars, with large starspots covering a substantial fraction of their surfaces. Many of these rapidly rotating close binaries have a non-uniform longitudinal distribution of spots, with the spots appearing preferentially at long-lived active longitudes. In the fastest rotators (periods less than one day), the active longitude coincides with the quadrature points on the stellar surface (Zeilik *et al.* 1994), suggesting that the preferred location of the spots is due to tidal effects (Holzwarth and Schüssler 2002).

9.2.4   AB Doradus

The single K0 dwarf star AB Doradus (HD 36705) is one of the brightest, most rapidly rotating, and most active lower main-sequence stars in the sky. Its estimated age of $10^6 - 3 \times 10^7$ years (e.g. Vilhu, Gustafsson and Edvardsson 1987) means it is in the

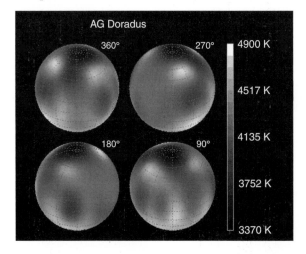

Fig. 9.9. Doppler maps of the BY Dra star AG Doradus. (Courtesy of K. G. Strassmeier.)

final stages of core contraction or just arriving on the main sequence. It shows many signs of stellar activity: it is photometrically variable in all wavelength bands, it is variable in Hα and other chromospheric lines, showing evidence of coronal prominences (Collier Cameron and Robinson 1989a,b), and it is also a variable radio source and soft X-ray source (see Kürster, Schmitt and Cutispoto 1994 for references). It has been closely monitored for stellar activity and starspots for a number of years. It is an ideal candidate for Doppler and Zeeman–Doppler imaging, having significant rotational broadening ($v_e \sin i = 90\,\mathrm{km\,s^{-1}}$) and a relatively short rotation period ($P_{rot} = 12.4\,\mathrm{h}$), which allows nearly 60% of its surface to be mapped in a single night.

Doppler images of AB Dor have been obtained at least once a year since 1992. The surface intensity pattern is generally very different from that on the Sun (see Fig. 9.10), with large dark spots at high latitudes and smaller, more isolated spots near the equator (Hussain *et al.* 2000; Donati *et al.* 2003). At times, however, it has a large number of smaller spots distributed over all latitudes but concentrated near an active latitude of about 25°. (Activity belts at low latitudes have been seen on several other stars, including the K1 giant YY Men: Piskunov 1991; Kürster *et al.* 1992.)

The surface magnetic field has a strong azimuthal component at all visible latitudes (Donati and Collier Cameron 1997; Donati *et al.* 1999). In December 1996, AB Dor showed a complex magnetic topology, with at least 12 different patches of radial field of both polarities at various locations, and also significant azimuthal fields in the form of a ring of negative polarity encircling the pole at high latitudes and several patches of positive polarity at intermediate latitudes (Donati *et al.* 1999). This arrangement of azimuthal-field polarity, which is very similar to that found a year earlier (in December 1995; Donati and Collier Cameron 1997), may well be indicative of the large-scale toroidal field configuration of the dynamo operating in this star. Hussain *et al.* (2000) showed that maps of surface magnetic fields of AB Dor produced by two independent inversion codes are nearly identical.

Photometric monitoring of AB Dor, which began in 1978, shows that the star went through an intensity minimum (maximum spottedness) in about 1988 and has been increasing in

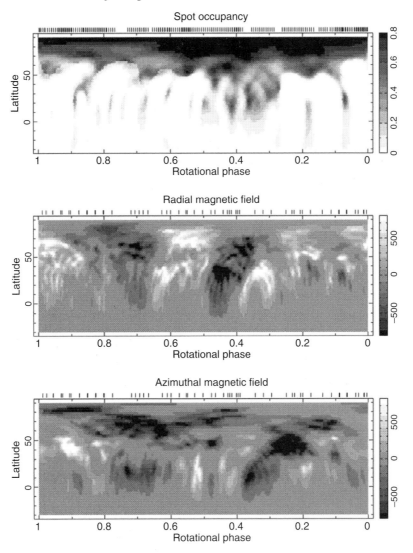

Fig. 9.10. Brightness and magnetic field map of AB Doradus. (From Donati *et al.* 2003.)

brightness ever since (Amado, Cutispoto and Lanza 2001). The first Doppler images of AB Dor, obtained near the time of minimum intensity (December 1988 and February 1989) showed a solar-like band of low-latitude spots but no polar spot (Kürster *et al.* 1992; Hussain 2002). As the star increased in brightness (decreased in spottedness) in subsequent years it developed a large polar spot, as shown in Figure 9.10. From the distortion of the brightness images, Donati *et al.* (1997) detected surface differential rotation on AB Dor, with the pole rotating more slowly than the equator by about one part in 120. A further study of its differential rotation (Collier Cameron and Donati 2002) indicates that it was rotating almost rigidly in December 1988 (140 day equator–pole lap time) but increased its shear as the polar spot developed (70 day equator–pole lap time in 1992).

9.2.5   *T Tauri stars*

T Tauri stars are a class of young, active stars that have not yet reached the main sequence. They acquire high rotation rates as a result of the conservation of angular momentum as they contract toward their main-sequence state. T Tauri stars are of interest because of what they can reveal about the early stages of stellar evolution and, in particular, about the role of a magnetic field in transferring angular momentum away from the star in order to allow contraction to occur in spite of spin-up (see Mestel 2003). They can be separated into two categories: the classical T Tauri stars, which are surrounded by a proto-planetary accretion disc, and the weak-line T Tauri stars, which have evolved closer to the main sequence and may have lost their accretion discs.

V410 Tau is a weak-line T Tauri star that exhibits large photometric variations (up to 1 mag in V) and has a large rotational velocity ($\sim$70 km s^{-1}), making it an ideal candidate for Doppler imaging. The first Doppler image of V410 Tau, by Strassmeier, Welty and Rice (1994), showed it to have a large, high-latitude spot that reached but did not straddle the pole. Subsequent Doppler images by Joncour, Bertout and Ménard (1994) and Hatzes (1995) confirmed this configuration (see Fig. 9.11).

Several other young, pre-main-sequence T Tauri stars have been found to have polar or near-polar spots. For example, in February 1993 the weak-line star HDE 283572 (V987 Tau), with rotation period 1.55 days, had one of the largest and coolest polar spots ever observed: a complete dark polar cap some 1600 K cooler than its surrounding photosphere (Joncour, Bertout and Bouvier 1994). When this star was Doppler imaged at higher resolution in October 1997 (Strassmeier and Rice 1998b), it again showed a large polar spot (covering about

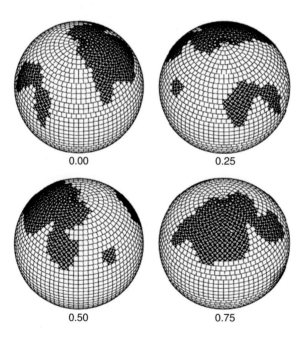

0.00   0.25

0.50   0.75

Fig. 9.11. Doppler image of V410 Tau shown at four phases of rotation ($\phi = 0.0, 0.25, 0.5, 0.75$). The dark spotted regions correspond to pixels that are at least 500 K cooler than the photospheric temperature. (From Hatzes 1995.)

6% of the star's surface) but with several appendages extending downward to latitudes as low as 40°. One possible interpretation of the appendages is that they correspond to lower-latitude spots that are migrating toward the pole where they merge with the polar cap, as suggested by Vogt and Hatzes (1996) for HR 1099, but such migration has not yet been observed.

Zeeman–Doppler imaging of the classical T Tauri star V2129 Oph reveals a strong radial magnetic field with a strength of 2 kG, associated with a pair of spots near the pole and reversed fields at lower latitudes (Donati *et al.* 2007). Potential field extrapolations show both closed field lines that return to the star and 'open' field lines that extend into its surrounding accretion disc, as illustrated already in Figure 1.9.

9.3 Starspots, differential rotation and dynamo patterns

9.3.1 *Differential rotation*

The Sun's surface differential rotation was discovered by tracking the motion of sunspots across the solar disc, so it is only natural to think that similar tracking of starspots might reveal surface differential rotation on stars. Analysis of the photometric variability of stars such as BY Dra led some authors to suggest that the surface features that caused this variability differed in their rotation rates because of surface differential rotation (Vogt 1975; Oskanyan *et al.* 1977). Other early indications of differential rotation were provided by pho-tometric studies of RS CVn binaries, in which the scatter in the rotation periods determined from several starspots, presumably formed at different latitudes, was taken to be a measure of differential rotation in the star (Hall and Busby 1990; Hall 1990). Adopting a simple solar-like form for the variation of rotation period P with colatitude θ,

$$P(\theta) = P_{eq}/(1 - k \cos^2\theta), \tag{9.1}$$

Hall (1990) found a linear dependence of the parameter k on rotation period P_{eq}, extending over three orders of magnitude of P_{eq}, for the 85 stars surveyed up to that time.

The development of Doppler imaging has provided a more direct means of measuring surface differential rotation (see Collier Cameron 2002). The unambiguous determination of the latitude and longitude of at least two spots on at least two occasions provides information on the surface rotation profile. The time interval between observations must be large enough to determine the rotation rate of the individual spots accurately, but not so large that some of the spots have decayed away. In binaries in which the rotation period of the component stars is synchronized with the orbital period, the orbital period itself provides the ephemeris 'clock' with which to determine the differential rotation. In this way, Hatzes and Vogt (1992) found that a large polar spot on EI Eri rotated with a period longer than the orbital period. If we assume that the rotation period at the equator is locked to the orbital period, this indicates a solar-like differential rotation for this star, with the equator rotating more rapidly than the pole. Barnes *et al.* (2005) combined earlier results on differential rotation from photometry (Henry *et al.* 1995) and chromospheric emission (Donahue, Saar and Baliunas 1996) with their own results from Doppler imaging to reveal a strong decrease in differential rotation with increasing depth of the convection zone along the main sequence.

Using Doppler imaging for the RS CVn binary HU Vir, Strassmeier (1994) and Hatzes (1998) found differential rotation in the opposite sense to the Sun's, with the equator rotating more slowly than the pole, but less extreme by a factor of ten. Doppler imaging of σ Gem (one of the stars originally studied by Eberhard and Schwarzschild 1913) has also revealed antisolar differential rotation, together with evidence of a poleward migration of the spots (Kővari *et al.* 2007b). Antisolar differential rotation was also inferred for the RS CVn system HR 1099 (Vogt *et al.* 1999); however, Petit *et al.* (2004a) later measured weak solar-like differential rotation for this same star in a more densely sampled sequence of images. Solar-like differential rotation has also been found for the RS CVn giant ζ Andromedae (Kővari *et al.* 2007a) and for the T Tauri star V410 Tau (Rice and Strassmeier 1996).

If enough small, well-imaged spots are found over a wide range of latitudes, cross-correlation techniques can yield a well-determined surface rotation profile. A good example is given in Figure 9.12, which shows the surface rotation profile for AB Doradus measured in December 1995 (Donati and Collier Cameron 1997; see also Petit, Donati and Collier Cameron 2004; Jeffers, Donati and Collier Cameron 2007). The differential rotation is very solar-like, with the 'lap time' required for the equatorial region to lap the polar region being 110 days as compared to the Sun's 120 days. Similar solar-like differential rotation was found for the young, late-type (K0 V), single star PZ Tel (Barnes *et al.* 2000). Another solar-type star, HD 171488 (a single G dwarf; cf. Fig. 9.6), also shows solar-like differential rotation, but seven times stronger than the Sun's (lap time of 16 days; Marsden *et al.* 2006). Rüdiger and Küker (2002) have proposed that the strong solar-like differential rotation in AB Dor and PZ Tel and similar single stars may be understood as a consequence of a deviation from spherical symmetry of the heat flux from the core due to rapid rotation, which causes a non-uniform heating of the base of the convection zone that drives a meridional circulation with equatorward flow at the surface, carrying angular momentum toward the equator.

Some of the results for differential rotation are puzzling. For example, for the FK Com star HD 199178 (cf. Fig. 9.8) Hackman and Jetsu (2003) found antisolar differential rotation in 1994–1996 whereas Petit *et al.* (2004b) found solar-like differential rotation in 2002–2003.

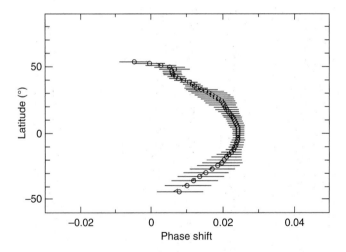

Fig. 9.12. Surface differential rotation on AB Dor in December 1995, as measured by cross-correlating constant-latitude slices of Doppler images and Zeeman–Doppler magnetic field maps taken at different times. The data points represent Gaussian fits to the peaks in the cross-correlation at each latitude. The dashed line is a fit to a $\cos^2\theta$ law for the variation of rotation with colatitude. (From Donati and Collier Cameron 1997.)

What about possible changes in differential rotation with time, possibly caused by more extreme versions of the Sun's torsional oscillations? Collier Cameron and Donati (2002) presented evidence for a change in the amplitude of the surface differential rotation on AB Dor by a factor of two over the period 1988 to 1996. However, they caution that this apparent change could instead be due to the restricted range of latitudes of spots in some of the years. Further analysis of the same data showed that the surface differential rotation varied on a time scale of at least one year, with the highest value occurring in late 1994 (Jeffers, Donati and Collier Cameron 2007).

9.3.2 *Starspots and surface patterns of activity*

Starspot tracking allows us to detect patterns associated with the stellar dynamo, such as the zone of spot appearance and the migration of this zone in latitude, or the existence of active longitudes.

The latitude distribution of starspots

Whereas sunspots are generally restricted to a latitude band within $30°$ of the equator, the latitudes of starspots on more rapidly rotating cool stars range from mid-latitudes (for stars near or on the main sequence) to the polar regions (for RS CVn systems and T Tauri stars). The surprising findings of polar spots on rapidly rotating stars, so unlike the spot distribution on the Sun, were questioned at first (e.g. Byrne 1992, 1996), but the cumulative evidence now strongly supports the existence of polar spots.

Not all fast rotators have high-latitude spots. For example, PW Andromedae (HD 1405), a very young pre-main-sequence star with rotation period 1.754 days, shows a solar-like distribution of several spots confined to $\pm 40°$ latitude (Strassmeier and Rice 2006). These spots are some 10 to 100 times larger than sunspot groups, however.

Attempts to explain the formation of polar spots on the basis of dynamo theory, including rising flux-tube and surface flux-transport models, are discussed in Section 11.4.2. In various papers, Strassmeier and his colleagues have pointed out several examples of stars on which the latitude distribution of spots disagrees with the predictions of the rising flux-tube models (see Strassmeier and Rice 2006 and references therein).

Active longitudes

Long-term photometric monitoring of RS CVn stars has shown that large spots live for years and are perhaps associated with one or two active longitudes, similar to the active longitudes observed on the Sun. Photometric data spanning more than 50 years showed that spots on the RS CVn star SV Camelopardis tended to be concentrated in two active longitude belts separated by about $140°$ (Zeilik, De Blasi and Rhodes 1988). A similar pattern, with two active longitudes separated by about $180°$, was found for the RS CVn star σ Geminorum (Strassmeier 1988; Oláh *et al.* 1989; Jetsu 1996).

Active longitudes have also been found in the spot pattern on FK Com stars and active young solar analogues. For example, for FK Comae Berenices itself, using 24 years of photometric data (1966–90), Jetsu, Pelt and Tuominen (1993) found that spot activity was concentrated at two active longitudes $180°$ apart, with the activity switching between the two longitudes three times during this time interval (see also Henry *et al.* 1995).

9.3.3   *Starspot cycles from long-term photometry*

Long-term photometric records of stars make it possible to detect cycles in the appearance of starspots (see Oláh and Strassmeier 2002), thus providing a useful supplement to the detection of stellar activity cycles by monitoring Ca II emission (described in Chapter 10). The earliest detections of starspot cycles were based on photographic plates. Using the Harvard plate collection, Phillips and Hartmann (1978) found apparently cyclic luminosity variations of amplitude 0.3 mag with periods of order 50–60 years on the dMe stars BY Dra and CC Eri, and Hartmann *et al.* (1981) found a 60-year spot cycle with amplitude 0.5 mag on the dK5e star BD +26°730. The cycle was especially apparent in the latter star because it is viewed nearly pole-on, which means that the long-term cyclic changes in spottedness are not swamped by the short-term rotational modulation of brightness due to spots.

Programs of long-term photoelectric photometry have revealed several other starspot cycles. Oláh and colleagues (Oláh, Kolláth and Strassmeier 2000; Oláh and Strassmeier 2002) have used data sets up to 34 years long to detect cycles in nine stars, with cycle periods generally in the range of 11 to 16 years. Most of these stars also show an additional shorter, weaker cycle with period in the range of about 2 to 5 years. The much-studied star HD 1099 (V711 Tau), for example, shows a 15.7-year cycle and a weaker 3.5-year cycle. The shortest fundamental cycle in their sample is the 6.4-year cycle of V833 Tau, which has rotation period 1.794 days.

In general, the photometric records used to determine starspot cycles are not long compared to the detected cycle periods, so the periods are not well determined and it is not known how regular the cycles are. (It should be kept in mind that the Sun's cycle is irregular: the times between successive sunspot maxima or minima vary from about 9 to 12 years.)

9.4   Properties of individual starspots

We now turn to the physical properties of individual starspots, including their temperatures and magnetic fields, and their lifetimes, as deduced from the available observations. Finally, we assess the similarities between starspots and sunspots.

9.4.1   *Temperatures and areas*

Measurements of starspot temperatures are made using several different techniques, the most elementary being simultaneous modelling of photometric brightness and colour variations. Doppler imaging also provides best-fit values of these temperatures. The most accurate methods are spectroscopic, involving the modelling of molecular bands or of atomic line-depth ratios. In all of these techniques, the surface area coverage (or filling factor) of the spots must be determined along with the temperature.

Spot temperatures and filling factors have been reported for almost all stars known to have spots. Berdyugina (2005) provides a useful table of values for a representative sample of spotted dwarfs, giants and supergiants. Figure 9.13 shows the dependence of spot temperature difference $\Delta T = T_{phot} - T_{spot}$ on the undisturbed photospheric temperature T_{phot} for these stars. There is a tendency for the temperature difference to increase with stellar temperature, with ΔT ranging from 200 K in M4 stars to 2000 K in G0 stars.

For some spotted stars, for example II Peg (see Fig. 9.1 and Section 9.1.2), the amplitude of the optical brightness variations is so great that a large fraction of the stellar surface must be covered by dark spots. The largest known brightness variations (e.g. $\Delta V = 0.65$ mag for

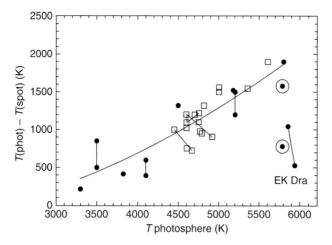

Fig. 9.13. Starspot temperature difference versus photospheric temperature for active dwarf (circles) and giant (squares) stars. Short lines connect symbols referring to the same star. The continuous curve is a second-order polynomial fit to the points, excluding EK Dra. The large circles with dots inside represent the umbra ($\Delta T = 1700$ K) and penumbra ($\Delta T = 750$ K) of a typical sunspot. (From Berdyugina 2005.)

the weak-line T Tauri star V410 Tau; Strassmeier *et al.* 1997), along with their associated colour variations, imply spotted areas covering more than 20% of the entire stellar surface. (An extreme value of 70% coverage for the contact binary VW Cep, reported by Hendry and Mochnacki 2000, should perhaps be interpreted as bright patches on a darker surface.)

The temperatures and filling factors of starspots are a possible issue in the determination of element abundances for a spotted star from the equivalent widths of its spectral lines. For example, for Li lines the equivalent width is greater in the cooler regions and, although the spots contribute less to the overall flux than the undisturbed photosphere, if there are enough spots then the net equivalent width for the star will be increased (see e.g. Soderblom *et al.* 1993c). If only a few large spots are present, then one might detect a rotational modulation of the equivalent width, but a more uniform distribution of spots would not produce a measurable variation.

9.4.2 Magnetic fields

Measurements of highly structured magnetic fields on stars other than the Sun are very difficult because of the cancellation effect of fields of opposite polarities. The technique of Zeeman–Doppler imaging (discussed in Section 9.1.4) can in principle detect the surface distribution of the vector magnetic field in rapidly rotating stars and has produced important results for a few stars (e.g. AB Dor; see Section 9.2.4). However, most of our knowledge of magnetic fields on cool stars comes from measurements of Zeeman broadening (Robinson, Worden and Harvey 1980; see Section 8.3), which work best for slowly rotating stars, for which the Zeeman broadening exceeds the rotational broadening. The method generally assumes that the magnetic field is perpendicular to the stellar surface and concentrated into patches of uniform strength, distributed uniformly across the stellar surface. Observed line profiles are then fitted with a combination of synthetic line profiles for the magnetic and non-magnetic regions, yielding values of the field strength and filling factor. Such measurements have been made for a number of G, K and M dwarfs and T Tauri stars (see the summary in Berdyugina 2005). The results show field strengths in the range 1.5–5 kG, with a tendency towards greater field strengths in cooler stars. The filling factors, however, are generally not in agreement with those determined from brightness and colour variations. The differences are most likely due to the assumption of a single field strength in the magnetic regions and the additional presence of strong fields in bright (plage) regions. Hence one should be suitably cautious in associating the measured field strengths with starspots.

9.4.3   Lifetimes

Both photometry and Doppler imaging have been used to study starspot lifetimes (see the review by Hussain 2002). Photometry provides less information than Doppler imaging but is possible from smaller telescopes and hence has provided many more of the long data sets required (see e.g. Strassmeier and Hall 1988; Strassmeier *et al.* 1989). Henry *et al.* (1995) studied photometric data sets ranging in length from 15 to 19 years for four active binaries (λ And, σ Gem, II Peg, and V711 Tau) and found lifetimes of individual starspots ranging from a few months (for II Peg) to more than six years. They found no significant trends with rotation period or spectral type.

Long-term Doppler-imaging data sets are available for a few RS CVn stars. A large polar spot on HR 1099 has been seen in all observations (one or two per year) taken over a span of more than 15 years, albeit with changes in shape between observations (Vogt *et al.* 1999; Donati 1999).

On the basis of a theoretical flux-transport model, Işik, Schüssler and Solanki (2007) find lifetimes of a few months for single bipolar spot pairs; the duration depends on the rotational shear and the latitude of emergence. In their model, polar spots are produced by tilted bipolar regions emerging at mid or high latitudes, and can be maintained for several years.

9.4.4   Sunspots as prototypes for starspots?

To what extent are starspots analogous to sunspots? Can our theoretical models of sunspots simply be scaled up in size to produce a model of a much larger starspot? Are the large starspots revealed by Doppler imaging actually close-packed groups of smaller spots? These are still very much open questions, but it is worth making some preliminary comments (see also Schrijver and Title 2001 and Schrijver 2002).

If spots cover a significant fraction of a star's surface, we can expect them to have a significant effect on the overall structure of the star. By analogy with sunspots, we can assume that the darkness of the starspots is caused by the inhibiting effect of the spot's strong magnetic field on convective heat transport. After the starspots first appear, the star will try to adjust the structure of its outer convection zone in order to carry the luminosity generated in its interior. The effect of the spots will be quite different on time scales short or long compared to the thermal time scale of the convection zone, and quite different in stars that are mostly convective or mostly radiative (Spruit 1992; Spruit and Weiss 1986).

The temperature contrasts of starspots are not widely different from those of sunspots. The starspot temperatures detected reliably by molecular-line diagnostics are in the range of 500 to 2000 K below that of the surrounding photosphere. For a sunspot, the temperature contrast is about 1800 K for the umbra alone but only about 600 K for the average temperature of the whole spot. Since the starspot temperatures are essentially averages over the spots, we see that the sunspot temperature contrast is in good agreement with that of the warmest starspots.

The most striking difference between sunspots and starspots is the much greater size that starspots can attain. Starspots sometimes cover a very large fraction of the stellar surface, 20% or more. What we don't know, of course, is whether the large spotted areas consist of a few very large spots or many smaller spots arranged in tight groups. This open question is a major impediment to understanding the relation between sunspots and starspots.

Another significant difference between sunspots and starspots is the occurrence of starspots at high latitudes, especially at the poles. This provides a challenge to dynamo theory as applied to rapidly rotating stars, where the pattern of differential rotation may be very different from that in the Sun (see Section 11.4.2 below).

The magnetic field strengths that have been measured for starspots are not significantly greater than those in sunspots. They are consistent with a balance between magnetic pressure and gas pressure near the stellar photosphere, as is the case for sunspots. However, the magnetic geometry – with a prevalence of strong azimuthal fields – that is revealed by Zeeman–Doppler imaging differs significantly from that in sunspot groups. We may conjecture that this provides evidence for bipolar spot groups, with unidirectional azimuthal fields but oppositely directed radial fields that cancel out in the measured signal. If so, we should expect the individual spots to have penumbrae resembling those in sunspots. If, on the other hand, there are only a few huge spots with monolithic structures, then the ratio of penumbral to umbral area may well be much less than in a sunspot. We may hope that this issue will be resolved by future imaging.

10

Solar and stellar activity cycles

So far, we have focused our attention mainly on the properties of individual sunspots and starspots, and their associated magnetic fields. Now we turn to systematic variations in magnetic activity and in the incidence of spots. In this chapter we shall only consider observable manifestations of activity. The magnetic fields that emerge through the surface of a star are actually generated in its interior, by dynamo processes which will be discussed in Chapter 11.

We begin by describing the well-known sunspot cycle, with an average period of about 11 years, which was first recognized by Schwabe (as explained in Section 2.2). This cycle is apparent in the record of telescopic observations, though it was interrupted during the Maunder Minimum in the seventeenth century. Fortunately, the record can be extended back through hundreds, thousands and tens of thousands of years by using measured abundances of cosmogenic isotopes as proxy data. These data confirm that similar grand minima are a regular feature of solar activity, and we can explore their statistical properties.

Next, we turn to other Sun-like stars. As expected, they can exhibit activity cycles too, although these are most apparent in middle-aged slow rotators, like the Sun itself. Younger, more rapidly rotating stars are much more active but their behaviour is erratic and less obviously periodic, as can be seen in Figure 1.7.

10.1 Cyclic activity in the Sun

Figure 2.3 shows how the area covered by sunspots has varied over the past 130 years (since the daily Greenwich photoheliographic record was initiated). The cyclic behaviour is readily apparent, as is the variability from one cycle to another: the maxima around 1958 and 1990 were abnormally high, while those prior to 1930 were relatively low. Also shown is the incidence of sunspots as a function of latitude and time – the well-known butterfly diagram. At the beginning of a new cycle, spots appear at latitudes around $\pm 30°$; thereafter the zones occupied by spots expand until they approach the equator. In the declining phase of the cycle, these zones retreat from higher latitudes until they eventually dwindle away at the equator, just as the next cycle starts. As can be seen, sunspot activity is very nearly symmetrical about the solar equator.

Solar activity is conventionally measured in terms of the Zurich sunspot number, $\mathcal{R}$, as arbitrarily defined by Wolf in 1848 (see Section 2.2). An alternative definition, of the group sunspot number introduced by Hoyt and Schatten (1998), is sometimes preferred. Fortunately, values of $\mathcal{R}$ are closely related to sunspot areas and to other, more physical, measures

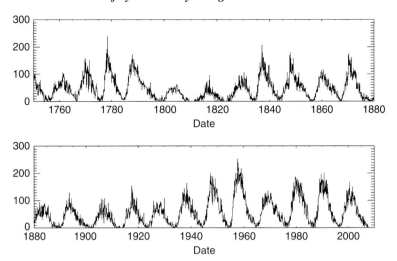

Fig. 10.1. Sunspot cycles since 1750: monthly averages of the Zurich sunspot number, $\mathcal{R}$. (Courtesy of D. H. Hathaway.)

of solar activity based on 10.7 cm radio emission or infrared emission in the He 1083 nm line. Sunspot numbers have been reconstructed from early observations going back to the seventeenth century, and the record in Figure 10.1 shows how $\mathcal{R}$ has varied with time through 23 cycles from 1750 to the present day.

As we have already explained in Sections 2.4.1 and 7.4.1, sunspots tend to appear in pairs oriented approximately parallel to the equator and with oppositely directed magnetic fields. The systematic properties of these fields are described by Hale's laws: in any cycle, the preceding spots have the same polarity in each hemisphere but the polarities of preceding and following spots are antisymmetric about the equator; moreover, these polarities reverse from one cycle to the next. Thus the magnetic field follows a 22-year cycle.

It is clear from Figure 10.1 that cycles have varied irregularly during the past 300 years. Peak values of $\mathcal{R}$ at sunspot maximum range from 60 in 1805 to 200 in 1958; the period (from minimum to minimum) has varied from 9.0 to 13.6 yr, although the mean period of 11.1 yr is well defined over this interval. Curiously, the phase drifted during the late eighteenth century but recovered in the early nineteenth century; however, this isolated incident is probably not significant (Gough 1981). Two robust features of the sunspot record are, first, the anticorrelation between the cycle period and the value of $\mathcal{R}$ at sunspot maximum and, second, the asymmetry of the cycle, with a short rise time followed by a slower decay. The record shows a tendency for the amplitudes of maxima to alternate from cycle to cycle, as if temporal symmetry were broken, but the phase of this 22-year periodicity is not maintained. There is also evidence of a longer-period variation with a period of 80–90 yr – the Gleissberg (1939, 1945) cycle – though the record is again too short to be certain whether this periodicity is real.

10.2 Modulation of cyclic activity and grand minima

Direct records of solar variability are limited to the era of the telescope, the past four centuries. Figure 1.4 shows group sunspot numbers from 1610 to 2000. As explained in Chapter 2, there was a dearth of sunspots from 1645 to 1715, and this episode is now referred to as the Maunder Minimum (Eddy 1976). Although sunspots did appear occasionally through most of this period, they were few and far between and hence individual cycles cannot be unambiguously identified from this record. Indeed, it was not until after the minimum in 1715 that a regular cycle, with spots in both hemispheres, reappeared.

Fortunately, there is another source of information. The magnetic fields that are transported outwards by the solar wind also vary with the sunspot cycle and these fields deflect galactic cosmic rays impinging on the heliosphere (McCracken *et al.* 2004). Thus the incidence of galactic cosmic rays on the Earth's atmosphere varies in antiphase with solar activity, as shown in Figure 10.2; these cosmic rays lead to the formation of radioactive isotopes, such as ^{14}C (with a half-life of 5730 yr) and ^{10}Be (with a half-life of 1.5×10^6 yr), which are deposited and stored in trees and polar ice, respectively. Hence the record of solar activity can be derived from measurements of the abundances of these cosmogenic isotopes in tree rings or ice cores, after correcting for changes in the geomagnetic field (Beer 2000). In Figure 10.3 we compare the ^{10}Be concentration from a Greenland ice core with the corresponding group sunspot numbers. The 11-year Schwabe cycle is clearly present in both records and there is a marked increase in ^{10}Be concentration at the end of the Maunder Minimum. It is apparent, however, that the 11-year activity cycle persisted, albeit at a reduced level, throughout the Maunder Minimum (Beer, Tobias and Weiss 1998). Since ^{14}C remains in the atmosphere as CO_2 for 6–7 yr and is well mixed, whereas ^{10}Be has a residence time of only 1–2 yr, evidence of the 11-year cycle is heavily damped in the ^{14}C record.

By measuring the abundances of cosmogenic isotopes it is possible to extend the record of solar activity back for tens of thousands of years. Since ^{14}C abundances are used for radiocarbon dating, the corresponding record has been carefully calibrated for the last 26 000 yr (Reimer *et al.* 2004), while the ^{10}Be records from Arctic and Antarctic ice cores extend back for up to 50 000 yr (Wagner *et al.* 2001a). These data sets provide evidence of recurrent grand minima going back into the past. The most recent examples have been given names, and they are labelled in Figure 10.4, which displays ^{10}Be abundances from two ice cores, one from Greenland and the other from the South Pole, after smoothing to give approximately

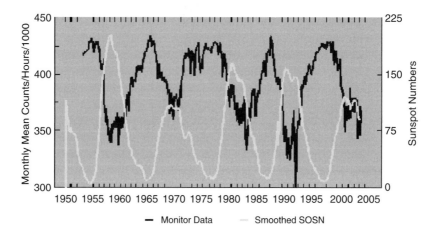

Fig. 10.2. Solar activity and galactic cosmic rays. Comparison between sunspot numbers and secondary neutrons counted at the University of Chicago Climax Neutron Monitor during the interval 1951–2005. (Courtesy of NOAA.)

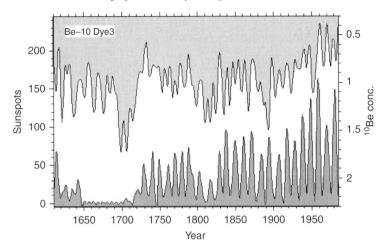

Fig. 10.3. Cyclic activity from sunspots and from the ^{10}Be concentration in the Dye 3 ice core from Greenland, determined by accelerator mass spectrometry. The 11-year cycle can be followed in both records from 1700 onwards, and it is also present in the ^{10}Be record through the Maunder Minimum. (Courtesy of J. Beer.)

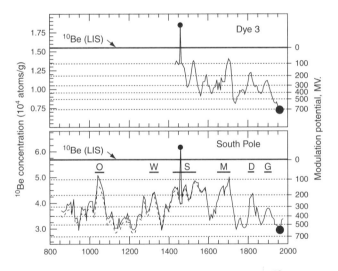

Fig. 10.4. Comparison between variations in smoothed ^{10}Be concentration in the Dye 3 ice core from Greenland (Beer *et al.* 1994) and in an ice core from the South Pole (Raisbeck *et al.* 1990). The ^{10}Be concentrations (which differ owing to different deposition rates) are indicated on the left, with the corresponding modulation potentials Φ on the right. The horizontal lines mark the level of zero modulation potential and the filled circles denote values of Φ in 1958. The outliers around AD 1460 are thought to be due to a nearby supernova. (From McCracken *et al.* 2004.)

22-year averages and thereby filtering out the Schwabe cycle (McCracken *et al.* 2004). Also indicated is an alternative measure of solar magnetic activity, in terms of the modulation potential Φ (Vonmoos, Beer and Muscheler 2006). During the interval of overlap the two records are in qualitative agreement: both show not only the Maunder Minimum but also the abortive Dalton Minimum around 1805. The South Pole data set spans an interval of 1100 yr, from AD 850 to 1958. During this period there were three more grand minima, namely the drawn-out Spörer Minimum (AD 1415–1535), the weaker Wolf Minimum (AD 1280–1340) and the deep Oort Minimum (AD 1010–1050). These minima alternated with grand maxima: high levels of activity comparable with that experienced at the end of the twentieth century were attained on five previous occasions.

This pattern persists in longer records. Figure 10.5 shows the modulation potential Φ reconstructed from part of the 3029 m GRIP ice core from Central Greenland (Vonmoos, Beer and Muscheler 2006). This record, extending over about 9000 yr, has been high-pass filtered (with a cut-off corresponding to a period of 2000 yr) to eliminate long-term trends, and it can be compared with the corresponding potential reconstructed from the ^{14}C production rate of Stuiver and Braziunas (1988). The Spörer, Wolf and Oort Minima are again present, along with many other grand minima and maxima going back into the past. Over the last half-century the mean value of Φ was about 700, a level that is high but by no means unprecedented in the long-term record.[1]

Inspection of the records in Figure 10.5 suggests that solar activity is aperiodically (and perhaps chaotically) modulated, with a characteristic time scale of around 200 yr. Frequency analysis of the ^{14}C and ^{10}Be abundance records reveals several persistent periodicities: for

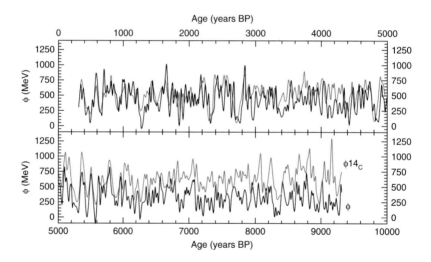

Fig. 10.5. Modulation of solar activity over a 9000 yr interval from 304 to 9315 BP (AD 1646 to 7365 BC): grand maxima and grand minima occur throughout this interval. The modulation potential Φ derived from ^{10}Be abundances in the GRIP ice core (thick line) can be compared with that derived from the ^{14}C production rate (thin line). Both records are high-pass filtered to reduce the effects of changes in the geomagnetic field and other long-term variations. The two records are initially in close agreement, though they gradually drift apart. (From Vonmoos, Beer and Muscheler 2006.)

---

[1] The modulation potential Φ is a measure of the Sun's open magnetic flux, corresponding to field lines that emerge from coronal holes and extend into the heliosphere, while the sunspot number $\mathcal{R}$ relates to the closed flux associated with active regions. Solanki *et al.* (2004; see also Usoskin, Solanki and Korte 2006) have reconstructed sunspot numbers going back to 11 400 BP from the ^{14}C record. They find that recent values of $\mathcal{R}$ are higher than at any time in the past 8000 yr.

both records there are peaks in the power spectra corresponding to periods of 88 yr (the Gleissberg cycle), 205 yr (the de Vries cycle) and 2300 yr (the Hallstatt cycle), as well as the obvious 11-year period of the Schwabe cycle (Damon and Sonett 1991; Stuiver and Braziunas 1993; Beer 2000; Wagner *et al.* 2001a). It seems likely that the prominent 205 yr periodicity, as well as the weaker Hallstatt cycle, has a solar origin.

The ^{14}C and ^{10}Be records are long enough to provide statistical information about the variation of Φ over the past 10 000 yr (Abreu *et al.* 2008). Values of Φ are normally distributed about a mean value around 480, with many episodes that can be classified as grand maxima or grand minima. Although the current grand maximum is not exceptionally high, it has already lasted much longer than the average. While it is safe to forecast that this maximum will terminate in the not too distant future, it is not possible to predict how deep the ensuing minimum will be.

10.3 Differential rotation in the Sun

Carrington (1863) had noticed that sunspots at the equator rotated more rapidly than those at higher latitudes, and this equatorial acceleration has been confirmed by Doppler measurements as well as tracking proper motions of magnetic features (Stix 2002). Different techniques give slightly different values for the surface rotation rate, with sunspots rotating slightly faster than the ambient plasma. Their sidereal angular velocity Ω can be represented by an expression of the form

$$\Omega = A - B\cos^2\theta \qquad (10.1)$$

with $A = 14.55°\,\mathrm{day}^{-1}$ and $B = 2.87°\,\mathrm{day}^{-1}$, where θ is the colatitude. The corresponding equatorial rotation period is 24.7 days.

Perhaps the greatest triumph of helioseismology has been the determination of the Sun's internal angular velocity profile from the rotational splitting of p-mode acoustic frequencies (Thompson *et al.* 2003; Christensen-Dalsgaard and Thompson 2007). Figure 10.6 shows how Ω varies with radius and latitude in the solar interior. At the surface, Ω matches the plasma rotation rate and in the bulk of the convection zone Ω is roughly independent of radius (and constant on conical surfaces). The radiative interior rotates more or less uniformly and there is an abrupt shear in a very narrow layer – the *tachocline* – whose thickness is less than 4% of the solar radius.

Superimposed on this mean rotation profile are variations associated with the sunspot cycle. Doppler measurements of surface rotation have revealed regions of anomalously rapid rotation that migrate from mid-latitudes towards the equator, tracking the zones of maximum activity (Howard and LaBonte 1980; Ulrich *et al.* 1988). At a fixed position on the solar surface, these 'torsional oscillations' have a period of 11 yr. They are also detected by helioseismology and are better described as *zonal shear flows*. Figure 10.7 shows the pattern of behaviour just below the solar surface, at a depth of 0.01 $R_{\odot}$: note that in each hemisphere there are two zones of slightly more rapid rotation, one migrating towards the equator while the other migrates towards the poles. Vorontsov *et al.* (2002; see also Basu and Antia 2003; Thompson *et al.* 2003; Howe *et al.* 2005) find that these zonal shear flows penetrate downwards for at least one-third of the depth of the convection zone and maybe as far as the tachocline itself.

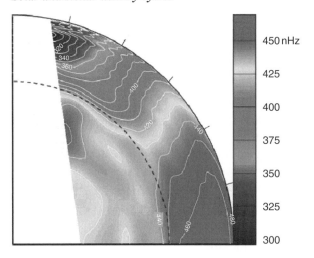

Fig. 10.6. Differential rotation in the solar interior. Rotation frequencies $\Omega/2\pi$ (in nHz) in the solar interior, as determined by helioseismology. The dashed line marks the base of the convection zone at $r = 0.713\,R_\odot$. In the convection zone Ω is predominantly a function of latitude; in the radiative zone Ω appears to be uniform. In between is the tachocline, which is the site of a strong rotational shear but is too thin to be resolved. (From Christensen-Dalsgaard and Thompson 2007.)

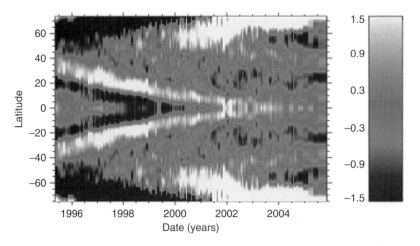

Fig. 10.7. Zonal shear flows ('torsional oscillations') just below the solar surface, at $r = 0.99\,R_\odot$, as determined by helioseismology from p-mode splittings measured by the GONG network. Note the two branches of slightly more rapid rotation in each hemisphere. (Courtesy of R. Howe.)

On a longer time scale, there is evidence that the solar rotation is affected by grand minima. The sunspot positions recorded by Hevelius in 1642–44, just before the Maunder Minimum, are consistent with the current rate of differential rotation (Abarbanell and Wohl 1981; Yallop *et al.* 1982).[2] However, the detailed series of sunspot observations made

---

[2] The rates derived by Eddy, Gilman and Trotter (1976, 1977) are apparently inaccurate.

at the Paris Observatory during the Maunder Minimum show that the equatorial rotation rate was about 2% lower during the period 1666–1719, and that the differential rotation was significantly enhanced (Ribes and Nesme-Ribes 1993).

10.4   Variable activity in stars

As we saw in Chapter 8, there are many stars that exhibit evidence of magnetic activity. In some of them the activity varies cyclically, as it does in the Sun. It is only by studying activity and activity cycles on stars other than the Sun that we can gain a full understanding of these phenomena, for it is essential to investigate their dependence on stellar mass, luminosity, age and – above all – rotation rate.

10.4.1   Stellar activity cycles

Enhanced chromospheric emission in the Ca II H and K lines provides one of the most effective indicators of magnetic activity in a star, as discussed in Section 8.1 above. Moreover, the star's rotation period can also be directly determined. In 1966 Olin Wilson and his colleagues at Mount Wilson Observatory began their important long-term study of calcium emission from 91 (since increased to 111) cool dwarf stars using the 100-inch Hooker telescope (Wilson 1978). In 1978 this 'H-K Project' was transferred to the 60-inch telescope on Mount Wilson, which was dedicated solely to these observations until the project was terminated in 2004 (Baliunas *et al.* 1995, 1998). Among these stars, there are about a dozen slow rotators that exhibit quasi-regular activity cycles similar to that of the Sun (Baliunas *et al.* 1995; Baliunas, Sokoloff and Soon 1996; Frick *et al.* 2004). One of the best examples, the K1 dwarf HD 10476 with a rotation period of 36 days and a cycle period of 9.6 yr, is displayed in Figure 10.8; another, the K4 dwarf HD 4628 with a rotation period of 38 days and a cycle period of 8.4 yr, was illustrated in Figure 1.7. Comparing these results with corresponding measurements for the Sun, which were shown in Figure 8.1, we conclude that main-sequence stars of spectral types G and K with ages and rotation rates similar to those of the Sun can exhibit similar magnetic activity. We may then infer that such stars also

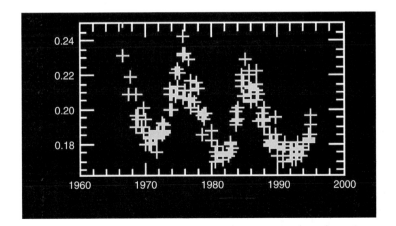

Fig. 10.8. Cyclic variation of Ca II H and K emission in a star. The lower main-sequence star HD 10476 shows an activity cycle, with a period of 9.6 yr, that is very similar to that in the Sun. (Courtesy of the Mount Wilson H-K Project.)

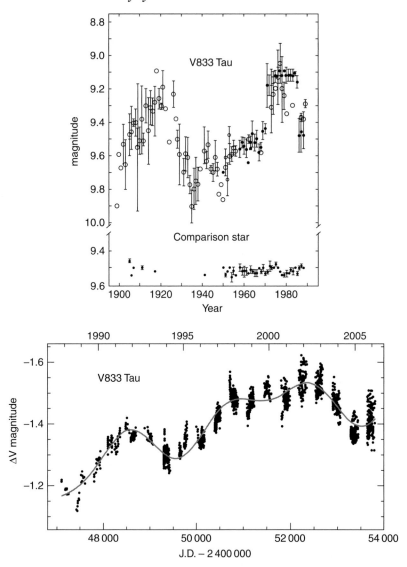

Fig. 10.9. Cyclic variability in the BY Dra star V833 Tau, a rapid rotator with a rotation period of 1.8 days. The long-term record (above) from the Harvard and Sternberg plate collections and more recent photoelectric measurements spans the interval from 1899 to 1990, and shows a very clear 60 yr periodicity (Hartmann *et al.* 1981; Bondar' 1995). The shorter, recent record (below) covers an interval of about 18 years, and shows a 6.4 yr cycle superimposed on the long-term trend; there are also suggestions of a further 2.5 yr cycle (Oláh, Kolláth and Strassmeier 2000; Oláh and Strassmeier 2002; Oláh *et al.* 2007).

harbour spots upon their surfaces, even though we cannot see them. There are other similar stars that show no evidence of activity, with low Ca II emission and no apparent variability, as though they were undergoing grand minima: the G2 dwarf HD 143761 is an example (see Figure 1.7). So far, however, no solar-type star has been reliably observed to enter into or come out of such a grand minimum, so this hypothesis remains untested.

Where Sun-like cycles are detected, the cycle period depends on the rotation period of the star: the more rapidly a star rotates, the shorter its cycle period is likely to be. There is, in addition, evidence of surface differential rotation (Donahue, Saar and Baliunas 1996). In order to compare stars with different masses and spectral types, it is convenient to introduce the Rossby number $Ro = P/\tau_c = 2\pi/\Omega\tau_c$, where τ_c is an estimate of the convective turnover time, as in Section 8.2; then Ro provides a normalized measure of the rotation period P, while Ro^{-1} is a corresponding dimensionless measure of the angular velocity Ω. The relationship between the cycle period $P_{cyc} = 2\pi/\omega_{cyc}$ and Ro can be represented by a power law of the form

$$P_{cyc}/P = \Omega/\omega_{cyc} = \text{const}.Ro^q, \tag{10.2}$$

where the observations are best fitted by an exponent $q > 0$ and various estimates have yielded $q = 0.25$ (Noyes, Vaughan and Weiss 1984), $q = 1.0$ (Ossendrijver 1997), $q = 0.5$ (Brandenburg, Saar and Turpin 1998; Saar and Brandenburg 1999) and, most recently, $q = 0.32$ (Saar 2002). It follows that $\omega_{cyc} \propto \Omega^{1+q}$ for stars of given spectral type.

10.4.2 Cyclic behaviour in more active stars

The stars that display Sun-like cycles are relatively inactive; they belong to the class of late-type stars with ages greater than 2–3 Gyr that are slow rotators and lie below the Vaughan–Preston gap in Figure 8.2 (Brandenburg, Saar and Turpin 1998). Other stars in this class exhibit less regular variability, though periodicities can still be extracted by frequency analysis. Cycles are not so easily recognizable for the more active stars in the Mount Wilson sample, which lie above the Vaughan–Preston gap: for example, the G0 dwarf HD 206860 in Figure 1.7 varies irregularly, though an underlying period of 6.2 yr is still present (Baliunas *et al.* 1995). Some of these active stars also show secondary periodicities. Their behaviour can be represented by a power law similar to that in Equation (10.2), with the same exponent but with periods that are about six times longer (Saar and Brandenburg 1999).

Chromospheric emission saturates for very rapidly rotating stars, with rotation periods of a day or less, as can be seen from Figure 8.3. Long-term photometric measurements have nevertheless revealed cyclic behaviour in some of these super-active stars (Saar and Brandenburg 1999; Oláh, Kolláth and Strassmeier 2000; Oláh and Strassmeier 2002). This method works best for stars whose rotation axis is aligned close to the line of sight (i.e. small inclination angle i) so that long-term secular changes are not masked by rotational modulation. The clearest example is the BY Dra star V833 Tau (see Fig. 10.9); this is a K4 dwarf with $P = 1.8$ days and cycle periods of 6.4 yr (Oláh, Kolláth and Strassmeier 2000; Oláh and Strassmeier 2002; Oláh *et al.* 2007) and 60 yr (Hartmann *et al.* 1981; Bondar' 1995). Cycle periods for these stars are much longer than would be expected for such rapid rotators if they obeyed the laws derived from chromospheric variations; indeed, their variability actually seems to follow a power law with $q < 0$ (Saar and Brandenburg 1999).

11

Solar and stellar dynamos

The magnetic field of a star with radius R is generated by electric currents flowing in its interior, and these currents are subject to Ohmic decay on a time scale $\tau_\eta \approx R^2/\eta_0$, where η_0 is a suitably weighted laminar magnetic diffusivity. For a star like the Sun, $\tau_\eta \approx 10^{10}$ yr, which is comparable to its lifetime on the main sequence; hence there would be no difficulty in regarding a steady field as a fossil relic.[1] The problem for the Sun is to explain the cyclic reversals of its magnetic field on a time scale $P_{\mathrm{cyc}} \ll \tau_\eta$. The simplest explanation would be to suppose that the Sun is a magnetic oscillator, and that its angular velocity oscillates with a 22-year period, acting on a steady poloidal field with dipole symmetry to generate a toroidal field that reverses every 11 years. This model faces three fatal difficulties: first, the observed dipole field itself reverses after 11 years; second, the only measured fluctuations in angular velocity (the zonal shear flows discussed in Section 10.3) have a period of 11 years; and, third, no credible mechanism for driving cyclic oscillations in Ω has ever been put forward. Hence it is generally accepted that the solar cycle is instead maintained by dynamo action in the interior of the Sun (Cowling 1981; Hughes 1992).

Stellar dynamos rely on the inductive effect of electrically conducting plasma flowing across a magnetic field. Here we first provide a brief mathematical outline of dynamo theory – which has developed into a rich and fascinating topic in its own right (see, for example, Dormy and Soward 2007) – followed by a more physical description of the appropriate mechanisms in stellar convection zones. Then we focus on the solar dynamo, which is most constrained by observations and has attracted the most attention from theorists; among recent reviews, see the brief surveys by Tobias (2002a) and Bushby and Mason (2004), the fuller discussions by Mestel (1999), Choudhuri (2003), Rüdiger and Hollerbach (2004), Solanki, Inhester and Schüssler (2006) and Tobias and Weiss (2007a), and the comprehensive accounts by Ossendrijver (2003) and Charbonneau (2005). Next we consider the status of dynamo models for more active stars, before concluding with a summary of progress towards direct numerical simulations.

11.1 Basic dynamo theory

Most studies of solar or stellar dynamos have been restricted to axisymmetric models, in which the magnetic field is averaged azimuthally to give a solenoidal mean field $\mathbf{B}$ that satisfies the induction equation

$$\partial\mathbf{B}/\partial t = \nabla \times (\mathbf{U} \times \mathbf{B}) + \eta_0\nabla^2\mathbf{B}, \tag{11.1}$$

---

[1] By contrast, although $\tau_\eta \approx 7000$ yr for the Earth, the geomagnetic field has existed for at least 3×10^9 yr, and therefore has to be maintained by a dynamo.

where the magnetic diffusivity η_0 is assumed to be uniform. Here $\mathbf{U} = \mathbf{u}_m + s\Omega(s, z)\mathbf{e}_\phi$ is the axisymmetric fluid velocity, where $\mathbf{u}_m$ is a meridional flow and Ω is the angular velocity, referred now to cylindrical polar co-ordinates (s, ϕ, z). Now $\mathbf{B}$ can be expressed as the sum of a toroidal field $\mathbf{B}_T = B_\phi(s, z)\mathbf{e}_\phi$ and a poloidal field $\mathbf{B}_P = \nabla \times A(s, z)\mathbf{e}_\phi$. From Equation (11.1),

$$\frac{\partial}{\partial t}\left(\frac{B_\phi}{s}\right) = \mathbf{B}_P \cdot \nabla\Omega - \nabla \cdot \left(\frac{B_\phi}{s}\mathbf{u}_m\right) + \left(\frac{\eta_0}{s}\right)\left(\nabla^2 - \frac{1}{s^2}\right)B_\phi \tag{11.2}$$

and

$$\frac{\partial}{\partial t}(sA) = -\mathbf{u}_m \cdot \nabla(sA) + s\eta_0\left(\nabla^2 - \frac{1}{s^2}\right)A. \tag{11.3}$$

11.1.1 Cowling's theorem

Equation (11.2) describes the generation of toroidal fields from poloidal fields by differential rotation (as expected in a highly conducting plasma) and their transport by meridional flows. In Equation (11.3) there is no source term, corresponding to $\mathbf{B}_P \cdot \nabla\Omega$, for the poloidal field. Cowling's theorem states that a steady axisymmetric magnetic field cannot be maintained (against Ohmic dissipation) by fluid motions. Equation (11.3) can be re-expressed in terms of the flux function $\chi = sA$ to give

$$\frac{\partial\chi}{\partial t} = -\mathbf{u}_m \cdot \nabla\chi - \mu_0\eta_0 j_\phi, \tag{11.4}$$

where the azimuthal current $j_\phi = -\mu_0^{-1}[\nabla^2 - (2/s)\partial/\partial s]\chi$. In its simplest form, the argument relies on the fact that the poloidal field must have closed field lines (contours of constant χ) enclosing a neutral point where $\mathbf{B}_P = 0$ and χ is an extremum. The source term vanishes at this neutral point but the current will normally be finite; hence the field cannot be maintained. This argument can be made rigorous for an incompressible flow with appropriate boundary conditions, for then Equation (11.4) yields

$$\frac{d}{dt}\int \chi^2 dV = -2\eta_0\int |\nabla\chi|^2 dV \le 0 \tag{11.5}$$

and so the poloidal field must decay (Moffatt 1978). The aim of mean-field dynamo theory is to remedy this defect by inserting an additional source term into Equation (11.3). This source term is provided by the turbulent α-effect.

11.1.2 Mean-field magnetohydrodynamics and the α-effect

We now suppose that the total velocity can be separated into an axisymmetric part $\mathbf{U}$ and a non-axisymmetric fluctuating part $\mathbf{u}$, and that the magnetic field can similarly be separated into a mean field $\mathbf{B}$ and a fluctuating field $\mathbf{b}$ (Parker 1955b; Steenbeck, Krause and Rädler 1965). Then the azimuthally averaged interactions between $\mathbf{u}$ and $\mathbf{b}$ generate a mean electromotive force $\mathcal{E} = \langle \mathbf{u} \times \mathbf{b} \rangle$ in the induction equation. Moreover, if we ignore any small-scale dynamo action and assume a separation of scales then it follows that $\mathbf{b}$ must be linearly related to $\mathbf{B}$ and that we can expand the components of $\mathcal{E}$ as

$$\mathcal{E}_i = \alpha_{ij}B_j + \beta_{ijk}\frac{\partial B_j}{\partial x_k} + \cdots. \tag{11.6}$$

If we assume that the turbulence is pseudo-isotropic (i.e. not mirror-symmetric) we can split α_{ij} into a symmetric part $\alpha \delta_{ij}$ and an antisymmetric part $\gamma_j \epsilon_{ijk}$, and set $\beta_{ijk} = \beta \epsilon_{ijk}$. Then we may truncate the expansion and rewrite Equation (11.6) as

$$\mathcal{E} = \alpha \mathbf{B} + \boldsymbol{\gamma} \times \mathbf{B} - \beta \nabla \times \mathbf{B}; \tag{11.7}$$

inserting this source term into the mean-field induction equation (11.1), we then have

$$\partial \mathbf{B}/\partial t = \nabla \times (\alpha \mathbf{B}) + \nabla \times [(\mathbf{U} + \boldsymbol{\gamma}) \times \mathbf{B}] + \eta \nabla^2 \mathbf{B}, \tag{11.8}$$

where $\eta = \eta_0 + \beta$ represents a combination of laminar and turbulent diffusivities; typically $\beta \gg \eta_0$. Correspondingly, Equation (11.3) acquires not only an enhanced diffusion term but also a novel source term αB_ϕ, while a corresponding source term is also inserted into Equation (11.2), in addition to that from differential rotation (the ω-effect). Kinematic mean-field dynamos are then governed by the equations

$$\frac{\partial}{\partial t}(sA) = s\alpha B_\phi - \mathbf{u}_\mathrm{m} \cdot \nabla(sA) + s\eta_0 \left(\nabla^2 - \frac{1}{s^2} \right) A \tag{11.9}$$

and

$$\frac{\partial}{\partial t}\left(\frac{B_\phi}{s} \right) = \frac{1}{s} \mathbf{e}_\phi \cdot \nabla \times (\alpha \mathbf{B}_\mathrm{P}) + \mathbf{B}_\mathrm{P} \cdot \nabla \Omega - \nabla \cdot \left(\frac{B_\phi}{s} \mathbf{u}_\mathrm{m} \right)$$
$$+ \left(\frac{\eta_0}{s} \right) \left(\nabla^2 - \frac{1}{s^2} \right) B_\phi. \tag{11.10}$$

Mean-field dynamos come in three different flavours: there are $\alpha^2 \omega$-dynamos, in which all three source terms are included; α^2-dynamos, in which differential rotation is omitted; and – more commonly – $\alpha\omega$-dynamos, where the α-effect is regarded as small compared with the ω-effect and is therefore omitted in the toroidal equation. This is an attractive and popular approach: it suffers, however, from the drawback that the presumed separation of scales is not obviously applicable in a star like the Sun.

It is natural to invoke rotation, acting through the Coriolis force, in order to generate turbulence that lacks mirror symmetry. In certain circumstances, where first-order smoothing (Roberts 1994; Ossendrijver 2003) is justified, it is possible to calculate the values of α and β. If the turbulent eddies have a length scale l and a characteristic velocity v, and either the magnetic Reynolds number $R_\mathrm{m} = v l/\eta_0 \ll 1$ or the correlation time $\tau_0 \ll \tau_\mathrm{c} = l/v$, then it is possible to relate α to the kinetic helicity $\langle \mathbf{u} \cdot \nabla \times \mathbf{u} \rangle$, while $\beta \sim v^2 \tau_\mathrm{c} = l^2/\tau_\mathrm{c}$ (Moffatt 1978; Krause and Rädler 1980). Unfortunately, neither of these conditions is satisfied in turbulent stellar convection and any simple dependence of α on helicity breaks down for flows at high R_m or for τ_0/τ_c of order unity (Courvoisier, Hughes and Tobias 2006). More generally, it would seem that first-order smoothing is only valid if $|\mathbf{b}| \ll |\mathbf{B}|$, which is not necessarily the case (Cowling 1981). Indeed, numerical studies of turbulent Boussinesq magnetoconvection in a rotating layer yield a very weak α-effect, $\alpha \approx \eta_0/l$, that does not even depend on the turbulent motion (Cattaneo and Hughes 2006, 2008).

In practice, therefore, we should regard the α-effect as a useful parametrization that captures the essential physics of the processes that regenerate the poloidal field – even if it is not possible to calculate a meaningful value of α from the statistics of the turbulent velocity field. Hence we need only adopt plausible distributions of α and Ω that are independent of the magnetic field strength and then solve Equation (11.8) in order to represent a kinematic (linear) dynamo. There are many examples of $\alpha\omega$-dynamo models, in Cartesian or spherical geometry, that display travelling wave or oscillatory solutions (e.g. Parker 1955b; Steenbeck and Krause 1969; Roberts and Stix 1972). These solutions vary exponentially with time and may grow if the dimensionless dynamo number $D = \alpha_0 \Delta\Omega L^3/\eta^2$ exceeds a critical value; here α_0 and $\Delta\Omega$ are measures of the α-effect and the rotational shear, respectively, and L is an appropriate length scale.

For nonlinear dynamo models it is necessary to model the back-reaction of the quadratic Lorentz force on the velocity field. Within the framework of mean-field dynamo theory this can be done by quenching the α-effect, for example by setting

$$\alpha = \alpha_0(1 + B^2/B_0^2)^{-1}, \tag{11.11}$$

where $B_0 = [\mu_0\langle\rho v^2\rangle]^{1/2}$ is the equipartition field (Jepps 1975); for consistency, η should be similarly quenched.[2] Alternatively, differential rotation may be modified or partially suppressed by the magnetic field, either macrodynamically through the mean field – the Malkus–Proctor (1975) effect – or microdynamically by the effect of the fluctuating fields on the small-scale turbulence (Rüdiger and Hollerbach 2004).

11.2 Phenomenology of the solar dynamo

The formation and orientation of sunspot groups and active regions, as described in Chapter 7, imply that they are caused by the emergence of toroidal flux tubes from the solar interior. Their scale and their systematic properties suggest that these flux tubes originate deep in the convection zone. Thus we are led to consider the behaviour of isolated flux tubes with predominantly azimuthal fields, embedded in a turbulent, convecting layer (Solanki, Inhester and Schüssler 2006). Such a flux tube can only maintain its identity if the magnetic energy density inside it is at least comparable with the external kinetic energy density; this requires field strengths of order 10^4 G near the base of the convection zone. It is well known that such flux tubes are magnetically buoyant (Parker 1955a; Jensen 1955; Hughes 2007a). If the total pressure (magnetic plus gas) inside balances the external gas pressure p_e, then the internal gas pressure $p_i < p_e$ and, if the flux tube reaches thermal equilibrium with its surroundings, it will be less dense than the ambient gas and therefore tend to rise. Moreover, the rise time is at most comparable to the convective turnover time τ_c, which is about a month in the deep convection zone.

The motion of isolated flux tubes within the convection zone can be treated in the thin flux tube approximation (Spruit 1981a) – see Fan (2004) for a comprehensive review, as well as the shorter accounts by Choudhuri (2003), Schüssler (2005) and Hughes (2007a). The picture here is of an initially axisymmetric flux tube, with a predominantly toroidal field, that becomes unstable to a non-axisymmetric mode, which develops into an Ω-shaped loop that rises through the convection zone until it emerges at the photosphere (e.g. Caligari, Moreno-Insertis and Schüssler 1995; see also Section 7.2.2). The axial field must be sufficiently strong for the tube to maintain its integrity and to avoid being disrupted by the ambient turbulence as it rises; in addition, the field has to be significantly twisted. The rising flux tube is subjected to Coriolis forces: unless the initial field within it is sufficiently strong, a tube will be carried polewards and emerge only at high latitudes (Choudhuri and

---

[2] Note, however, that the form and magnitude of the quenching effects are subject to debate (e.g. Brandenburg and Subramaniam 2005); see Section 11.3.1 below.

Gilman 1987; Choudhuri 1989). Furthermore, the tube should be only slightly twisted by Coriolis forces, so as to produce inclinations that satisfy Joy's law at the solar surface (D'Silva and Choudhuri 1993). These considerations all imply that the initial toroidal field should be as strong as 10^5 G. There is still a problem, for as an adiabatically rising tube approaches the surface the internal gas pressure within it is likely to exceed the external pressure, so that the tube 'explodes' (Moreno-Insertis, Caligari and Schüssler 1995; Rempel and Schüssler 2001). Such an explosion may allow the upper portion of the flux loop to become detached from its footpoints at the base of the convection zone.

Magnetic buoyancy can be counteracted by the downward expulsion of magnetic flux from the convection zone. Diamagnetic flux expulsion by a layer of two-dimensional eddies, whether persistent or turbulent, can easily be demonstrated (Weiss 1966; Tao, Proctor and Weiss 1998). In three dimensions, magnetic fields are pumped down the gradient of turbulent intensity with a velocity γ that can be calculated (Rädler 1968; Zeldovich, Ruzmaikin and Sokoloff 1983; Moffatt 1983). Up–down symmetry is broken in a stratified compressible layer, where convection is dominated by slender, rapidly sinking plumes, and magnetic fields are pumped preferentially downwards. In a configuration with a turbulent convecting layer lying above a stably stratified layer, magnetic flux will then end up in the lower, stable region (Nordlund *et al.* 1992; Tobias *et al.* 1998, 2001; Dorch and Nordlund 2001; Ossendrijver *et al.* 2002). Hence it is to be expected that toroidal magnetic fields will accumulate in the stably stratified layer of convective overshoot immediately below the base of the solar convection zone (Spiegel and Weiss 1980; Golub *et al.* 1981; van Ballegooijen 1982). Galloway and Weiss (1981) estimated that this magnetic layer might contain a total flux of 10^{24} Mx, with an *average* field strength of 10^4 G. Within such a layer there may be localized flux tubes with fields that are ten times stronger and therefore able to escape and reach the solar surface.

This layer is, of course, the location of the tachocline (see Section 10.3) and the steep radial gradient in angular velocity provides the most obvious source of the ω-effect, though the latitudinal gradient is of comparable importance. The latter is present throughout the convection zone and there is also a weaker radial gradient in the outermost 35 000 km. Hence there is no difficulty in modelling the generation of toroidal flux, although the source of the α-effect is far less clear. The simplest picture relies on the combined effects of a density stratification and the Coriolis force, which lead to a net helicity in the convective motion, and so to the presence of cyclonic eddies. They in turn distort the strong toroidal field, and so (with the aid of turbulent diffusion) create a net poloidal component, thereby providing an α-effect and completing the dynamo cycle (Parker 1955b, 1979a). Babcock (1961) associated this process with the systematic tilts of active regions, and his phenomenological picture underlies many subsequent models of the solar dynamo.

11.3   The solar dynamo

We are now ready to consider some of the mean-field dynamo models that have been developed in order to explain the origins of sunspots and of the Sun's magnetic cycle. After enumerating the various physical mechanisms that are involved, we contrast the two locations that have been proposed for the α-effect – at the tachocline (for interface dynamos) or at the photosphere (for flux transport dynamos) – and summarize the differences between them. Then we illustrate the results obtained with selected models and go on to show how nonlinear dynamos can explain the chaotic modulation of cyclic activity in the Sun.

11.3.1 Physical mechanisms

The total velocity $(\mathbf{u}_m + s\Omega\mathbf{e}_\phi + \boldsymbol{\gamma})$ in Equation (11.8) includes three components. We have already discussed the differential rotation $\Omega(s, z)$ and the downward pumping represented by $\boldsymbol{\gamma}$, but the meridional velocity $\mathbf{u}_m$ remains to be considered. Doppler measurements of surface motion reveal a somewhat erratic poleward flow in each hemisphere (Hathaway 1996) and helioseismology has confirmed the presence of a quadrupolar circulation with a peak velocity of about $20\,\mathrm{m\,s^{-1}}$ directed towards the poles (Thompson *et al.* 2003), which may extend downwards through much of the convection zone (Braun and Fan 1998; Duvall and Kosovichev 2001). Mass conservation demands that there should be a return flow at greater depths,[3] with an estimated equatorward velocity of $1\,\mathrm{m\,s^{-1}}$ at the base of the convection zone – enough to cover 40 degrees in 11 years (Hathaway *et al.* 2003; Dikpati *et al.* 2004).

In Parker's (1955b, 1979a) original dynamo model buoyant fluid elements rise and expand, carrying stitches of toroidal field with them; the action of the Coriolis force makes them rotate (but not too far) and so creates a meridional component of the field. Thus the α-effect relies on the generation by the Coriolis force of helicity and some form of cyclonic motion. It follows that α itself must be antisymmetric about the equator. As Parker (1955b) showed, the direction in which dynamo waves travel depends on the sign of the product $\alpha\partial\Omega/\partial r$, where (r, θ, ϕ) are spherical polar co-ordinates. Waves will travel towards the equator if $\alpha\partial U_\phi/\partial r < 0$ in the northern hemisphere. Given the measured velocity gradients near the tachocline, we may expect dynamo waves in each hemisphere to travel equatorwards at low latitudes, and polewards at high latitudes, provided that α is negative in the northern hemisphere.

The classical α-effect is distributed throughout a turbulent region, corresponding, in the solar context, to a large part of the convection zone. This raises two grave difficulties. First of all, numerical studies of turbulent, rotating convection show that, even in the kinematic regime, α is extremely small, as mentioned in Section 11.1 above. The problem here is that $\tau_0 \geq \tau_c$ and weak toroidal fields are twisted round through large, randomly distributed angles, so that the average value of α dwindles away (Cattaneo and Hughes 2008). In the nonlinear regime, the small-scale field $\mathbf{b}$ is likely to be much stronger than the mean field $\mathbf{B}$ and the α-effect is severely quenched as a result. We should therefore replace Equation (11.11) by an expression of the form

$$\alpha = \frac{\alpha_0}{1 + R_m^q\, B^2/B_0^2}\,, \tag{11.12}$$

with $0 < q \leq 2$ (Vainshtein and Cattaneo 1992; Diamond, Hughes and Kim 2005; Hughes 2007b); numerical experiments suggest that $q = 1$ (Cattaneo and Hughes 1996; Ossendrijver, Stix and Brandenburg 2001). Since $R_m \gg 1$ in the Sun, this implies that α is catastrophically quenched when B is negligibly small.

An alternative is to rely on magnetically driven instabilities as the source of the α-effect. It has long been known that a stratified magnetic field whose strength decreases upward is liable to instabilities driven by magnetic buoyancy (Newcomb 1961; Thomas and Nye 1975; Tobias 2005; Hughes 2007a).[4] In a rotating system these instabilities can give

---

[3] Estimates of the flow speed are sensitive to assumptions about its form; the meridional velocity might even reverse direction several times.

[4] Note that these instabilities are quite distinct from the lack of equilibrium of an isolated flux tube.

rise to magnetostrophic waves with a net helicity that provides an α-effect (Moffatt 1978; Schmitt 1987; Brandenburg and Schmitt 1998; Thelen 2000a), and a rotational shear adds further complications (Gilman and Cally 2007). Three-dimensional computations demonstrate how isolated flux tubes can emerge from a magnetic layer in the nonlinear regime (Kersalé, Hughes and Tobias 2007); once liberated, they can rise upwards to intersect the photosphere in active regions (Fan 2004). This process (Thelen 2000b), or a variant relying on undular instabilities of thin flux tubes (Ferriz-Mas, Schmitt and Schüssler 1994), can be incorporated into an $\alpha\omega$-dynamo model.

The magnetic field acts back on the angular velocity not only through the macrodynamic Lorentz force $\mu_0^{-1}(\nabla \times \mathbf{B}) \times \mathbf{B}$ (the Malkus–Proctor effect) but also through turbulent Maxwell stresses. These back-reactions limit the growth of the field in the nonlinear regime. Since the large-scale Lorentz force is quadratic in $\mathbf{B}$, it does not depend on the sign of the magnetic field and therefore produces fluctuations in angular velocity with twice the frequency of the magnetic cycle, i.e. with an 11-year rather than a 22-year period in the Sun, thus providing a natural explanation for the zonal shear flows that appear as 'torsional oscillations' at the solar surface and are displayed in Figure 10.7 (Schüssler 1981; Yoshimura 1981). This link with observations makes the Malkus–Proctor effect an attractive choice as the key nonlinear process in model calculations.

11.3.2   Location

Ever since helioseismology revealed the presence of the tachocline it has been generally assumed that the strong toroidal fields are formed at the base of the convection zone and stored in a region of convective overshoot beneath it. Various locations have been suggested for the α-effect at different times. It was originally thought that turbulent eddies provided a significant contribution to α throughout the convection zone but (as explained above) it has now been shown that α is negligible in the kinematic regime (Cattaneo and Hughes 2006, 2008), and it appears also that α is drastically quenched in the nonlinear regime. Hence most – but not all (see Brandenburg 2005) – recent models assume that the α-effect is concentrated either at the top or bottom of the convection zone.

Leighton (1969) extended Babcock's (1961) phenomenological model and developed what was essentially a mean-field dynamo operating at the solar surface (Stix 1974). He assumed that toroidal fields are formed by subsurface differential rotation and emerge, owing to magnetic buoyancy, to form active regions, containing sunspots, that are oriented according to Joy's law. These active regions are then subjected to differential rotation and to turbulent diffusion caused by supergranular motion, so that the preceding fields in a sunspot pair drift towards the equator, where they cancel out, while the following fields drift towards the poles. There they cancel and reverse the pre-existing dipolar field, and so provide the necessary input for the next cycle. This model was later extended to include a poleward meridional flow at the surface (Wang and Sheeley 1991; Sheeley 2005). By adopting the measured differential rotation and meridional flow, together with a supergranular diffusivity that is consistent with results of kinematic modelling (Simon, Title and Weiss 1995), it is possible to reproduce the observed two-dimensional behaviour of magnetic fields at the solar surface, including reversals of the polar fields at sunspot maximum (Dikpati *et al.* 2004; Durrant, Turner and Wilson 2004).

This success has led to the formulation of a family of flux-transport dynamo models, in which the α-effect acts near the solar surface at mid-latitudes, while the ω-effect is concentrated near the base of the convection zone. These models can be grouped into two subfamilies. In one, radial diffusion competes effectively with meridional transport: most estimates of the turbulent diffusivity β give values of 10^8 to 10^9 m^2 s^{-1} throughout the convection zone, with a corresponding radial diffusion time of about a year, which is smaller than the time taken for meridional motion to transport flux to high latitudes and to reverse the polar fields (Choudhuri, Schüssler and Dikpati 1995; Jiang, Chatterjee and Choudhuri 2007). Thus the poloidal field produced at the surface feeds the ω-effect below and leads also to a reversal of the polar fields – but those reversed fields do not directly cause the toroidal field to reverse. In the other subfamily, the turbulent diffusivity is assumed to be 50 times smaller, and the meridional flow acts as a conveyor belt, transporting the poloidal field from mid-latitudes at the surface to the poles, thence to the base of the convection zone and back again to mid-latitudes at the tachocline, a process that takes about 20 years (Dikpati and Charbonneau 1999; Dikpati *et al.* 2004).

The alternative is a family of interface dynamo models (Parker 1993; Tobias 2005), where the α-effect is concentrated at, or just above, the tachocline and may be produced by turbulent convection where the large-scale toroidal field is weak, or else as a result of instabilities powered by magnetic buoyancy. It can be shown that the α-effect is much more potent when α is localized at the base of the convection zone rather than at the top (Mason, Hughes and Tobias 2002; Moss and Sokoloff 2007). In these interface models, magnetic features observed at the solar surface – active regions, sunspots and even the reversal of the polar fields – are all epiphenomena (Cowling 1975) or secondary manifestations of the real action, involving the α- and ω-effects plus a slow meridional flow, which is concentrated near the base of the convection zone. The differences between these families of models can be expressed in terms of the lag between the emergence of sunspot groups at mid-latitudes and the time when the poloidal field in the tachocline begins to reverse (Jiang, Chatterjee and Choudhuri 2007). In the interface dynamos the two events are almost simultaneous; if anything, the lag is negative. In the highly diffusive flux-transport model the lag is only a year or two but in the weakly diffusive models there is a lag of 20 years, so that the new cycle is related not to its predecessor but to its predecessor but two (Dikpati and Gilman 2006).

11.3.3 Models of the solar cycle

Any reliable dynamo model should include a consistent representation of the two key mechanisms that are accurately known from observations, namely the Sun's differential rotation and the meridional flow in the upper convection zone.[5] The remaining ingredients, including the distributions of α, η and γ (which represents both pumping and magnetic buoyancy), as well as the choice of nonlinear quenching processes, can be mixed according to taste (Rempel 2006). There are many possible recipes that are able to produce similar, and equally plausible, butterfly diagrams. It turns out, however, that certain crucial properties – whether the dynamo is steady or oscillatory, whether the magnetic field has dipole or quadrupole symmetry (i.e. whether $\mathbf{B}_T$ is antisymmetric or symmetric about the equator) and

---

[5] Earlier models, which assumed that Ω was approximately constant on cylindrical surfaces (e.g. Ivanova and Ruzmaikin 1977; Yoshimura 1978a), are more relevant to rapidly rotating stars – see Section 11.4.2 below.

whether dynamo waves travel polewards or equatorwards at low latitudes – do depend sensitively on the assumptions that are made. It follows therefore that, although the successful models are instructive, they should be regarded as illustrative rather than predictive.

In this subsection, we shall confine our attention to a few examples of nonlinear interface dynamos in spherical geometry; for more comprehensive accounts of solar dynamo models, see the reviews by Choudhuri (2003), Ossendrijver (2003), Rüdiger and Hollerbach (2004) and Charbonneau (2005), who also consider flux-transport models (Küker, Rüdiger and Schultz 2001; Chatterjee, Nandy and Choudhuri 2004; Dikpati *et al.* 2004). Here we focus first on models with algebraic quenching of the α-effect and then go on to consider examples with a dynamically varying ω-effect.

Charbonneau and MacGregor (1997) found that kinematic dynamos yielded fields with dipole symmetry, and that dynamo waves propagated towards the equator provided that the α-effect was concentrated at low latitudes. Nonlinear interface dynamos with straightforward α-quenching, such as given by Equation (11.11), preserve this desirable feature (Covas *et al.* 1998; Markiel and Thomas 1999) as does a variant that invokes mean-field hydrodynamics (Rüdiger and Brandenburg 1995). Figure 11.1 shows a butterfly diagram calculated for an idealized model of this type (Markiel and Thomas 1999), which clearly reproduces the most essential features of the observed pattern in Figure 2.3.

A more interesting option is to allow α or Ω to vary dynamically, under the effects of Maxwell stresses or the Lorentz force. Guided by observations of zonal shear flows, the most obvious procedure is to add an equation describing the time-dependent evolution of perturbations to the azimuthal velocity, driven by the macrodynamic Lorentz force and moderated by a turbulent viscous diffusivity $\nu = P_{\mathrm{m}}\eta$ (Covas *et al.* 2000; Bushby 2005). Figure 11.2 shows results from a carefully tuned model. At the tachocline, the product $\alpha\partial\Omega/\partial r$ changes sign at mid-latitudes and, as might be expected, this sign reversal leads to the appearance of two dynamo branches: in addition to the equatorward-travelling waves at low latitudes

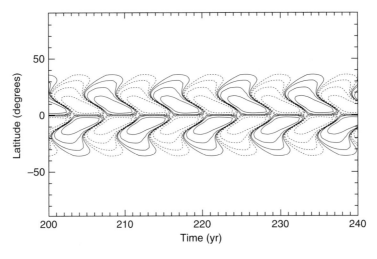

Fig. 11.1. Butterfly diagram showing cyclic activity for an interface dynamo model with the α-effect concentrated near the equator and limited by nonlinear quenching. Contours of the toroidal field just below the interface show antisymmetric dynamo waves propagating towards the equator, with a weak poleward branch. (From Markiel and Thomas 1999.)

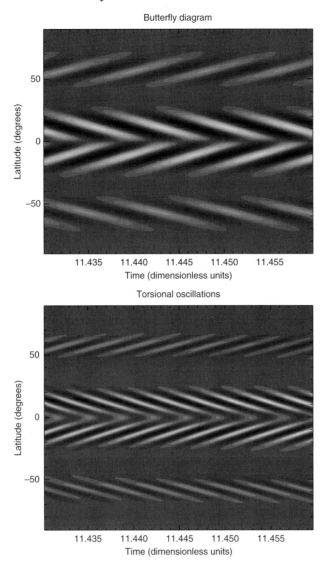

Fig. 11.2. Upper panel: butterfly diagram showing cyclic activity for an interface dynamo model with zonal shear flows driven by the macrodynamic Malkus–Proctor effect. Note the appearance of a weaker poleward branch as well as the normal equatorward branch. Lower panel: the corresponding zonal shear flows at the base of the convection zone. Both branches are again evident. (Courtesy of P. J. Bushby.)

there is also a weaker poleward-travelling branch at higher latitudes. Zonal shear flows consequently show two corresponding branches, in agreement with the observed pattern of 'torsional oscillations' in Figure 10.7. The effect of the radial density stratification is to enhance the zonal shear flows near the surface, so that the strongest effect may still appear at the photosphere even if the flows are driven near the base of the convection zone (Kleeorin and Ruzmaikin 1991; Covas, Moss and Tavakol 2004; Bushby 2005).

11.3.4   Modulation and grand minima

Dynamo models can also represent the modulation of cyclic activity that was described in Section 10.2 (Tobias 2002b). The most obvious explanation is to invoke stochastic effects that are excluded from the normal mean-field formalism; these can be ascribed to major fluctuations either in the turbulent convection (which modify the α-effect) or in the meridional circulation (Schmitt, Schüssler and Ferriz-Mas 1996; Ossendrijver 2000; Charbonneau, Blais-Laurier and St-Jean 2004). Such stochastic inputs can, with care, be tuned so as to lead to plausible representations of the long-term variation of solar activity; indeed, the sunspot record can be mimicked by a stochastic oscillator (Barnes, Sargent and Tryon 1980). The weakness of these stochastic models is that, of their nature, they cannot reproduce the 205-year periodicity that is such a prominent feature of the record derived from cosmogenic isotopes.

The alternative is that the apparently chaotic modulation of solar activity has a deterministic origin.[6] Periodic and aperiodic modulation of cyclic behaviour has been successfully demonstrated for spherical dynamo models with the nonlinearity provided by either macrodynamic or microdynamic quenching of the ω-effect (Kitchatinov *et al.* 1999; Pipin 1999; Küker, Arlt and Rüdiger 1999; Moss and Brooke 2000; Bushby 2006).

The origin of this modulation can be explained by reference to the underlying bifurcation structure (Tobias and Weiss 2007b), which is revealed most clearly in low-order models involving coupled nonlinear ordinary differential equations. Tobias, Weiss and Kirk (1995) showed for a canonical third-order system[7] that, as a control parameter corresponding to the dynamo number is increased, there is a transition from a solution in which the magnetic field decays to one in which the field varies cyclically and then a second transition to a doubly periodic solution in which the cycle is periodically modulated, followed by a more complicated transition, involving frequency locking and period-doubling, that leads to chaotically modulated behaviour.[8] (In the phase space of the system, trajectories are attracted first to a fixed point, then to a limit cycle, then to a two-torus and finally to a chaotic attractor after the torus is destroyed. From a mathematical point of view, this bifurcation sequence is generic and robust.) Subsequently, Tobias (1996) identified the same bifurcation sequence in a nonlinear mean-field dynamo model in Cartesian geometry, with dipole symmetry imposed. Once this symmetry restriction is relaxed, a further range of bifurcations can occur, allowing quadrupole and mixed-mode solutions to appear (Tobias 1997b).[9] This behaviour too is present in low-order models (Knobloch, Tobias and Weiss 1998).

Figure 11.3 shows chaotically modulated behaviour in a nonlinear Cartesian model (Beer, Tobias and Weiss 1998). The cycles persist through the grand minima, as they do in the ^{10}Be record. It is still possible to extract longer periodicities from such a chaotic record; the modulation period then corresponds to that on a 'ghost' of the destroyed attracting torus (Tobias,

---

[6] Deterministic modulation of cyclic activity in a mean-field dynamo model was first demonstrated by Yoshimura (1978b), who introduced a time delay into the nonlinear quenching process, thereby increasing the effective order of the system.

[7] Actually the normal form for a saddle-node/Hopf bifurcation (Guckenheimer and Holmes 1986; Kuznetsov 1998), whose properties are structurally stable.

[8] This bifurcation sequence was originally established for plane dynamo waves, using a truncated low-order model (Weiss, Cattaneo and Jones 1984; Jones, Weiss and Cattaneo 1985), and later confirmed for the corresponding nonlinear partial differential equations (Tobias 1997a).

[9] Transitions between dipole and quadrupole solutions in spherical geometry had already been found by Brandenburg *et al.* (1989).

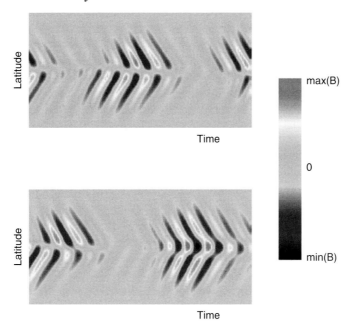

Fig. 11.3. Chaotic modulation of cyclic activity for a mean-field dynamo model in Cartesian geometry. These butterfly diagrams show the toroidal field as a function of latitude and time, with cyclic activity interrupted by grand minima. Upper panel: symmetry breaking in a grand minimum. During the grand maxima the toroidal fields are almost antisymmetric about the equator but this dipole symmetry is broken as they emerge from a grand minimum, just as it was at the end of the Maunder Minimum. Lower panel: a later sequence, showing flipping from dipole to quadrupole symmetry during a grand minimum. (From Beer, Tobias and Weiss 1998.)

Weiss and Kirk 1995). The ratio of this modulation period to that of the cycle depends on the magnetic Prandtl number P_m, the ratio of the viscous to the magnetic diffusivity, and becomes large for $P_m \ll 1$. Thus these model calculations can provide an explanation for the persistent 205-year periodicity in the proxy records.

In the upper panel of Figure 11.3, the nearly perfect dipole symmetry that prevails during grand maxima is broken during the grand minima and this loss of symmetry is most apparent as the solution emerges from a grand minimum, when all the activity is concentrated in one hemisphere, precisely as observed at the end of the Maunder Minimum in 1705. The symmetry occasionally changes during a grand minimum and the lower panel shows an example where the solution flips from dipole to quadrupole symmetry. This result implies that the Sun itself may have possessed fields with quadrupole symmetry from time to time in the past, although dipole symmetry has prevailed since telescopic observations began.

11.4 Stellar dynamos

While it is reasonable to adapt solar dynamo models to describe cyclic magnetic activity in other slowly rotating stars, it is dangerous to assume that younger, rapidly rotating and more active stars can be treated in the same way. Indeed, the observations suggest that the relationships between P_{cyc} and Ro differ depending on whether stars lie below or above the Vaughan–Preston gap in Figure 8.2 (see Section 10.4.1). Furthermore, the magnetic behaviour of late M stars, which are fully (or almost fully) convective, is bound to differ from that of the Sun, where the tachocline is all-important.

If we were able to follow the Sun's evolution backwards, as its angular velocity increased, we should expect to find significant changes in the patterns both of convection and of differential rotation. In any star that rotates sufficiently rapidly, the dominance of the motion by the Coriolis force is expressed by the Taylor–Proudman constraint: in the anelastic approximation, where sound waves are suppressed and $\rho_0(r)$ is the mean density stratification, $\nabla \cdot (\rho_0 \mathbf{u}) = 0$, and so $\mathbf{\Omega} \cdot \nabla(\rho_0 \mathbf{u}) = 0$ (e.g. Thompson 2006a). As a result, convection takes the form of 'banana cells', elongated parallel to the rotation axis, and Ω tends to be constant on cylindrical, rather than on conical, surfaces; these structures have been found in many numerical models (Gilman 1979; Glatzmaier 1985; Miesch *et al.* 2000; Miesch 2005).[10] It seems likely that the jump in magnetic activity across the Vaughan–Preston gap corresponds to such a change in the pattern of convection (Knobloch, Rosner and Weiss 1981).

11.4.1   Slow rotators

We expect middle-aged, slowly rotating stars, if they possess deep convection zones and tachoclines, to exhibit behaviour similar to the Sun's, with cyclic activity interrupted occasionally by grand minima, and fields that may have either dipole or quadrupole symmetry. Properties such as the cycle period, P_{cyc}, should then depend both on the rotation period, P, and on stellar structure. Early hopes (based on weakly nonlinear dynamo waves) that the exponent q in Equation (10.2) might serve as a diagnostic to distinguish between different quenching mechanisms have not been fulfilled. Tobias (1998) found, for a nonlinear model of a cyclic dynamo in Cartesian geometry, that all simple forms of quenching gave rather similar results, with $P_{cyc} \propto D^{-q_1}$ and $0.4 \leq q_1 \leq 0.7$. If we assume that $D \propto Ro^{-2} \propto \Omega^2$ (Durney and Latour 1978), then these values are all more or less consistent with the scalings derived from observations, $P_{cyc} \propto \Omega^{-(1+q)}$ with $1.25 \leq (1+q) \leq 2.0$ (see Section 10.4.1).

In all nonlinear dynamo models, magnetic activity increases monotonically with D, in agreement with the observational result that Ca II emission is a monotonically decreasing function of the Rossby number Ro (Noyes *et al.* 1984). Montesinos *et al.* (2001; see also Lorente and Montesinos 2005) have endeavoured to make this relationship more precise for interface dynamos, by introducing an extended definition of the Rossby number; they conclude that the proportional variation in Ω should scale inversely as the ratio of the radius of the convection zone to the pressure scale-height near its base.

11.4.2   Rapid rotators and polar spots

Fast rotators differ from the Sun not only in being much more active but also in the prevalence of polar spots. As already explained, the distribution of angular velocity in their interiors is likely to be very different from that in the Sun, with consequences for the extent – or even the presence – of a tachocline. The best studied example of a rapid rotator is AB Doradus, with a rotation period of 0.5 days and a pole–equator difference in angular velocity that is similar to the Sun's (but proportionately much less); moreover, this star shows

---

[10] The transition from a tesselated pattern to banana cells as the rotation rate is increased was nicely demonstrated in a series of experiments on convection driven by electrostatic forces in a rotating system, under zero-gravity conditions in space (Hart *et al.* 1986; Hart, Glatzmaier and Toomre 1986).

significant variations on time scales of a year or more (Donati and Collier Cameron 1997; Jeffers, Donati and Collier Cameron 2007).

Studies of convection in rapidly rotating spherical shells confirm the important role played by the tangent cylinder that encloses the uniformly rotating interior (e.g. Rüdiger and Hollerbach 2004).[11] Both in the Sun (at its present age) and in AB Doradus (a K0 star that has just reached the main sequence) the tangent cylinder intersects the surface at latitudes around $\pm 45°$. In the illustrative model shown in Figure 11.4, Bushby (2003) assumed that, outside the tangent cylinder, Ω depended mainly on the distance s from the rotation axis, while inside the cylinder, Ω tended to be a function of colatitude, so that the tachocline was only present at high latitudes; thus there was no net frictional stress applied to the radiative interior, which rotated at an intermediate rate. Figure 11.4 also shows the solution found for a mean-field dynamo model limited by α-quenching, with poleward propagating dynamo waves at high latitudes only, as might be expected. In a different model, with Ω increasing with increasing s throughout the convection zone, a tachocline at all latitudes and macrodynamic quenching of differential rotation, Covas, Moss and Tavakol (2005) found examples of dynamo waves propagating towards the equator at low latitudes.

If buoyant magnetic flux tubes emerge from the base of the convection zone and rise towards the surface they will be deflected polewards by the strong Coriolis force in rapidly rotating stars. Schüssler and Solanki (1992) suggested that this effect might be responsible for the appearance of spots at higher latitudes in these stars. Several detailed numerical studies have since been carried out (Schüssler *et al.* 1996; DeLuca, Fan and Saar 1997; Granzer *et al.* 2000) for pre-main-sequence and main-sequence stars with differing rotation rates. These studies model the development and rise of undulatory instabilities in thin flux tubes until they emerge at the surface. A variety of behaviour appears, depending on the strength of the toroidal field, the rotation rate, the depth of the convection zone and the latitude at which the flux tube is injected. Granzer *et al.* (2000) found a strong increase of the latitude of emergence of the flux tube with increasing rotation rate, a moderate decrease with increasing stellar mass, and a stronger decrease with increasing stellar age. High-latitude emergence was indicated for the more evolved pre-main-sequence stars and young main-sequence stars. However, the latitude of emergence never reached the poles themselves. Apparently some other mechanism is needed to transport the magnetic flux to the poles, such as a poleward migration of the foot of the anchored flux tube at the base of the convection zone or a meridional circulation. Schrijver and Title (2001) have indeed suggested that tilted magnetic bipoles are advected to the poles, much as they are on the Sun, resulting in a small excess accumulation of the trailing polarity at each pole. For a star ten times more active than the Sun, they propose that this process could lead to flux densities of some 300–500 G in the polar caps which might condense into dark spots; this process requires that the bipoles emerge at latitudes of 50–70° and that the meridional velocity reaches $100 \, \mathrm{m \, s^{-1}}$ (Mackay *et al.* 2004).

11.4.3 Fully convective stars

Doppler and Zeeman–Doppler imaging of AB Doradus and other active stars (e.g. Donati *et al.* 2003; Hussain *et al.* 2007) show that spots appear and disappear at low latitudes as well as at the poles. If the interface dynamo only operates within the tangent cylinder, as

---

[11] This is especially relevant in the Earth's core.

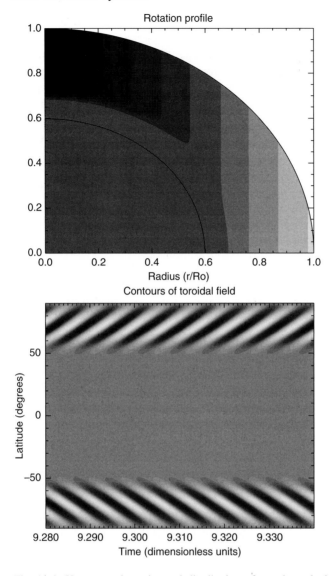

Fig. 11.4. Upper panel: conjectural distribution of angular velocity in a rapidly rotating star, with Ω constant on cylindrical surfaces outside the tangent cylinder and a tachocline confined to high latitudes. Lower panel: the corresponding butterfly diagram for a nonlinear α-quenched interface dynamo, showing cyclic activity near the poles. (From Bushby 2003.)

suggested above, then there must be a very effective distributed dynamo at low latitudes as well. It is interesting therefore that the fully convective dwarf M star V374 Peg displays a strong, predominantly axisymmetric, radial field, without any significant latitudinal variation of its angular velocity (Donati *et al.* 2006b). Unless this field is a fossil remnant, which is unlikely with such vigorous convection, there must be an efficient large-scale dynamo operating in the interior of this star. Moreover, the dynamo must be distributed throughout the stellar volume, since no interface is present. At the moment there are no satisfactory models for this process.

11.5   Small-scale dynamo action

So far, we have only considered large-scale dynamo action, responsible for global magnetic fields in a star. Turbulent motion at high magnetic Reynolds numbers can also maintain small-scale fluctuating fields (such that $\langle \mathbf{b} \rangle = 0$) indefinitely against Ohmic decay. This was first demonstrated for turbulent convection in the Boussinesq approximation, with $R_m \approx 1000$, by Cattaneo (1999). The resulting fields are highly intermittent and concentrated into slender downdrafts (updrafts) at the upper (lower) boundaries of the convecting layer. Careful inspection of the results shows that these long-lived flux concentrations lie at the vertices of a mesocellular pattern (Cattaneo, Lenz and Weiss 2001). Similar behaviour has since been found for fully compressible convection, including the effects of ionization and radiative transfer, for $R_m \approx 2600$ (Vögler and Schüssler 2007). Cattaneo and Hughes (2006, 2008) have also established that small-scale dynamo action is present, with increased efficiency, in a rotating convecting layer. The critical value of R_m for dynamo action to occur depends on the magnetic Prandtl number P_m, and increases as P_m decreases, but calculations for forced turbulence suggest that it tends to a limit as $P_m \to 0$ (Boldyrev and Cattaneo 2004; Isakov *et al.* 2007).

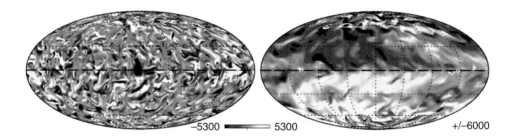

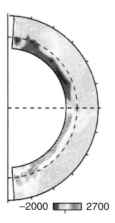

Fig. 11.5. The toroidal field in an ambitious attempt to simulate the solar dynamo. Snapshots showing (left) the toroidal field in the middle of the convection zone and (right) in the region of convective overshoot, both as Mollweide projections, with (below) the temporally averaged axisymmetric component of the toroidal field, which is approximately antisymmetric about the equator. (From Browning *et al.* 2006.)

Bipolar fields emerge into the solar photosphere on a variety of small scales, forming a magnetic carpet, as explained in Section 7.6. Ephemeral active regions, with fluxes of 10^{18}–10^{19} Mx, are closely associated with supergranules: they emerge near the centres of supergranules and shredded flux elements migrate towards the supergranular network, where their fields reconnect and cancel out. This suggests that they are generated by small-scale dynamo action associated with supergranular convection (Hagenaar, Schrijver and Title 2003), influenced perhaps by the velocity shear in the upper part of the convection zone. More recently, the Hinode results have revealed much smaller flux concentrations, of order 10^{15} Mx, nestling between individual granules. These fields are omnipresent outside active regions and appear to be products of small-scale dynamo action driven on the scale of the granulation. Thus we are left with a picture of dynamos on three scales (one large and two small), which are only loosely coupled to each other.

11.6   Numerical simulation of solar and stellar dynamos

Our discussion of large-scale dynamos has shown the limitations of mean-field models, with parameters that can be arbitrarily distributed. The way ahead has to rely on massive numerical computation. In a pioneering calculation, Gilman (1983) obtained self-consistent nonlinear solutions, in the Boussinesq approximation, that demonstrated dynamo action in a rapidly rotating spherical annulus (see also Glatzmaier 1985). With a rotation pattern dominated by the Taylor–Proudman constraint, he found cyclic activity, with dynamo waves that propagated towards the poles (Gilman 1983; Weiss 1994); although this is not what happens in the Sun, it is clearly relevant to behaviour in rapidly rotating stars.

Despite the enormous development in computing power since then, it remains difficult to model motion in the solar convection zone, and it is even harder to produce a large-scale dynamo. Anelastic calculations have succeeded in representing turbulent convection that leads to a solar-like angular velocity profile (Thompson *et al.* 2003; Miesch 2005). So far, however, simulations have not yet yielded a fully convincing large-scale dynamo (Brun, Miesch and Toomre 2004; Browning *et al.* 2006). Figure 11.5 shows details of the toroidal fields that appear in the most recent computational model. The field within the convection zone looks like the product of a small-scale dynamo. There is, however, a large-scale field pumped downwards into the underlying region of convective overshoot, with an axisymmetric component that is antisymmetric about the equator, although it does not reverse. We may expect that further, yet more ambitious, simulations will succeed in reproducing most of the salient features of the solar cycle in the not too distant future.

12

Solar activity, space weather and climate change

Viewed as a star, the Sun exhibits very mild variability. Its total luminosity fluctuates by only 0.1%, following the sunspot cycle. Although most of the energy radiated is in the visible and infrared ranges, corresponding to the peak in the Planck spectrum for the photosphere, the temperature rises through the chromosphere, and then undergoes an abrupt transition to reach millions of degrees in the corona. Since the structure of the solar atmosphere is dominated by its magnetic field, it is not surprising that the Sun's ultraviolet, extreme ultraviolet and X-ray emission vary much more drastically as a result of solar activity. This activity also gives rise to flares and coronal mass ejections, generating energetic particles and enhanced magnetic fields that are carried outwards into interplanetary space by the solar wind. These then impinge upon the Earth, producing aurorae and magnetic storms.

In this chapter we describe the solar irradiance variations and the influence of solar activity on the heliosphere, which gives rise to 'space weather'. We also explain how the incidence of geomagnetic storms, which have been studied since the time of Gauss, provides a proxy measure of the Sun's open magnetic flux, which is responsible for deflecting galactic cosmic rays. Finally, we consider the controversial topic of the influence of solar variability on terrestrial climate (Friis-Christensen *et al.* 2000; Haigh 2003, 2007); here we emphasize that this influence, though it is clearly present, remains small compared with the recent global warming caused by anthropogenic greenhouse gases.

12.1 The variable solar irradiance

The *total solar irradiance*, or 'solar constant', S is the total amount of solar radiative energy at all wavelengths received per unit time and unit area at the top of the Earth's atmosphere at the mean Sun–Earth distance, expressed in SI units of watts per square metre. The term solar constant is misleading because S is known to vary with time. The theory of stellar evolution tells us that the Sun's luminosity has increased by nearly 40% over its lifetime of 4.6×10^9 yr. Measurements from space over the past three decades have shown that S varies by a few tenths of one per cent over time scales ranging from a week (due to the passage of large sunspots and facular regions across the disc) to several years. The most highly variable parts of the spectrum of solar radiation are at the shortest (UV and X-ray) and longest (radio) wavelengths. The currently quoted average value of S, based on measurements from space, is $S = 1366\,\mathrm{W\,m^{-2}}$, but there is an uncertainty of $4\,\mathrm{W\,m^{-2}}$ in the absolute value of S.

Early attempts to monitor changes in the solar irradiance from the ground were confounded by variations in transmission by the Earth's atmosphere, and it was not until radiometers could be flown in space that we had any trustworthy measurements. Monitoring of the total solar irradiance from space began in November 1978 with the HR radiometer on the NIMBUS 7 spacecraft and has continued with a sequence of missions: the ACRIM I radiometer on the Solar Maximum Mission (SMM, launched in 1980) which provided the first convincing evidence of variations in S (Willson and Hudson 1991); the Earth Radiation Budget Satellite (ERBS); ACRIM II on the Upper Atmosphere Research Satellite (UARS); the Solar and Heliospheric Observatory (SOHO), with two radiometers; and ACRIM III on ACRIM-SAT (Fröhlich 2006). Since 1980 there have been continuous measurements by at least two electrically calibrated solar radiometers in space. The resulting data sets are displayed in Figure 12.1a. It is apparent that the records differ both in absolute calibration and in sensitivity; moreover, the radiometers are progressively degraded owing to exposure to solar radiation. Splicing these overlapping records together to construct a composite is therefore not straightforward; Figure 12.1b shows the latest normalized record assembled by Fröhlich (2006). The total irradiance S is greatest at sunspot maximum and least at sunspot minimum,

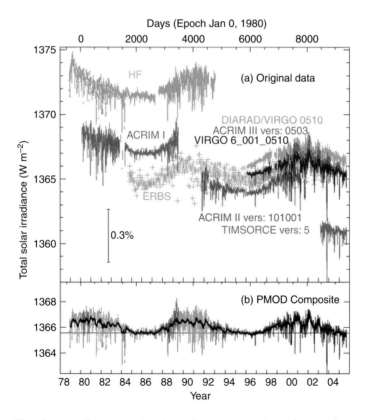

Fig. 12.1. (a) Time series showing daily averaged values of the total solar irradiance S as measured by radiometers on different spacecraft since 1978. The individual series do not match owing to differences in absolute calibration and sensitivity, as well as instrumental degradation. (b) Composite time series showing daily averaged values of the total solar irradiance S and an 81-day smoothed record from 1978 to 2006. (Courtesy of C. Fröhlich.)

and varies between them by about $1.3\,\mathrm{W\,m^{-2}}$, which is about 0.1% of S itself. This variation is well reproduced by a simple empirical model involving only the deficit in radiation from sunspots and the excess radiation from faculae, effects that we discuss below. These variations produce a climate-forcing input of $0.2\,\mathrm{W\,m^{-2}}$, assuming the average albedo of the Earth is 0.7 and correcting for the difference between the cross-sectional area and surface area of a sphere (a factor of $\frac{1}{4}$).

12.1.1 Sunspots, faculae and the solar irradiance

The effects of sunspots and faculae on the total solar irradiance may usefully be separated into short-term and long-term effects. The short-term effects are associated with the passage of individual features across the solar disc as the Sun rotates, and are dominated by the passage of large sunspots, which cause significant dips in solar irradiance. The upper-left panel of Figure 12.2 shows several such dips during a year around sunspot maximum; their timing and magnitude can be accounted for by the size and intensity of large sunspot groups crossing the disc (Foukal 2004; Foukal *et al.* 2006). As we have seen (in Section 3.2),

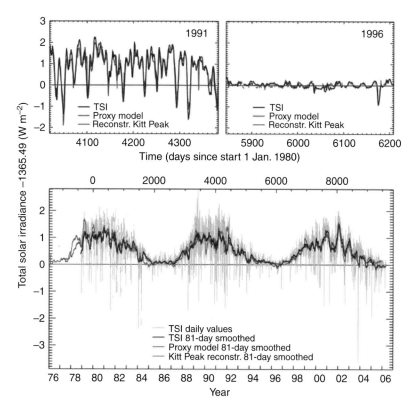

Fig. 12.2. Lower panel: smoothed time series of the total solar irradiance S for the period 1978–2006, from space measurements, with the daily variations in a lighter format. Also shown are two reconstructions, one from the separate contributions of sunspots, faculae and network, and the other derived from Kitt Peak magnetograms. The upper panels compare the measured and modelled variations in S at finer time resolution during periods of high (left) and low (right) overall magnetic activity levels. (From Foukal *et al.* 2006.)

the radiation deficit in a sunspot is not simultaneously balanced by an excess of radiation in the surrounding photosphere; rather, the diverted energy is stored in the convection zone and released slowly over a very much longer period (Foukal *et al.* 2006). Thus a sunspot by itself produces a short-term drop in irradiance. On the other hand, a large facular region by itself can produce a short-term increase in irradiance, and such increases can also be seen in the irradiance time series in the same panel. At sunspot minimum, as shown in the right-hand panel, scarcely any variations can be recognized.

Over a longer term, the time series in the lower panel of Figure 12.2 clearly shows that the solar irradiance varies (by $\pm 0.05\%$) in phase with the sunspot cycle, with maximum irradiance occurring near sunspot maximum, the opposite of what might be expected based on the short-term effect of sunspots alone. The positive effects of faculae, which have lower photometric contrast than sunspots but typically cover areas of the solar surface up to 20 times that of sunspots (Chapman, Cookson and Dobias 1997), more than compensate for the negative effects of sunspots. Indeed, as shown in Figure 12.2, both the short-term and long-term variations in the measured total solar irradiance are well reproduced by a simple empirical model involving only the deficit from sunspots and the excess from faculae.

It is perhaps surprising that the variations in S are so well reproduced by a model involving only sunspots and faculae. Other mechanisms can be imagined, such as fluctuations in the net heat transport across the turbulent convection zone due to random changes in the pattern or intensity of convective cells, or effects of magnetic fields on the efficiency of the convection. It is conceivable that the large thermal inertia of the convection zone, which prevents the sunspot deficit or facular excess from being compensated simultaneously by changes over the rest of the solar surface, will smooth out any variations due to local changes in the turbulent convection. In any case, the time record of measured variations in S is too short to rule out all alternative mechanisms.

Reliable measurements of solar irradiance extend only over the past 30 years. The success of models involving only sunspots and faculae in reproducing these measurements has encouraged researchers to attempt to reconstruct the variations in S over a much longer period in the past based either on the historical sunspot record or on the proxy record from abundances of cosmogenic isotopes, or even on models of cyclic activity in the solar photosphere (e.g. Lean 2000; Fröhlich and Lean 2004; Wang, Lean and Sheeley 2005; Krivova, Balmaceda and Solanki 2007). The upper panel of Figure 12.3 shows the most straightforward reconstruction, relying only on the measured correlation between sunspot numbers and irradiance since 1978 (Fröhlich and Lean 2004). Other reconstructions (e.g. Lean 2000; Wang, Lean and Sheeley 2005) differ in the inclusion or omission of an arbitrarily varying contribution from ephemeral active regions, or on the basis of a questionable difference in Ca II emission between active and inactive stars, in assuming that there was a long-term increase in S, as shown in the lower panel of Figure 12.3. In reality, since we know that cycles persisted through the Maunder Minimum (Beer, Tobias and Weiss 1998), it seems unlikely that the average value of S could have dropped significantly below its level at a normal sunspot minimum.

As we have seen, the positive contribution to S from faculae exceeds the negative contribution from sunspots. We should note, however, that this does not hold for younger and more active stars. As the overall level of activity increases, the contribution from starspots predominates (Foukal 1993, 1998); as a result, the apparent magnitude of a vigorously active

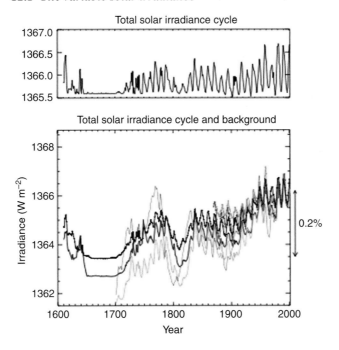

Fig. 12.3. Reconstructions of solar irradiance variations over the period 1600–2000. The upper panel shows a conservative reconstruction derived from surface activity, using the correlation of recent means of total solar irradiance and sunspot numbers. The change since the end of the Maunder Minimum in the mean value of S, averaged over a cycle, is estimated to be about $0.5\,\mathrm{W\,m}^{-2}$. This reconstruction does not include the long-term trends that have been assumed by many authors, as illustrated by the differing versions in the lower panel. (From Fröhlich and Lean 2004.)

star is indeed reduced when its magnetic activity is greatest. Thus photometric changes are anticorrelated with starspot coverage, as was assumed in Chapter 9.

12.1.2 Spectral variability

Figure 12.4 shows how the solar irradiance varies with wavelength across the spectrum, from $10\,\mathrm{nm}$ to $10\,\mu\mathrm{m}$. Across the visible and infrared ranges this variation closely matches the Planck spectrum of a black body at the photospheric temperature of $5770\,\mathrm{K}$. In the ultraviolet and extreme ultraviolet the spectrum is complicated by the presence of strong emission lines. Also shown is the proportional range of variation during an 11-year activity cycle, given by the ratio of the difference between maximum and minimum spectral radiances to the minimum value. In the visible range this ratio has a value around 10^{-3}, as expected from Figure 12.2, but in the EUV the ratio rises to unity. Such significant variations in UV and EUV emission influence the production of ozone in the upper stratosphere, which in turn affects the lower stratosphere and then the troposphere below it. Since ultraviolet radiation is emitted from the corona, whose structure is dominated by magnetic fields in active regions, the mean level of UV and EUV emission during the Maunder Minimum is likely to have been close to that at sunspot minimum and therefore significantly less than it is now.

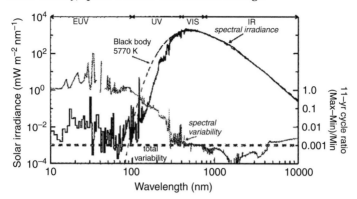

Fig. 12.4. Solar spectral irradiance: the thick line shows the irradiance per unit wavelength interval (in mW m^{-2} nm^{-1}, on the left-hand scale), as a function of wavelength (in nm). For comparison, the dashed line shows the spectrum of a black body at 5770 K. The thin line shows the proportional variability during a typical 11-year activity cycle (on the right-hand scale). (From Lean 2001.)

12.2   Interplanetary effects of solar activity

12.2.1   *Solar flares and coronal mass ejections*

A *solar flare* is an explosive event that produces a sudden brightening in the chromosphere, best seen in Hα through a narrow-band filter (see Fig. 12.5). The brightness typically increases for several minutes (the *flash phase*) and then decreases more slowly over an hour or more (the *main* or *decay phase*). The Hα brightness can increase by as much as a factor of ten in the strongest flares, which are also visible in white light, and the line can broaden by as much as 2 nm, indicating plasma velocities as high as 1000 km s^{-1} in the flare.

A flare produces a burst of energetic particles (mostly protons and electrons) and radiation at essentially all wavelengths in the electromagnetic spectrum. The total energy released by a solar flare can be as high as 10^{32} erg. Flares are readily detected as bursts of noise at radio wavelengths. With the advent of space observations, flares are routinely observed in the EUV and X-ray regions of the spectrum. Strong X-ray emission is produced by the high-energy electrons accelerated and trapped in the flare by its magnetic field. Flares are classified according to their brightness in the X-ray spectrum in the wavelength range 0.1–0.8 nm. In order of increasing energy flux I_f at the Sun–Earth distance, the classes are B ($I_f < 10^{-6}$ W m^{-2}), C ($10^{-6} \leq I_f < 10^{-5}$ W m^{-2}), M ($10^{-5} \leq I_f < 10^{-4}$ W m^{-2}), and X ($I_f > 10^{-4}$ W m^{-2}), and each class is divided into nine gradations in order of increasing intensity (e.g. C1 through C9). X-class flares can trigger major radiation storms and radio blackouts on Earth, M-class flares typically cause only brief radio blackouts at polar latitudes, and C-class flares usually have no significant terrestrial consequences.

All solar flares occur in or near active regions, but not all active regions produce flares. Flares tend to occur above or near sunspots, especially those that are highly asymmetric. Many flares occur immediately after the sudden disappearance of a chromospheric filament lying along a magnetic neutral line (i.e. line of vanishing longitudinal magnetic field), in which case the flare is usually in the form of two bright ribbons lying on either side of and parallel to the vanished filament. EUV observations show that these ribbons lie along the

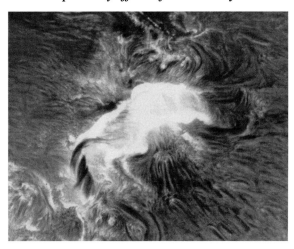

Fig. 12.5. Hα image of a solar flare. (Courtesy of NASA Marshall SFC.)

footpoints of an arcade of coronal loops that straddle the magnetic neutral line. Often a new filament will form in the position of the old one several hours, or even a day or more, after the flare, and the new filament will sometimes produce another flare.

The energy released in a flare comes from the energy stored in a highly sheared or twisted magnetic field. The onset of a flare corresponds to an instability of the magnetic field configuration, after which the field lines rapidly tear and reconnect. In the case of a two-ribbon flare, the horizontal component of the magnetic field is observed to be nearly parallel to the filament before the flare but nearly perpendicular to the filament in its relaxed state after the flare.

Coronal mass ejections (CMEs) are massive (10^{14} to 10^{17} g) bursts of coronal plasma ejected from the Sun into interplanetary space at speeds of 200 to 1000 km s^{-1}. They appear above the solar limb as expanding loops or bubbles and are most likely associated with rising and expanding helical magnetic flux ropes, triggered by an instability of the magnetic field configuration in the solar atmosphere. CMEs of various sizes occur at an average rate of several per day and their combined mass flux is an appreciable fraction (up to 10%) of the total mass flux associated with the solar wind.

Thousands of CMEs have been observed with coronagraphs aboard satellites, most notably with the Large Angle and Spectrometric Coronagraph (LASCO) aboard the SOHO satellite, which has been imaging the solar corona from 1.1 to 32 solar radii since 1995. Figure 12.6 shows an example of a CME imaged by LASCO. In addition to recording CMEs, LASCO has also recorded more than 1000 comets, many of which were found by amateur astronomers searching through the images.

12.2.2 Space weather

The term *space weather* refers to variations in the electromagnetic and particle radiation emanating from the Sun and passing through the near-Earth space environment (Schwenn 2006; Pulkkinen 2007). Solar flares and coronal mass ejections cause large increases in the flux of solar particles impinging on the Earth's magnetosphere, by factors

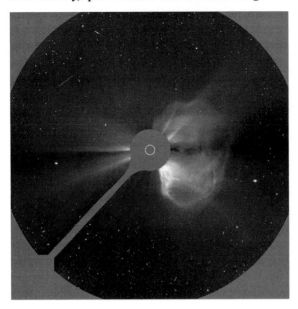

Fig. 12.6. The large coronal mass ejection (CME) of 20 April 1998, imaged by the LASCO coronagraph aboard the SOHO satellite. (Courtesy of NASA.)

of up to 10 000 times the normal flux. These sudden surges in the solar wind stream reach the Earth in some 30 to 70 hours and cause *geomagnetic storms* involving distortions of the Earth's magnetic field, which normally shields us from most of the charged-particle emission from the Sun. Among the effects of these magnetospheric distortions are the injection of many electrons into the Van Allen belts and the appearance of aurorae round the magnetic poles.

There are practical needs for being able to predict space weather on time scales from a few hours up to a sunspot cycle period of 11 years or longer. Before the advent of artificial Earth satellites and human space travel, the primary need was to predict disruptions in long-range radio communication caused by solar-induced distortions of the radio reflecting layers in the Earth's ionosphere. X-ray and UV radiation from solar flares hits the dayside of the Earth and is absorbed by atoms in the ionosphere, raising them to excited states and freeing electrons which in turn causes increased absorption of short-wave (HF) radio waves, producing at times a complete fadeout of short-wave communications. Local heating of the ionosphere by solar radiation, especially at tropical latitudes, causes ascending bubbles of gas that distort the normal reflective layering. In the era of satellite communications, the concern is with even shorter wavelength radio waves (VHF, UHF and microwaves) that penetrate the ionosphere. Solar-induced inhomogeneities in the ionosphere cause phase and amplitude fluctuations in the signals to and from satellites, sometimes disrupting communications. For example, such distortions can cause Global Positioning Satellite (GPS) receivers to lose the lock on their signals.

During geomagnetic storms the fluctuating currents in the ionosphere produce fluctuating magnetic fields that reach the Earth's surface and induce currents in large-scale electrical conductors such as the ocean, large rock formations, and man-made structures, for instance electrical power lines and pipelines. The induced EMFs can be as strong as six volts per kilometre. The man-made structures are most vulnerable at high latitudes, especially in regions (such as in North America) where they are built above mostly igneous rock (a relatively poor electrical conductor) and hence offer the path of least resistance to the induced currents. In large-scale electrical power networks the induced currents can disrupt the distribution of electricity by causing failures in the transformers that step the voltage up and down throughout the grid. On 13 March 1989, a transformer on one of the main transmission lines of the HydroQuebec power network failed in response to currents induced by a very large geomagnetic storm, causing a chain of events that produced a catastrophic collapse of the entire system within 90 seconds, leaving some six million people across eastern North America without power for nine hours or more. Similar induced currents can occur in gas and oil pipelines at high latitudes, causing increased corrosion of the pipes.

In the era of artificial satellites, space weather has become increasingly important. Solar-induced turbulence in the ionosphere affects the radio communication links with satellites. Increases in the solar EUV radiation cause increased heating of the upper atmosphere, making it expand and thereby causing increased drag on satellites in low orbit; hence, estimates of the useful lifetime of a satellite depend on predictions of the level of solar activity many years ahead. Solar radiation can also affect satellites directly, damaging sensitive electronic components. Solar power cells are particularly vulnerable to high-speed solar protons and ions; for example, in October 1989 the GOES satellite lost six years of normal lifetime of its solar panels during a single solar proton event lasting several days.

With the advent of human space travel, space weather has taken on significantly greater importance. The energetic protons and other ions emitted by solar flares and coronal mass ejections, as well as galactic cosmic rays, pose a significant hazard for astronauts in space (and to a lesser extent even to passengers and crew in aircraft flying at high altitudes over polar regions). Solar energetic particles are most prevalent around sunspot maximum, while galactic cosmic rays are more common at sunspot minimum. Astronauts who ventured outside the Earth's magnetosphere during the Apollo missions were lucky that they happened to avoid the most serious solar events (Lockwood and Hapgood 2007). Although it should be possible to shield future astronauts even from extreme solar events, provided they are warned in time, it is not clear how they can be fully protected from galactic cosmic rays. Thus a 600-day mission to Mars would involve significant risks.

Short-term forecasts of space weather concentrate on predictions of solar flares, prominence eruptions, and other features of active-region evolution. These predictions are based largely on the complexity of the magnetic field configuration, especially the amount of twist or shear in the field. Once an active region has produced a flare, it receives more attention and the success rate for predicting further flares goes way up. Predictions of CMEs are more difficult and rely on observed changes in arcades of loops observed in X-rays. Direct detections of CMEs by coronagraphs in orbit, such as the LASCO instrument on SOHO, indicate which ones might impact the Earth and provide at least a few days' warning, although the delays are hard to predict. Predictions of the level of solar activity several years in advance are based mostly on empirical methods. One method is to use previous cycles to predict into the future, whether by extrapolating Fourier series, by constructing the attractor in phase space or by using cellular automata. Other methods are based on using some features of the current cycle to predict the next cycle. Such methods have at least some physical basis in the context of dynamo models: for example, in most models, the strength of the next cycle depends to some extent on the strength of the polar fields in the declining phase of the current cycle. The danger in all these predictions is that solar activity appears to be chaotic and, moreover, that there is always the risk of an unexpected descent into a grand minimum.

12.2.3   Variations in the open solar magnetic flux

Magnetic field lines emerging from the Sun can be divided into two classes: there are those closed lines – often associated with active regions – that return to the solar surface without extending far into space, and the open field lines that become almost radial and are carried out to a great distance by the solar wind. It is the open magnetic flux that extends into interplanetary space, deflecting galactic cosmic rays and ultimately impinging on the Earth's magnetosphere. There it gives rise to magnetic storms, which have been carefully recorded for almost 140 years.

The open solar flux at the Earth can be computed in three ways: the interplanetary field can be measured directly, or it can be derived from measurements of photospheric fields on the Sun, or it can be estimated from a quantitative measure of geomagnetic activity, the aa index (Lockwood 2003). Since the Ulysses space mission confirmed that the meridional part of the interplanetary magnetic field is, on average, purely radial (and approaches that of a split monopole at sunspot minimum), the total open flux can be obtained from local measurements. Figure 12.7 shows the variation with time of the open solar magnetic flux, as derived from the aa index, over the last 50 years of the twentieth century, compared with a scaled record of cosmic ray counts and with the sunspot number over the same period. As expected, the aa index is closely related to the cosmic ray count (and therefore to the abundances of cosmogenic isotopes) and both vary with the sunspot number. Thus the longer record of the

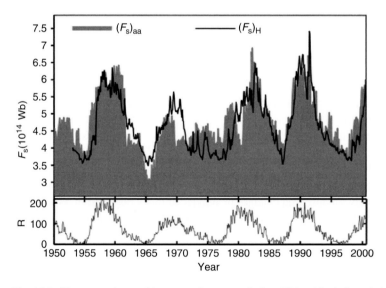

Fig. 12.7. Upper panel: monthly open solar magnetic flux $(F_s)_{aa}$ (shaded grey) derived from the aa geomagnetic index averaged over 12 months, compared with values $(F_s)_H$ derived using linear regression of cosmic ray counts at Huancayo and Hawaii with $(F_s)_{aa}$. Lower panel: monthly means of the sunspot number $\mathcal{R}$. (From Lockwood 2003.)

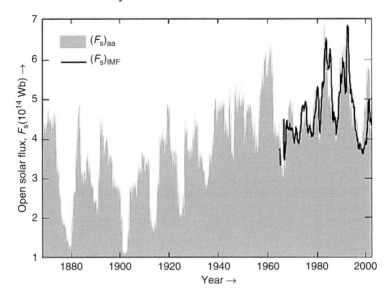

Fig. 12.8. Variation with time of the open solar magnetic flux $(F_s)_{aa}$, as derived from the aa geomagnetic index, since 1868. The solid line shows the flux $(F_s)_{IMF}$ derived from directly measured monthly values of the interplanetary magnetic field. (From Lockwood 2003.)

aa index in Figure 12.8 provides a reliable measure of solar activity ever since 1868. The current level is certainly higher than at any previous time during this interval.

12.3 Solar variability and the Earth's climate

The Sun's unfailing radiation and the Earth's atmospheric blanket, with its 'greenhouse' gases, are responsible for the warm climate that harbours life on Earth. It follows that variations in the solar irradiance must have some effect on the Earth's climate. The Earth's climate system is complicated, however, and determining the sensitivity of this system to variations in the solar input has proved to be very difficult. Understanding the climatic effects of solar variability has become increasingly important in recent years because of the concern about global warming caused by increasing concentrations of greenhouse gases in the atmosphere due to industrial processes. It is imperative that we be able to identify, understand and distinguish between the solar, anthropogenic, and other influences on climate.

12.3.1 Changes in the Earth's orbit

The amount of solar radiation received by the Earth varies not only because of changes on the Sun itself but also because of slow and predictable variations in the Earth's orbit about the Sun and in the tilt of the Earth's axis to its orbital plane. Those changes, caused by the gravitational pull of the Moon and the other planets, produce cyclic variations in the amount and distribution of sunlight received by the Earth, with periods of roughly 23 000, 41 000, and 100 000 years. These cycles are known as the *Milankovitch cycles*, after the Serbian astronomer Milutin Milanković who was the first to calculate their amplitude and point out their significance for the Earth's climate. The 100 000-year cycle corresponds to a small change in the eccentricity of the Earth's orbit, which in turn affects the total

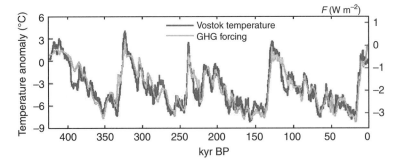

Fig. 12.9. Variation of temperature (half the difference from the 1880–99 mean) in an Antarctic (Vostok) ice core over the past 420 000 years. Also shown is the calculated greenhouse gas forcing derived from the measured concentrations of CO_2, N_2 and methane in the core. The records show four ice ages and five warm periods, including the present. (From Hansen *et al.* 2007.)

amount of sunlight received by the Earth. The 23 000-year cycle corresponds to the precession, or wobble, in the orientation of the Earth's rotation axis (which causes the time of year at which the Earth is closest to the Sun – its perihelion – to progress through the seasons), and the 41 000-year cycle corresponds to oscillations in the tilt of the Earth's rotation axis to its orbital plane, which varies from 21.5° to 24.5° (the current value being 23.44°). These two cyclic variations affect the distribution of sunlight in latitude over the Earth. As a result of the three Milankovitch cycles, the total annual amount of sunlight averaged over the entire Earth can vary by as much as 0.1%; seasonal variations at mid-latitudes can reach a few percent, while the summer sunshine in polar regions may vary by as much as 10%. The impact of these variations depends on the asymmetrical distribution of continental land masses with respect to the equator, and it is widely accepted that the Milankovitch cycles are responsible for the Earth's ice ages. Temperature measurements, based on the $^{18}O/^{16}O$ isotope ratio, first in deep sea sediments (Hays, Imbrie and Shackleton 1976) and subsequently in polar ice cores, show that these ice ages have recurred at intervals of about 100 000 years, and this periodicity has now been traced back for tens of millions of years.

Figure 12.9 shows how Antarctic temperature has varied over the past 420 000 years, based on measurements of oxygen isotope and deuterium abundances in a Vostok ice core (Vimeux, Cuffey and Jouzel 2002). This record demonstrates the sensitivity of the Earth's climate, for the temperature difference of 8 K between ice ages and warm periods is more than can be explained by changes in orbital parameters alone. Their effect is amplified by alterations in the Earth's albedo (as incoming radiation is reflected from icecaps) and by the gradual release of carbon dioxide and methane dissolved in the ocean as its temperature is increased. These greenhouse gases absorb infrared radiation emitted from the Earth's surface and then re-emit it partly back. The variation of their concentrations, which can be measured in air bubbles included in the ice core, does seem to follow, rather than to precede, the changes in temperature (Hansen *et al.* 2007).

12.3.2 *Effects of solar variability*

The solar variability that influences the Earth's climate occurs on two different time scales, arising from distinct causes: very slow changes, occurring over many millions of years, caused by solar evolution, and shorter-term changes, occurring over weeks to hundreds of thousands of years, caused by solar magnetic activity. The long-term evolutionary changes are well understood: when the Sun arrived on the main sequence, 4.6 billion years ago, its luminosity was about 30% lower than its present value; since then its luminosity has been increasing steadily and will continue to increase until it reaches roughly double its present value at the end of its main-sequence lifetime, at an age of about 10 billion years. Then, as it leaves the main sequence, the Sun will evolve much more rapidly and its luminosity (and radius) will increase significantly and undergo large variations. The slow evolutionary changes in solar luminosity have important implications for the history of the Earth's climate. For example, the early Sun was not bright enough to prevent the Earth (assuming it had its present atmosphere) from being covered in ice during its early years, and the high albedo of the ice would have kept it from melting for about 2 billion years. Yet we know from the fossil record that flowing water and life were present on the Earth at least 3.5 billion years ago. This apparent paradox is resolved if the Earth's early atmosphere had a greater concentration of greenhouse gases than it does now, thereby offsetting the reduced solar energy flux.

Of more immediate concern to humanity is the shorter-term solar variability associated with the Sun's magnetic activity. The forms of this variability that might induce climate changes include variations in total and spectral irradiance, which directly affect the amount of energy received by the Earth, and variations in the strength and configuration of the Sun's 'open' magnetic field (in the heliosphere), which modulates the flux of galactic cosmic rays reaching the Earth. The irradiance variations, described in Section 12.1, are very small but they can in principle cause changes in the Earth's climate, though the nature and magnitude of these changes are not yet reliably understood.

The 11-year activity cycle

The sunspot record, being the longest direct record of solar activity, presents an almost irresistible invitation to compare it with records of climate and of other variables affected by the climate. Numerous claims of correlations with terrestrial records have been made, including those with the price of wheat and the value of the stock market (see Chapter 2); most of these claims rest on very shaky grounds. In trying to assess the impact of solar activity on climate it is important to separate out any internal variations of the atmosphere and ocean. Together they form an extremely complex dynamical system, with its own rich natural behaviour, including the North Atlantic Oscillation, with a characteristic decadal time scale, and the El Niño–Southern Oscillation (in the South Pacific but with world-wide ramifications) which recurs at intervals of 3–8 years.

Effects of the 11-year activity cycle are clearly apparent in sea-surface temperatures in the Indian, Pacific and Atlantic Oceans, whether taken separately or combined (Reid 1991, 2000; White *et al.* 1997). These fluctuations have an amplitude of about ±0.05 K, which is consistent with the measured changes in solar irradiance (cf. Fig. 12.2). Unfortunately, signals of the 11-year cycle in atmospheric temperature records are less straightforward; a recent estimate of the globally averaged surface warming gives an amplitude of ±0.08 K (Camp and Tung 2007). There is also an apparent connection between solar activity and variations in summer rainfall, associated with tropospheric temperature fluctuations, in tropical regions (van Loon, Meehl and Arblaster 2004). These variations are in turn linked to changes in the height of surfaces of constant pressure in the lower stratosphere, which are found to vary with the solar cycle in a manner that depends on the phase of the so-called quasi-biennial oscillation in the atmosphere (van Loon and Labitzke 2000).

Longer-term climatic change

As we have seen in Chapter 10, the 11-year cycle is modulated on a longer time scale by the appearance of grand minima and associated maxima. Eddy (1976) drew attention to an apparent link between solar activity and climate, claiming that the Maunder and Spörer Minima coincided with a Little Ice Age that was preceded by a Medieval Warm Period associated with a maximum of activity. There is certainly evidence of local climatic effects, for example from advances and retreats of Alpine glaciers (Holzhauser, Magny and Zumbühl 2005), that are apparently correlated with solar activity (Hormes, Beer and Schlüchter 2006). However, the story is less simple than that: the Little Ice Age was a Northern Hemisphere phenomenon only, running from the fifteenth to the nineteenth century, with the coldest temperatures in the 1690s, when sunspots had indeed disappeared; similarly, the Warm Period – which made it possible for Vikings to settle in Greenland – was not globally important. Nevertheless, there is convincing evidence of large-scale correlations between surface temperature and solar activity, both recently (as indicated in Fig. 12.10) and over past millennia.

Figure 12.11 shows the global temperature variation over the interval from AD 200 to 2000, as well as that in the Northern and Southern Hemispheres separately, as reconstructed by Mann and Jones (2003) and Jones and Mann (2004). The expanded detail for the last millennium displays the differences between various attempts to reconstruct the Northern

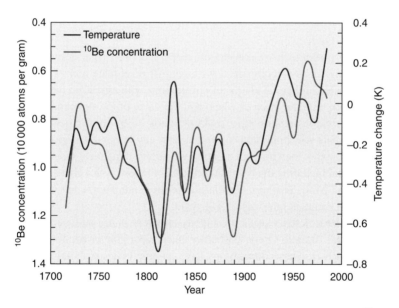

Fig. 12.10. Comparison between solar magnetic activity, as measured by ^{10}Be concentration in the Dye 3 ice core from Greenland, and a reconstruction of Northern Hemisphere temperature, from 1715 to 1985. The data are filtered to eliminate time scales shorter than 20 years. Note, in particular, the minima in both solar activity and temperature around 1815 and 1890, and the increase in both between 1900 and 1940. (After Beer 2001.)

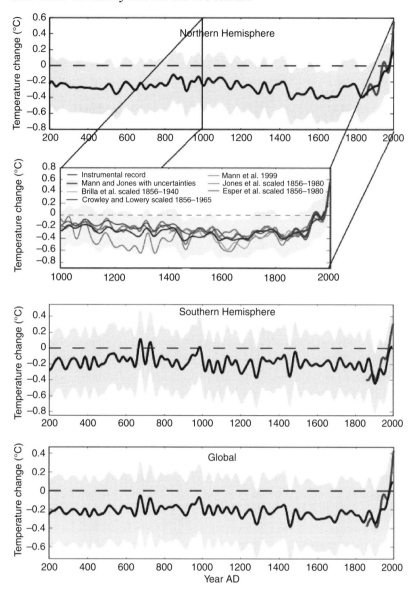

Fig. 12.11. Reconstruction of the temperature variation over almost two millennia in (top panels) the Northern Hemisphere, (middle panel) the Southern Hemisphere and (bottom panel) globally. The heavy lines follow the smoothed reconstructions of Mann and Jones (2003) with 2σ annual variations shaded. Also shown are instrumental measurements for the last 150 years. (From Jones and Mann 2004.)

Hemisphere record; some such attempts show significantly greater variations (e.g. Moberg *et al.* 2005; Juckes *et al.* 2007). There is general agreement, however, on three points: the coolest period was in the sixteenth and seventeenth centuries – Juckes *et al.* (2007) estimate a drop of almost 0.6 K; the warmest episode was around AD 1000; and the rise in the latter half of the twentieth century is unprecedented in this interval. The temperature changes

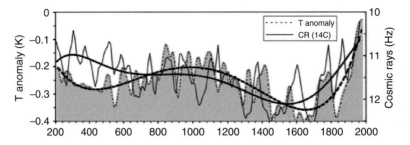

Fig. 12.12. Comparison between a measure of cosmic ray flux (measured downward), derived from cosmogenic ^{14}C abundances (Solanki *et al.* 2004), and the Northern Hemisphere temperature record (grey-shaded) of Mann and Jones (2003). Long-term trends in temperature and cosmic rays are shown by heavy broken and full lines, respectively. (From Usoskin *et al.* 2005.)

are influenced both by internal climatic dynamics and by external forcing, including not only solar activity but also volcanoes and changes in land use. Volcanoes emit sulphate aerosols, which remain in the atmosphere for several years and lead to temporary cooling that is relatively severe (Jones and Mann 2004): the most recent example was the eruption of Pinatubo in 1991, but that of Tambora in 1815, followed in 1816 by 'the year without a summer', was much more serious. Averaged over a long time, however, volcanic aerosols cause a reduction of only $0.2\,W\,m^{-2}$ in climate forcing (Hansen 2000).

There is convincing evidence for a correlation between hemispheric variations in temperature and long-term solar activity. In Figure 12.12 we compare the smoothed Northern Hemisphere record with a measure of cosmic ray fluxes derived from ^{14}C abundances (Usoskin *et al.* 2005). The persistent anticorrelation indicates a significant solar influence on climate, though the long-term increase in cosmic rays is also affected by a slow decay of the Earth's magnetic dipole moment. On a yet longer time scale, Bond *et al.* (2001) used deposits from melting drift ice in deep ocean sediments from the North Atlantic as proxy measures of temperature that could be compared with ^{10}Be abundances over a period of 8000 years from the beginning of the Holocene postglacial era. As can be seen in Figure 12.13 there is a striking correlation, with a characteristic periodicity of around 2300 years.

Amplification mechanisms
    Prior to 1900, the temperature record in Figure 12.11 shows fluctuations of 0.3 K, which are significantly larger than would be expected from changes of 0.1% in total solar radiance alone. This suggests that some amplification process is acting in the Earth's atmosphere. The most obvious source is the effect of large variations in ultraviolet and extreme ultraviolet radiation impinging on the upper stratosphere and consequent changes in ozone production (van Loon and Labitzke 2000; Haigh 2003, 2007; Baldwin and Dunkerton 2005). The resulting disturbances can then interact with Rossby waves and penetrate first to the lower stratosphere and thence into the troposphere below. This process seems to involve interactions with the North Annular Mode (or Arctic Oscillation) as well as with the quasi-biennial oscillation (Ruzmaikin and Feynman 2002; Ruzmaikin *et al.* 2004; Labitzke 2005;

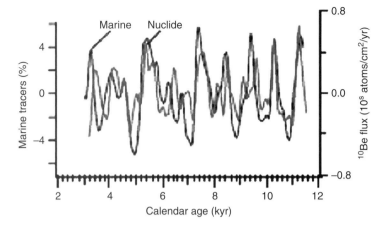

Fig. 12.13. Comparison between fluctuations in ^{10}Be flux, from a Greenland ice core, and mineral tracers deposited from drift ice and preserved in marine sediments. (From Bond *et al.* 2001.)

Baldwin and Dunkerton 2005). A further possibility is that there might be nonlinear resonant coupling of processes driven by solar variability with natural oscillations of the Earth's climatic system, as suggested by Tobias and Weiss (2000).

An alternative proposal, which has attracted much attention recently, is that galactic cosmic rays may have a direct effect on cloud formation, which would therefore be modulated by changes in solar magnetic activity (Svensmark and Friis-Christensen 1997; Svensmark 1998, 2007; Marsh and Svensmark 2000). The claim here is that secondary particles such as muons ionize the air at low altitudes, so encouraging the formation of low clouds, which then lead to global cooling. The physics of cloud formation still remains controversial (see Harrison 2000; Harrison *et al.* 2007 and references therein). To be sure, it is well known that energetic particles lead to condensation of droplets in supersaturated air – that is the principle of the Wilson cloud chamber. The general belief, however, is that there are enough aerosol clusters present to act as condensation nuclei already. One way of checking Svensmark's hypothesis is to look for other changes in the incidence of cosmic rays and then to seek a corresponding effect on cloud formation. For instance, coronal mass ejections distort the Earth's magnetosphere and reduce the incidence of cosmic rays (the so-called Forbush decreases); Čalagović *et al.* (2008) have investigated the effect of these Forbush decreases on cloud formation without obtaining any significant correlation. Again, there was a sudden drop in the Earth's magnetic field 40 000 years ago (the Laschamp event), which certainly increased the incidence of cosmic rays but did not produce any significant change in temperature (Wagner *et al.* 2001b). At the moment, Svensmark's claims still lack credibility.

12.4 Global warming

The temperature records in Figure 12.11 show a characteristic pattern of low-amplitude fluctuations (over a range of 0.3 K) up to the mid twentieth century. Thereafter, the temperature takes off and rises through 0.7 K in 50 years, at a rate that is far greater than those associated with earlier grand maxima in solar activity. It is clear from the past record that the Sun's magnetic variability cannot be responsible for such a large and rapid change. Even if the present grand maximum were to give way to a grand minimum, any consequent cooling would be insufficient to cancel out the recent increase in global temperature. Furthermore, the sudden rapid increases in concentrations of carbon dioxide and methane in the last 150 years have no parallel in the previous 400 millennia. They are the result of man's activities, in burning fossil fuel and changing patterns of agriculture. A range of large-scale computational climate models have confirmed that the current rise in temperature is to be expected as a result of pumping carbon dioxide into the atmosphere at a faster rate than the oceans can absorb it, and that this effect is exacerbated by the release of dissolved gases as sea temperature rises – see the detailed discussion in the latest report of the Intergovernmental Panel on Climate Change (2007) or the more approachable survey by Houghton (2004). The potential consequences of this climatic change are extremely serious, especially for developing countries. Global warming by anthropogenic greenhouse gases does indeed raise important challenges for all mankind – but such issues lie beyond the scope of this book.

13

The way ahead

L'imagination ... se lassera plutôt de concevoir que la nature de fournir.[1]

Blaise Pascal (*Pensées*, 15° – 390)

The many results that we have described amply demonstrate the rapidly accelerating rate of progress in our knowledge of the properties of sunspots and starspots, as well as the profound gaps in our understanding of some aspects of their behaviour. Our purpose in this concluding chapter is to identify the major unsolved problems involving the physics of sunspots, starspots and stellar magnetic activity, and to indicate those areas where we expect to see significant progress in the future, as techniques and facilities develop.

It is already apparent that solar observations, from the ground and from space, will continue to achieve higher resolution and increased precision. Meanwhile, stellar observations will attain greater resolution through improved spectroscopy and the introduction of interferometry, and stellar activity cycles will be followed using dedicated telescopes. Theory will also progress, depending not only on physical insight but also on the ever increasing power of high-performance computers, and the possibility of carrying out ever more realistic simulations.

13.1 The structure and dynamics of a sunspot

The Hinode satellite has only recently begun to deliver results, and will continue for a good many years, to be joined by the Solar Dynamics Orbiter in 2009. On the ground, the 1-m Swedish Solar Telescope will continue to provide important new results, and the 1.5-m German GREGOR solar telescope in Tenerife is due to become operational in 2008, to be followed in the USA by the 4-m Advanced Technology Solar Telescope in Hawaii by about 2014. These facilities will refine and extend our knowledge of the structure of a sunspot and the patterns of intensity, velocity and magnetic fields within it.

We can anticipate that helioseismology will help to reveal the overall subphotospheric structure of a sunspot, once the magneto-acoustic nature of its modes of oscillation have been properly incorporated into the modelling process. It may also be possible to establish how much of the radiant energy emitted from the umbra and penumbra is transmitted from deep within the underlying flux concentration, and how much is transported inwards from the near-surface surroundings of the spot. Detailed modelling will also help to explain the

---

[1] The imagination tires sooner of conceiving than Nature does of providing.

sharpness of the umbra–penumbra boundary, as well as the relationship between a spot and the moat cell that surrounds it.

Concerning the umbra itself, we can expect a closer correspondence between observations and computational models of the fine structure of umbral dots, which will settle the issue of whether external, relatively field-free plasma actually penetrates into the core of a sunspot. The interlocking-sheet structure of the penumbral magnetic field raises more fundamental problems: we need to establish the vertical extent of the steeply inclined component, and the relationship (if any) between the dark cores within bright filaments and the Evershed flow. Above all, theoreticians have to develop a sequence of increasingly realistic numerical models that can explain the magnetoconvective structure of the penumbra itself, and its connections to external granular convection in the moat. So far as sunspot oscillations are concerned, the relationship between umbral oscillations and penumbral waves has yet to be established, and there is a need to study oscillations in more realistic axisymmetric and three-dimensional models of the sunspot.

13.2   Solar and stellar activity cycles

From the observational side, the greatest need is to exploit the solar–stellar connection by continuing and extending the Mount Wilson survey of Ca II emission from nearby bright stars. Such a synoptic survey is a long-term project that can best be carried out with robotic telescopes (see below). In due course it should be possible to record cyclic behaviour in a significantly larger sample of slowly rotating Sun-like stars and to pin down the relationship between cycle period and rotation period more precisely, thereby providing an essential check on dynamo models of solar and stellar magnetic cycles.

So far as dynamo theory itself is concerned, it is time to leave mean-field dynamos behind and to focus on developing fully nonlinear, three-dimensional numerical models that can represent the interactions between convection, rotation and magnetic fields with reasonable precision. Such models will still have to employ the anelastic approximation, along with physically realistic but exaggerated diffusion coefficients. As a first stage, it is necessary to reproduce the observed differential rotation in the convection zone, and this is already feasible. The next stages might be to include the ultra-thin tachocline at the base of the convection zone and then to add magnetic fields. The problem becomes distinctly more complicated if, as seems increasingly likely, the structure of the tachocline itself depends on the magnetic fields within it, so that all three physical effects have to be included together. That will remain a challenge for some time to come.

13.3   Starspots

Photometry remains the most common observational method for studying starspots. Routine, nightly photometric measurements, for a large sample of stars, are being made by a number of dedicated robotic telescopes (listed by Berdyugina 2005). For example, the wide-field STELLA Imaging Photometer on Tenerife is being used to study stars in open clusters of different ages up to 2 Gyr in order to investigate stellar properties and activity. The future International Concordia Explorer Telescope (ICE-T) at Dome C in Antarctica, which will search for transiting extrasolar planets during the continuous 2000-hour winter night, will also yield copious data on starspots. Space missions designed to study stellar oscillations and detect Earth-sized extrasolar planets, such as the existing MOST (launched in 2003) and COROT (launched in late 2006) satellites and the future Kepler and Gaia missions, due to launch in 2009 and 2011, will also provide an immense amount of photometric data related to starspots.

Direct imaging of stellar surfaces using interferometry is a promising approach to observing starspots and stellar activity in general. Earth-based interferometers have already begun to resolve the surfaces of some of the largest stars, but the resolution required to sample activity on many stars will only be possible with space-based instruments. The proposed NASA Stellar Imager mission, for example, would put a large UV–optical interferometer in orbit to image the surfaces of nearby stars, detect surface activity patterns, and probe their interiors through asteroseismology. Another, more remote possibility of detecting starspots is offered by gravitational microlensing, which can in principle measure the radial distribution of a star's surface brightness as a function of wavelength (Hendry, Bryce and Valls-Gabaud 2002).

Meanwhile, we can expect improvements in the spectroscopic techniques of Doppler imaging and Zeeman–Doppler imaging to allow us to better determine the structure of individual starspots and their associated magnetic fields. As the spatial resolution of these techniques improves, we may be able to address the important question of whether a large starspot is a single entity or a close-packed assembly of smaller spots, which is closely related to the issue of the extent to which a sunspot is a suitable prototype for a starspot. Once this is settled, theoreticians can begin constructing detailed models of spots on various types of stars.

13.4 Prospect for the future

The problems we have outlined range in scale from the smallest distances that can be resolved on the Sun (limited only by the mean free path of photons) to the dimensions of a post-main-sequence giant star. All ultimately involve the rich nonlinear interactions between fluid motions and magnetic fields in the stellar plasma, and the observations can only be explained by developing adequate theoretical models. To some astrophysicists it may seem surprising that, 400 years after Galileo and a century after Hale, there is still so much to be explained. That, however, reflects the burst of discoveries in the last two decades, which still continues, and in turn provides a host of opportunities for a new generation of researchers.

Appendix 1 Observing techniques for sunspots

Over the past four decades, remarkable advances in high-resolution studies of sunspots and active regions have been made through the use of new ground-based telescopes, space missions and associated instrumentation, and of new observing techniques. In this appendix we give a brief summary of these facilities and techniques. (Techniques for observing starspots are described separately in Chapter 9.) An excellent general introduction to modern solar telescopes and their instrumentation is given in Chapter 3 of Stix (2002). References to more thorough treatments of particular topics are provided in the sections below.

A1.1   High-resolution solar telescopes

Although there are more than 50 professional ground-based solar telescopes in regular operation around the world, we focus here only on the few large telescopes that are best suited to high-resolution studies of sunspots. These existing telescopes are listed in Table A1.1 along with two future telescopes, one nearing completion (GREGOR) and the other recently through its design phase and awaiting construction (ATST). There are also plans for a large European Solar Telescope, to come into operation around 2020. A comprehensive list of solar telescopes and their specifications has been compiled by Fleck and Keller (2003).

Ground-based observations are limited to electromagnetic radiation in the visible and near-infrared ranges, at wavelengths between about 300 nm and 2200 nm, and in a range of radio wavelengths. Space missions have allowed us to observe at shorter wavelengths (UV, EUV, X-ray and gamma-ray), revealing the properties of the higher-temperature chromospheric and coronal layers of the solar atmosphere, and have also provided long time series of seeing-free measurements in the visible for studying solar oscillations (helioseismology) and the evolution of solar magnetic fields. Table A1.2 lists current and planned space missions relevant to high-resolution studies of sunspots. Closely related to space missions is the Sunrise project, in which a balloon-borne 1-m solar telescope will make long-duration flights at an altitude of 120 000 feet (37 000 m) in the polar stratosphere.

A1.2   Correcting for atmospheric seeing and stray light

For ground-based telescopes, the limiting factor in high-resolution observations is usually atmospheric 'seeing', the distortion of the wavefronts of light passing through the Earth's turbulent atmosphere on its way to the telescope. Much of this seeing originates in the free atmosphere and is beyond control other than through careful selection of the

Table A1.1 *High-resolution ground-based solar telescopes*

Telescope	Aperture	Location	Comments
McMath–Pierce Solar Telescope	1.6 m	Kitt Peak, Arizona	
Swedish Solar Telescope	97 cm	La Palma, Canary Islands	vacuum tower
THEMIS	90 cm	Tenerife, Canary Islands	helium-filled
Dunn Solar Telescope	76 cm	Sacramento Peak, New Mexico	vacuum tower
German Vacuum Tower Telescope	70 cm	Tenerife, Canary Islands	vacuum tower
Big Bear Solar Telescope	65 cm	Big Bear Lake, California	
Dutch Open Telescope	45 cm	La Palma, Canary Islands	open structure
GREGOR	1.5 m	Tenerife, Canary Islands	open structure 2008
Big Bear Solar Telescope	1.6 m	Big Bear Lake, California	2008
Advanced Technology Solar Telescope	4 m	Maui, Hawaii	open structure ~2014

Table A1.2 *High-resolution solar space missions*

Satellite	Agency	Launch date	Comments
Solar and Heliospheric Observatory (SOHO)	NASA	1995	visible, UV, EUV
Transition Region and Coronal Explorer (TRACE)	NASA	1998	visible, UV, EUV
Hinode	JAXA	2006	visible, EUV, X-ray
Solar Dynamics Observatory	NASA	2008	visible, EUV
Sunrise	NCAR, NASA, MPI	2009	UV

telescope site, informed by measurements of seeing properties at candidate sites. 'Local' seeing caused by ground-level turbulence can be minimized to a considerable extent by careful design of the telescope structure: approaches include elevated towers (placing optics above the ground-generated turbulence) and open structures allowing moderate winds to clear away local convective turbulence. Internal seeing, due to turbulence generated within the telescope by solar heating, can be minimized either by evacuating the optical path or by using an open design.

If a solar telescope is to produce diffraction-limited images over a substantial fraction of its observing time, measures must be taken to correct for the remaining seeing. Methods in current use include frame selection, post-processing image reconstruction, and adaptive optics. *Frame selection* relies on the fact that even during average seeing conditions there are brief moments of excellent seeing. To exploit this, many short-exposure images are taken sequentially but only the best frames are selected for storage, based on some criterion (such as maximum rms contrast) that is evaluated in real time on a high-speed computer.

The most commonly used post-processing methods for image reconstruction are those based on *speckle interferometry* or *phase diversity*. Because the field of view of the detector is usually larger than the isoplanatic angle (the angular extent over which wavefronts suffer the same distortion, typically around $5''$), the reconstruction must be applied separately to different segments of the image. In speckle interferometry, image details smaller than the limit imposed by seeing are recovered from the speckle patterns in several sequential, short-exposure images. For images of the solar surface, where no point source is available as a reference, it is necessary to measure both the Fourier phase as well as amplitude. Most of the techniques for solar speckle imaging are based on the method of Knox and Thompson (1974) in which the autocorrelation of the image transform is computed as an average over many short-exposure images. In the phase-diversity method, two images are recorded simultaneously using a beam splitter: a conventional focused image and a 'diversity' image intentionally defocused by a known amount. The two images are subject to the same seeing aberrations, but their point-spread functions differ because the complex wave field in the diversity image is altered by a known amount. This allows a simultaneous solution of the two image convolutions in order to obtain an object function (the undistorted image) and an aberration function that are consistent with the two images. The method can be extended to include more than two simultaneous images and to different techniques for diversifying the phase.

A more efficient way to correct for seeing is to make the correction in real time using *adaptive optics* (AO). The simplest form of AO is an *image-motion compensator* that guides on a solar feature (such as a pore) and employs a single tip-tilt mirror to remove the overall image motion due to the varying average tilt of the wavefront (i.e. the lowest mode of wavefront distortion). A *correlation tracker* (von der Lühe 1983), which senses image motion by cross-correlating successive images, allows an image-motion compensator to guide on the granulation pattern. A full AO system employs a multi-element deformable mirror to correct for higher-order modes of wavefront distortion, using the granulation pattern as a reference. Several different techniques of wavefront sensing have been employed, including curvature sensing and phase diversity. The standard technique is the Shack–Hartmann method in which an image of the entrance pupil is formed on an array of lenslets. The several subimages (or subapertures) produced by the lenslets are displaced relative to each other because of the differing local tilts of the wavefront. These displacements are measured (usually by correlation tracking) and used as input to control the deformable mirror.

The first operating AO system on a solar telescope was a 24-element (i.e. 24-subaperture) system installed on the Dunn Solar Telescope in 1998, followed shortly thereafter by a system on the Swedish 50-cm telescope (now replaced by the 1-m SST). More recently, higher-order 76-element AO systems were put in place at the Dunn telescope and the Big

Bear telescope, demonstrating the feasibility of scaling an AO system up to match the requirements of new-generation large telescopes such as the 4-m ATST. A 36-element AO system has been in place at the German VTT since 2002, and an 80-element system is ready for use in the new GREGOR telescope.

A1.3 Imaging and narrow-band filters

In contrast to stellar observations, the availability of sufficient sunlight allows the use of filters with very narrow band passes. Filters of bandwidth around 1 nm are routinely used to isolate certain spectral regions, such as the broad chromospheric Hα and Ca II H and K lines, and for the photosphere most notably the G-band at wavelength 430 ± 1 nm which is dominated by many absorption lines of the CH molecule. The intensity in this band is highly temperature sensitive, due to the high temperature sensitivity of the dissociation of the CH molecule; hence images made in the G-band show very high contrast, and bright points associated with intense magnetic fields show up very clearly. Narrow-band filters have played a prominent role in solar astronomy ever since Bernard Lyot (1933) introduced the *birefringent filter*. This type of filter consists of an alternating sequence of polarizers and birefringent crystals whose net transmission profile consists of a row of narrow windows separated in wavelength. An additional broader-band filter can be used to select one of these windows and close the others. The filter can be made tunable by adding a quarter-wave plate to one or more of the birefringent crystals in the sequence and allowing the subsequent polarizer(s) to be rotated, which shifts the transmission windows in wavelength. A simple quarter-wave plate (itself a birefringent crystal with optical axis at 45° to the adjacent crystal) only allows tuning over a narrow range around one wavelength, but the development of layered achromatic quarter-wave plates (Beckers 1971) allows the construction of a *universal birefringent filter* (UBF), tunable over a large wavelength range. UBFs with narrow bandwidth, 0.025 nm or less, are available, and because their bandwidth is comparable to that of a typical photospheric absorption line, they can be used to measure Doppler shifts by placing the bandpass in the sloping wings of the line.

Other modern narrow-band filters for solar observations are based on Fabry–Perot and Michelson interferometers. In a Fabry–Perot interferometer, the incoming beam passes through two parallel, partially reflecting optical surfaces. The beam undergoes multiple reflections between these surfaces, but with each reflection a fraction of the beam is transmitted and the several transmitted fractions interfere in the outgoing beam to form a periodic sequence of narrow transition windows whose wavelength locations depend on the separation distance of the optical surfaces. The wavelength spacing between these windows (the 'free spectral range') is rather narrow, but because the maximum transmission is relatively high (compared to other types of filter), two or more Fabry–Perot filters can be used in combination to isolate a particular spectral window.

A Michelson interferometer can be used to produce a tunable, narrow-band, wide-field filter capable of imaging the whole solar disc in a narrow wavelength band. Inserting glass blocks of different index of refraction into the two arms of the interferometer produces a phase difference in the beams traversing the two arms. By using a polarizing beam splitter and inserting a quarter-wave plate in front of each mirror of the interferometer, the two output beams can also be made to be linearly polarized in orthogonal directions, and then tuning is possible much as it is for a birefringent filter. Tunable Michelson interferometers have been employed most notably in the six stations of the Global Oscillation Network

Group (GONG) and the Michelson Doppler Imager (MDI) aboard SOHO, producing full-disc velocity images for use in helioseismology.

A1.4   Spectroscopy

The high intensity of sunlight allows the use of solar spectrographs of high dispersion and hence high spectral resolution. Resolving the detailed shape of photospheric spectral lines, whose width is of order 0.01 nm, requires a spectral resolution of order 0.001 nm = 1 pm, or $\lambda/\delta\lambda \sim 500\,000$ in the visible. The use of highly efficient gratings, such as an echelle grating, allows measurements at high spectral orders, with unwanted orders eliminated through the use of a predisperser or masks and filters. Because the height of formation of an absorption line decreases with increasing distance (in wavelength) from the line centre, high spectral resolution allows resolution in height (along the line of sight) by measuring spectral properties such as line width, Doppler shift, line bisector shape and polarization as functions of line depth across the line profile. Height resolution can also be achieved by using spectral lines of different strength, formed over different ranges of height in the solar atmosphere.

Spectral lines of diatomic molecules, formed only in the cooler sunspot umbra and absent in the quiet photosphere, are particularly useful in sunspot (and starspot) observations: they serve as good diagnostics of temperature, pressure, and element abundances in the umbra and are relatively unaffected by scattered light from the surroundings of the spot. Telluric spectral lines (formed in the Earth's atmosphere) are useful as a wavelength reference, since they are unaffected by motions or magnetic fields on the Sun.

A1.5   Polarimetry

A solar polarimeter consists of a spectrograph with additional optical elements designed to distinguish the various polarization states of the sunlight. Knowing the state of polarization, caused by the Zeeman effect, of the light in a magnetically sensitive absorption line enables us to determine the magnetic field in the part of the solar atmosphere in which the line is formed. The measurement of solar magnetic fields was revolutionized in 1940 by the Babcocks' development of the *magnetograph* (see Babcock 1953), a photoelectric device designed to measure weak magnetic fields from their Zeeman effect. In its original form, the magnetograph measures only the longitudinal (line-of-sight) component of the magnetic field. More recent developments are *vector magnetographs* and *Stokes polarimeters*, which can determine the vector magnetic field by measuring the full polarization state of light in the spectral line, as represented by the four Stokes parameters I, Q, U and V. A more detailed discussion of the Zeeman effect and the measurement of vector magnetic fields on the Sun is given in Section 3.4.

Polarimetry can also be employed to reduce the problem of stray light in sunspot observations: for example, velocity measurements based on the Doppler shift of Stokes V profiles are largely uncontaminated by stray light from the much less magnetic surroundings. Accurate polarimetry requires telescopes that are polarization-free themselves, or at least have well-determined polarization characteristics. The longer exposure times required for polarimetric measurements make them especially susceptible to seeing effects, and hence new polarimeters being developed to take advantage of adaptive optics systems will significantly increase the accuracy of vector magnetic field measurements.

A1.6   Inversion methods

Information about the distribution of temperature, velocity and magnetic fields on the Sun must be extracted from measurements of the spectrum of sunlight and its polarization state. Various techniques are employed: direct methods, such as the determination of Doppler shifts of a line profile; forward modelling by constructing a synthetic spectrum; and 'inverse' methods that seek to invert the equations of radiative transfer. Especially important for the study of sunspots are techniques for inverting the Stokes parameters of line profiles to determine the vector magnetic field (see Section 3.4). All inverse methods assume some physical model of the solar atmosphere, and the resulting model dependence is often the greatest source of uncertainty in the derived quantity.

Appendix 2 Essentials of magnetohydrodynamic theory

In this appendix we offer a very brief account of the basic equations of magnetohydrodynamics (MHD) and of some fundamental concepts of MHD that are relevant to sunspots and starspots. For a fuller account of MHD theory see the books by Roberts (1967), Cowling (1976a) and Priest (1982), as well as relevant material in the volumes by Moffatt (1978), Parker (1979a), Choudhuri (1998), Mestel (1999) and Thompson (2006a).

A2.1 Basic equations

We begin with Maxwell's equations, in SI units, for the electric field $\mathbf{E}$ and the magnetic field $\mathbf{B}$,

$$\nabla \times \mathbf{E} = -\frac{\partial \mathbf{B}}{\partial t}, \qquad \nabla \times \mathbf{B} = \mu_0 \mathbf{j} + \frac{1}{c^2}\frac{\partial \mathbf{E}}{\partial t}, \qquad \nabla \cdot \mathbf{B} = 0, \qquad \nabla \cdot \mathbf{E} = \rho_{\mathrm{e}}/\epsilon_0, \tag{A2.1}$$

where $\mathbf{j}$ is the electric current, ρ_{e} is the electric charge density, c is the velocity of light, ϵ_0 and μ_0 are the permittivity and permeability of free space, and $\epsilon_0\mu_0 = 1/c^2$. To these equations we add Ohm's law, in its simplest form, for a fluid medium moving with a non-relativistic local velocity $\mathbf{u}$:

$$\mathbf{j} = \sigma_{\mathrm{e}}(\mathbf{E} + \mathbf{u} \times \mathbf{B}), \tag{A2.2}$$

where σ_{e} is the electrical conductivity. (Note that Ohm's law becomes more complicated in the solar photosphere, where the electron gyro-frequency is higher than the collision frequency, and the gas is only partially ionized: see Mestel 1999.) If we assume that all motion is non-relativistic, so that $u^2/c^2 \ll 1$, it then follows that the displacement current can be neglected in Equation (A2.1b) and so we recover Ampère's equation $\nabla \times \mathbf{B} = \mu_0 \mathbf{j}$. Substitution into Faraday's law, Equation (A2.1a), then yields the induction equation

$$\partial \mathbf{B}/\partial t = \nabla \times (\mathbf{u} \times \mathbf{B}) - \nabla \times (\eta \nabla \times \mathbf{B}), \tag{A2.3}$$

where the magnetic diffusivity $\eta = 1/(\mu_0\sigma_{\mathrm{e}})$. If η is uniform, this reduces to

$$\partial \mathbf{B}/\partial t = \nabla \times (\mathbf{u} \times \mathbf{B}) + \eta \nabla^2 \mathbf{B} \tag{A2.4}$$

(cf. Section 11.1).

In this approximation, the electrostatic force is negligible compared with the Lorentz force, $\mathbf{j} \times \mathbf{B}$, and so the equation of motion takes the form

$$\rho \left(\frac{\partial \mathbf{u}}{\partial t} + (\mathbf{u} \cdot \nabla) \mathbf{u} \right) = -\nabla p + \mathbf{j} \times \mathbf{B}, \tag{A2.5}$$

where ρ is the density, p is the pressure and viscous terms have been neglected. The Lorentz force can be written as the divergence of the Maxwell stress tensor M_{ik}, so that

$$(\mathbf{j} \times \mathbf{B})_i = \frac{\partial M_{ik}}{\partial x_k} , \text{ where } M_{ik} = \frac{1}{\mu_0} \left(B_i B_k - \tfrac{1}{2} B^2 \delta_{ik} \right). \tag{A2.6}$$

Thus the stress can be decomposed into an isotropic magnetic pressure $B^2/2\mu_0$ and a tension B^2/μ_0 along the lines of force, while the Lorentz force

$$\mathbf{j} \times \mathbf{B} = \frac{1}{\mu_0} \left((\mathbf{B} \cdot \nabla) \mathbf{B} - \nabla(\tfrac{1}{2} B^2) \right) \tag{A2.7}$$

is itself a combination of a curvature force (caused by the tension) and a pressure gradient: this provides a convenient means of estimating its effects. Correspondingly, the magnetic energy density is $B^2/2\mu_0$.

In general, the induction equation (A2.3) and momentum equation (A2.5) must be supplemented by the continuity equation (expressing conservation of mass),

$$\partial \rho / \partial t + \nabla \cdot (\rho \mathbf{u}) = 0, \tag{A2.8}$$

and a suitable energy equation, which can vary in form from a simple isentropic or polytropic relation to a complex equation including viscous and Ohmic dissipation, heat conduction and perhaps radiative transfer.

In SI units, magnetic fields are measured in teslas and $\mu_0 = 4\pi \times 10^{-7}$. In this book, however, as in most astrophysical literature, we choose to use electromagnetic units (gauss) rather than teslas when measuring magnetic fields ($1\,\mathrm{T} = 10^4\,\mathrm{G}$). Then the Maxwell stress corresponds to an isotropic pressure $B^2/8\pi$ plus a tension $B^2/4\pi$ along the field, in cgs units. The corresponding magnetic energy density is likewise given by $B^2/8\pi$.

A2.2 Kinematic MHD: flux freezing, flux concentration and flux expulsion

In the limit of perfect electrical conductivity ($\eta = 0$), the induction equation reduces to

$$\partial \mathbf{B}/\partial t = \nabla \times (\mathbf{u} \times \mathbf{B}) = (\mathbf{B} \cdot \nabla)\mathbf{u} - (\mathbf{u} \cdot \nabla)\mathbf{B} - \mathbf{B}\nabla \cdot \mathbf{u}, \tag{A2.9}$$

whence it follows that the magnetic flux through a circuit moving with the fluid remains constant (Alfvén's theorem). Alternatively, combining Equations (A2.9) and (A2.8), we find that

$$\frac{\partial}{\partial t} \left(\frac{\mathbf{B}}{\rho} \right) + (\mathbf{u} \cdot \nabla) \left(\frac{\mathbf{B}}{\rho} \right) = \left(\frac{\mathbf{B}}{\rho} \cdot \nabla \right) \mathbf{u}. \tag{A2.10}$$

This can be interpreted as implying that the magnetic field moves with the fluid, i.e. that magnetic field lines are *frozen* into the perfectly conducting fluid. It follows that the field lines are stretched in any flow with transverse shear and that the field strength will be correspondingly increased.

More generally, the ratio of the two terms on the right-hand side of Equation (A2.4) is given by the magnetic Reynolds number, $R_\mathrm{m} = UL/\eta$, where U and L are a characteristic speed and length scale, respectively. In stellar interiors, $R_m \gg 1$ – near the solar surface $R_m \approx 10^6$ – but nevertheless the effects of a small but finite diffusivity cannot be ignored. Two important examples are the concentration of magnetic flux by converging flows and the expulsion of magnetic flux from a persistent eddy (Proctor and Weiss 1982).

In a two-dimensional, incompressible stagnation point flow, with a velocity given by $\mathbf{u} = U(-x/L, 0, z/L)$ and $\mathbf{B} = B(x)\hat{\mathbf{e}}_z$, referred to Cartesian co-ordinates with the z-axis pointing upwards, there is a steady solution given by

$$B(x) = B_0 \left(\frac{2R_m}{\pi}\right)^{1/2} \exp\left(-\frac{R_m x^2}{2L^2}\right), \tag{A2.11}$$

corresponding to a flux sheet of thickness $R_m^{-1/2}L$, containing a total flux $2B_0 L$, with a peak field strength of order $R_m^{1/2}B_0$. The corresponding solution for an axisymmetric three-dimensional flow, referred to cylindrical polars (s, ϕ, z), with $\mathbf{u} = U(-s/L, 0, 2z/L)$ and $\mathbf{B} = B(s)\hat{\mathbf{e}}_z$, is

$$B(s) = \frac{1}{2}B_0 R_m \exp\left(-\frac{R_m s^2}{2L^2}\right); \tag{A2.12}$$

thus the total flux $\pi L^2 B_0$ is confined to a tube or rope with a peak field of order $R_m B_0$ and a diameter of order $R_m^{-1/2}L$. In this kinematic approximation, the velocity $\mathbf{u}$ is assumed to be prescribed and unaffected by the Lorentz force. In reality, unless B_0 is very small, the Lorentz force will act to limit flux concentration by excluding motion from the flux rope.

Flux expulsion is most simply illustrated by considering the effect of a band of two-dimensional eddies on an initially uniform field, with $R_m \gg 1$, as shown by the numerical results in Figure A2.1. To start with, the sheared motion, acting through the $(\mathbf{B} \cdot \nabla)\mathbf{u}$ term in Equation (A2.9), winds up the magnetic field, increasing the total magnetic energy but decreasing the scale on which the field varies, until diffusion bites and the field lines reconnect. On a time scale given by $\tau_{\text{crit}} \approx R_m^{1/3}\tau_c$, where the eddy turnover time $\tau_c = L/U$, magnetic flux is expelled from the cores of the eddies and concentrated by the converging flows between them into narrow regions like those described above. Note that this reconnection occurs in a time much shorter than the characteristic resistive time scale $\tau_\eta = R_m\tau_c$. Similar effects occur in three dimensions, for example in a tesselated pattern of hexagonal cells, with fluid rising at the centre of each cell and sinking at its edges. Near the top of the layer, magnetic flux is swept outwards and concentrated at the vertices of each cell, while at the bottom flux is concentrated near the central axis of a cell.

A2.3 MHD waves

Consider a uniform, inviscid, perfectly conducting ($\eta = 0$), isothermal gas at rest ($\mathbf{u} = 0$), permeated by a uniform magnetic field $\mathbf{B}_0$ and in the absence of gravity. If we introduce a small adiabatic (and hence isentropic) perturbation to this equilibrium configuration, and linearize the governing Equations (A2.3), (A2.5) and (A2.8) by neglecting products of perturbation quantities, we can then eliminate the perturbations in pressure, density, temperature and magnetic field to obtain a single equation in the velocity perturbation $\mathbf{u}$:

$$\partial^2\mathbf{u}/\partial t^2 = c_s^2\nabla(\nabla \cdot \mathbf{u}) + v_A^2[\nabla \times \nabla \times (\mathbf{u} \times \mathbf{b})] \times \mathbf{b}, \tag{A2.13}$$

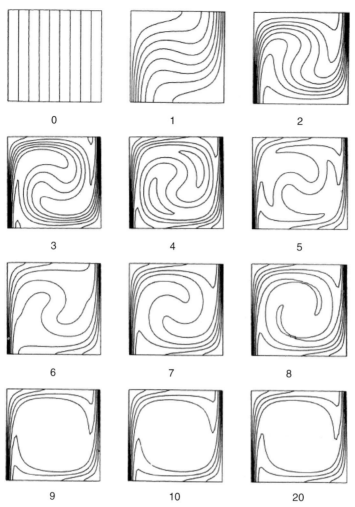

Fig. A2.1. Flux expulsion illustrated in a numerical experiment, with a velocity given by $\mathbf{u} = U(-\sin \pi x/L \cos \pi z/L, \ 0, \ \cos \pi x/L \sin \pi z/L)$ and $R_m = 250$. The time is measured in units of $5\tau_c/8$. (From Galloway and Weiss 1981.)

where c_s is the isentropic sound speed, $v_A = (B_0^2/\mu_0 \rho_0)^{1/2}$ is the Alfvén speed (with ρ_0 being the undisturbed density), and $\mathbf{b}$ is a unit vector along $\mathbf{B}_0$. If we then assume a plane-wave solution in the form $\mathbf{u} = \mathbf{u}' \exp[i(\mathbf{k} \cdot \mathbf{x} - \omega t)]$, Equation (A2.13) becomes

$$\omega^2 \mathbf{u}' = [(c_s^2 + v_A^2)(\mathbf{u}' \cdot \mathbf{k}) - v_A^2 (\mathbf{k} \cdot \mathbf{b})(\mathbf{u}' \cdot \mathbf{b})]\mathbf{k}$$
$$+ v_A^2 [(\mathbf{k} \cdot \mathbf{b})^2 \mathbf{u}' - (\mathbf{k} \cdot \mathbf{b})(\mathbf{u}' \cdot \mathbf{k})\mathbf{b}]. \tag{A2.14}$$

Taking the scalar product of Equation (A2.14) with the vector $\mathbf{k} \times \mathbf{b}$, we obtain

$$[\omega^2 - (\mathbf{k} \cdot \mathbf{b})^2 v_A^2][\mathbf{u}' \cdot (\mathbf{k} \times \mathbf{b})]. \tag{A2.15}$$

Hence the component $\mathbf{u}' \cdot (\mathbf{k} \times \mathbf{b})$ of the velocity perturbation, which is perpendicular to both $\mathbf{k}$ and $\mathbf{B}_0$, propagates along $\mathbf{B}_0$ as a pure *Alfvén wave* obeying the dispersion relation

$$\omega^2 = v_A^2 k^2 \cos^2 \theta, \tag{A2.16}$$

where θ is the angle between the wavenumber vector $\mathbf{k}$ and $\mathbf{B}_0$. In this Alfvén mode, the motion is incompressible and transverse to both $\mathbf{k}$ and $\mathbf{B}_0$, and the phase velocity is $v_A \cos \theta$ in the $\mathbf{k}$-direction.

Taking the scalar product of Equation (A2.14) with $\mathbf{k}$ and with $\mathbf{b}$, we obtain a pair of equations which can be written in matrix form as

$$\begin{bmatrix} \omega^2 - (c_s^2 + v_A^2)k^2 & v_A^2 k^2 (\mathbf{k} \cdot \mathbf{b}) \\ -c_s^2 (\mathbf{k} \cdot \mathbf{b}) & \omega^2 \end{bmatrix} \begin{bmatrix} (\mathbf{u}' \cdot \mathbf{k}) \\ (\mathbf{u}' \cdot \mathbf{b}) \end{bmatrix} = 0. \tag{A2.17}$$

For a non-zero solution, the determinant of the coefficient matrix must vanish, which yields the dispersion relation for the compressive *magneto-acoustic waves*,

$$\omega^4 - (c_s^2 + v_A^2)k^2 \omega^2 + c_s^2 v_A^2 k^4 \cos^2 \theta = 0. \tag{A2.18}$$

This quadratic equation in ω^2 has two real, positive roots

$$\omega_{1,2}^2 = \tfrac{1}{2} k^2 \left[(c_s^2 + v_A^2) \pm \sqrt{(c_s^2 + v_A^2)^2 - 4c_s^2 v_A^2 \cos^2 \theta} \right], \tag{A2.19}$$

where ω_1 corresponds to the plus sign and the 'fast' mode, while ω_2 corresponds to the minus sign and the 'slow' mode. The fast and slow modes have motions confined to the plane defined by $\mathbf{k}$ and $\mathbf{B}_0$. From the form of Equation (A2.19), we see that the phase velocities ω/k of the fast and slow modes are independent of the magnitude k of the wavenumber vector (but not of its direction), and hence these small-amplitude waves are non-dispersive.

For $\theta = 0$ (i.e. for $\mathbf{k} \| \mathbf{B}_0$), we have $\omega_1^2 = k^2 \max(c_s^2, v_A^2)$ and $\omega_2^2 = k^2 \min(c_s^2, v_A^2)$, i.e. for propagation parallel to the magnetic field the fast and slow modes travel at the sound speed and the Alfvén speed (for $c_s > v_A$) or vice versa (for $c_s < v_A$). Here one of the modes is a pure, longitudinal acoustic mode with motions directed along $\mathbf{B}_0$, and the other mode is a pure Alfvén mode with motions transverse to $\mathbf{B}_0$. For $\theta = \pi/2$ (i.e. for $\mathbf{k} \perp \mathbf{B}_0$), we have $\omega_1^2 = k^2(c_s^2 + v_A^2)$ and $\omega_2^2 = 0$, i.e. for propagation perpendicular to the undisturbed magnetic field, the slow mode disappears while the fast mode travels at the 'fast speed' $(c_s^2 + v_A^2)^{1/2}$. In the limit of vanishing magnetic field ($\mathbf{B}_0 = 0$), the pure Alfvén mode and the slow mode disappear and the fast mode becomes a pure acoustic wave (in any propagation direction).

The analysis of linearized MHD waves can be extended to the case in which the gas is stably stratified in a uniform gravitational field, which provides a more realistic model of the solar atmosphere. In that case we have the so-called *magneto-atmospheric waves* (or *magneto-acoustic-gravity waves*), for which, in addition to the pressure and magnetic force, the buoyancy force contributes to the total restoring force. A qualitative discussion of these waves is presented in Section 6.1; for a more detailed discussion, see the review by Thomas (1983). In general, the wave modes can still be classified as 'fast' or 'slow', but because of the stratification, the sound speed and Alfvén speed vary with height and the classification only applies locally. Indeed, a wave of fixed frequency might be classified as 'slow' at one height in the atmosphere but 'fast' at another height.

A2.4 Thin flux tubes

A number of theoretical studies of solar magnetic fields are based on the dynamics of an individual, thin magnetic flux tube, embedded in a surrounding medium which may or may not itself contain a magnetic field. In the so-called *thin flux tube approximation*, it is assumed that the diameter of the flux tube is small compared to both the local density scale height of the surrounding atmosphere and the radius of curvature of the flux tube axis, and all thermodynamic variables (and the velocity, if there is a flow along the tube) are assumed to be uniform across the tube.

Consider a thin, isolated flux tube, with internal density ρ_i, embedded in field-free surroundings of density ρ_e. Such a flux tube can support three different kinds of waves: (i) a pure, torsional Alfvén wave travelling along the tube at speed $v_A = (B^2/\mu_0\rho_i)^{1/2}$, with purely azimuthal, incompressible motions that do not distort the tube radius; (ii) 'kink' or sinuous modes, with sinusoidal displacements of the flux tube axis, travelling at the 'modified' Alfvén speed $v_m = [B^2/\mu_0(\rho_i + \rho_e)]^{1/2}$; and (iii) compressive 'sausage' or varicose modes, with sinusoidal variations in the tube radius, propagating at the 'tube speed' $c_t = [c^2 v_A^2/(c^2 + v_A^2)]^{1/2}$.

For studies of steady or slow flows along a thin flux tube, or slow motions of the tube through the surrounding medium, it may be assumed that exact pressure equilibrium with the surroundings is maintained all along the tube,

$$p_i + \frac{B_i^2}{8\pi} = p_e + \frac{B_e^2}{8\pi}, \tag{A2.20}$$

where the subscripts i and e denote internal and external values.

For modelling the buoyant rise of a thin flux tube through the convection zone (discussed in Section 7.2.2), one can derive an equation of motion for the flux tube in a frame rotating at uniform angular velocity Ω, in the form (e.g. Fisher *et al.* 2000)

$$\rho_i \left(\frac{\partial \mathbf{v}}{\partial t} + (\mathbf{u} \cdot \nabla)\mathbf{v} \right) = \mathbf{F_B} + \mathbf{F_T} + \mathbf{F_C} + \mathbf{F_D}, \tag{A2.21}$$

where $\mathbf{v}$ is the velocity of the flux tube relative to the surrounding medium. Here $\mathbf{F_B} = g(\rho_e - \rho_i)\mathbf{r}$ is the buoyancy force, where g is the local acceleration of gravity and $\mathbf{r}$ is the spherical radius vector; $\mathbf{F_T} = (B^2/2\mu_0)\kappa$ is the magnetic tension force, where $\kappa = \partial^2\mathbf{r}(s)/\partial s^2$ is the curvature vector, with s being the co-ordinate along the tube axis; $\mathbf{F_C} = -2\rho_i\Omega \times \mathbf{v}$ is the Coriolis force; and $\mathbf{F_D}$ is the drag force exerted on the tube as it moves relative to its surroundings, usually expressed in terms of a drag coefficient C_D in the form $\mathbf{F_D} = \rho_e(C_D/\pi a)|\mathbf{v_\perp}|\mathbf{v_\perp}$, where $\mathbf{v_\perp}$ is the transverse component of the velocity (normal to the tube axis) and $a = (\Phi/\pi B)^{1/2}$ is the radius of the flux tube, expressed in terms of the constant total magnetic flux Φ along the tube. The drag coefficient C_D is usually assumed to have a value near unity, appropriate for the drag on a circular cylinder at high Reynolds numbers.

A2.5 Fundamentals of magnetoconvection

Sunspots provided the original motivation for studying the interactions between convection and magnetic fields; since then, this topic has developed into a significant subject in its own right, and it is covered in a number of reviews (e.g. Proctor and Weiss 1982;

Proctor 1992, 2005; Schüssler 2001; Weiss 2003). Here we need only outline some basic aspects of magnetoconvection.

Convective motion is driven by thermally induced variations in density, and impeded by the effects of thermal and viscous dissipation. Within the bulk of a stellar convection zone – but not at the level where sunspots are observed – the convective motion is subsonic. If the relative density perturbation $\hat{\rho}$ is small ($\hat{\rho} = \mathcal{O}(u^2/c_s^2) \ll 1$), the density fluctuations can be dropped from the continuity equation (thereby eliminating sound waves), and Equation (A2.8) simplifies to become $\nabla \cdot (\rho_0 \mathbf{u}) = 0$, where ρ_0 is the unperturbed density (the *anelastic* approximation). If the convecting layer is also shallow, so that its depth d is much less than the density or pressure scale height, the continuity equation reduces to $\nabla \cdot \mathbf{u} = 0$ and the flow is incompressible; in this *Boussinesq* approximation, the equation of state takes the form $\rho = \rho_0[1 - \alpha_T(T - T_0)]$, where ρ_0 and T_0 are reference values of the density and temperature, and the coefficient of thermal expansion $\alpha_T = -(\partial \ln \rho/\partial T)_p$, so that $\alpha_T = 1/T$ for an ideal gas.

We shall confine our attention here to Boussinesq magnetoconvection (Chandrasekhar 1961; Proctor and Weiss 1982). Consider then a static, horizontal fluid layer $\{0 \leq z \leq d\}$ with an imposed vertical temperature gradient such that $T(z) = T_0 - \Delta T z/d$, and an imposed uniform vertical magnetic field $\mathbf{B}_0$. For simplicity, suppose that the temperature is fixed and the field is constrained to be vertical at the upper and lower boundaries, which are assumed to be impermeable and stress-free. The momentum equation (A2.5) must be augmented by adding a buoyancy force and a viscous term, so that

$$\frac{\partial \mathbf{u}}{\partial t} + (\mathbf{u} \cdot \nabla)\mathbf{u} = -\frac{1}{\rho_0}\nabla p' - \alpha_T T'\mathbf{g} + \frac{1}{\mu_0\rho_0}(\nabla \times \mathbf{B}') \times \mathbf{B} + \nu\nabla^2\mathbf{u}, \qquad \text{(A2.22)}$$

where the primes denote perturbations, $\mathbf{g}$ is the gravitational acceleration and ν is the viscous diffusivity. The energy equation takes the form

$$\frac{\partial T'}{\partial t} - \mathbf{u} \cdot (\beta\hat{\mathbf{e}}_z) = -\mathbf{u} \cdot \nabla T' + \rho_0 c_p \kappa \nabla^2 T, \qquad \text{(A2.23)}$$

where the superadiabatic gradient $\beta = \Delta T/d - g/c_p$, with c_p the specific heat at constant pressure, and κ is the thermal diffusivity. This configuration is characterized by four dimensionless numbers: these are the Rayleigh number $Ra = g\alpha_T\beta d^4/(\kappa\nu)$, the Chandrasekhar number $Q = B_0^2 d^2/(\mu_0\rho_0\eta\nu)$, the Prandtl number $\sigma = \nu/\kappa$ and the diffusivity ratio $\zeta = \eta/\kappa$.

With these illustrative boundary conditions, the eigenfunctions describing linearized perturbations are sines and cosines, and without loss of generality we can set

$$\mathbf{u} = U'e^{st}\left(-\sin\frac{ax}{d}\cos\frac{\pi z}{d},\ 0,\ \cos\frac{ax}{d}\sin\frac{\pi z}{d}\right), \qquad \text{(A2.24)}$$

giving convective rolls with axes in the y-direction. In the absence of a magnetic field, convection sets in as a monotonically growing mode, corresponding to overturning motion, and the layer is unstable for $Ra > Ra_0 = (\pi^2 + a^2)^3/a^2$; thus Ra_0 is least for $a = \pi/\sqrt{2}$. In the presence of a magnetic field, the dimensionless growth rate $\hat{s} = sL^2/[\kappa(\pi^2 + a^2)]$ satisfies the cubic characteristic equation

$$\hat{s}^3 + (1 + \sigma + \zeta)\hat{s}^2 + [\sigma + \zeta + \sigma\zeta - \sigma(r - \zeta q)]\hat{s} - \sigma\zeta[r - (1 + q)] = 0, \qquad \text{(A2.25)}$$

where $r = Ra/Ra_0$ and $q = \pi^2 Q/(\pi^2 + a^2)^2$. Equation (A2.25) has three roots, of which one is always real and negative, while the other pair may be real or complex conjugates. If $\zeta > 1$ overturning convection sets in at a stationary bifurcation when $r = r^{(e)} = 1 + q$. However, if $\zeta < 1$ and $q > \zeta(1 + \sigma)/[\sigma(1 - \zeta)]$, then convection sets in at an oscillatory (Hopf) bifurcation when

$$r = r^{(o)} = 1 + \frac{\zeta}{\sigma}(1 + \sigma + \zeta) + \frac{\zeta(\sigma + \zeta)}{1 + \sigma}q < r^{(e)}, \tag{A2.26}$$

giving rise to oscillatory convection (corresponding to trapped slow magneto-acoustic oscillations that are thermally excited). The preferred horizontal scale is also much reduced when $Q \gg 1$: in dimensional terms, the critical values of Ra are least when $a \sim (\pi^4 Q/2)^{1/6}$.

The development of nonlinear Boussinesq magnetoconvection has also received considerable attention – see the reviews by Proctor and Weiss (1982), Weiss (1991) and Proctor (1992) – while three-dimensional behaviour has been explored by Cattaneo, Emonet and Weiss (2003). Compressible magnetoconvection is discussed by Schüssler and Knölker (2001), Weiss (2002) and Proctor (2005).

References

Abarbanell, C., and Wohl, H. 1981. *Sol. Phys.*, **70**, 197.

Abbett, W. P., and Fisher, G. H. 2003. *Astrophys. J.*, **592**, 475.

Abbot, C. G. 1929. *The Sun* (New York: Appleton).

Abdelatif, T. E. 1990. *Sol. Phys.*, **129**, 201.

Abdelatif, T. E., Lites, B. W., and Thomas, J. H. 1986. *Astrophys. J.*, **311**, 1015.

Abdelatif, T. E., and Thomas, J. H. 1987. *Astrophys. J.*, **320**, 884.

Abdussamatov, H. I. 1971. *Sol. Phys.*, **16**, 384.

Abell, G. 1964. *Exploration of the Universe* (New York: Holt, Rinehart and Winston).

Abetti, G. 1932. *Publ. R. Osserv. Arcetri*, **50**, 47.

Abetti, G. 1957. *The Sun* (London: Faber & Faber).

Abreu, J. A., Beer, J., Steinhilber, F., Tobias, S. M., and Weiss, N. O. 2008. *Geophys. Res. Lett.*, submitted.

Adam, M. G. 1990. *Sol. Phys.*, **125**, 37.

Adam, M. G., and Petford, A. D. 1991. *Sol. Phys.*, **135**, 319.

Albregtsen, F., Jorås, P. B., and Maltby, P. 1984. *Sol. Phys.*, **90**, 17.

Albregtsen, F., and Maltby, P. 1978. *Nature*, **274**, 41.

Albregtsen, F., and Maltby, P. 1981. *Sol. Phys.*, **71**, 269.

Alissandrakis, C. E., Georgakilas, A. A., and Dialetis, D. 1992. *Sol. Phys.*, **138**, 93.

Amado, P., Cutispoto, G., and Lanza, A. F. 2001. In ASP Conf. Ser. 223, *Cool Stars, Stellar Systems and the Sun*, ed. R. J. Garcia Lopez, R. Rebolo and M. R. Zapaterio Osorio (San Francisco: Astron. Soc. Pacific), p. 895.

Antia, H. M., and Chitre, S. M. 1979. *Sol. Phys.*, **63**, 67.

Antia, H. M., Chitre, S. M., and Gokhale, M. H. 1978. *Sol. Phys.*, **60**, 31.

Archontis, V., Hood, A., and Brady, C. 2007. *Astron. Astrophys.*, **466**, 367.

Archontis, V., Moreno-Insertis, F., Galsgaard, K., Hood, A., and O'Shea, E. 2004. *Astron. Astrophys.*, **426**, 1047.

Arena, P., Landi degl'Innocenti, E., and Noci, G. 1990. *Sol. Phys.*, **129**, 259.

Avrett, E. H. 1981. In *The Physics of Sunspots*, ed. L. E. Cram and J. H. Thomas (Sunspot, NM: Sacramento Peak Observatory), p. 235.

Ayres, T. R. 1996. In IAU Symp. 176, *Stellar Surface Structure*, ed. K. G. Strassmeier and J. L. Linsky (Dordrecht: Kluwer), p. 371.

Babcock, H. D. 1959. *Astrophys. J.*, **130**, 364.

Babcock, H. W. 1947. *Astrophys. J.*, **105**, 105.

Babcock, H. W. 1953. *Astrophys. J.*, **118**, 387.

Babcock, H. W. 1960. *Astrophys. J.*, **132**, 521.

Babcock, H. W. 1961. *Astrophys. J.*, **133**, 572.

Babcock, H. W., and Babcock, H. D. 1955. *Astrophys. J.*, **121**, 349.

Bagnulo, S., Hensberge, H., Landstreet, J. D., Szeifert, T., and Wade, G. A. 2004. *Astron. Astrophys.*, **416**, 1149.

Bagnulo, S., Landstreet, J. D., Lo Curto, G., Szeifert, T., and Wade, G. A. 2003. *Astron. Astrophys.*, **403**, 645.

Bahng, J., and Schwarzschild, M. 1961. *Astrophys. J.*, **134**, 312.

Balasubramaniam, K. S., Pevtsov, A., and Rogers, J. 2004. *Astrophys. J.*, **608**, 1148.

Baldwin, M. P., and Dunkerton, T. J. 2005. *J. Atmos. Sol.-Terr. Phys.*, **67**, 71.

Baliunas, S. L., Donahue, R. A., Soon, W. H., *et al.* 1995. *Astrophys. J.*, **438**, 269.

Baliunas, S. L., Donahue, R. A., Soon, W. H., and Henry, G. W. 1998. In ASP Conf. Ser. 254, *The Tenth Cambridge Conference on Cool Stars, Stellar Systems and the Sun*, ed. R. A. Donahue and J. A. Bookbinder (San Francisco: Astron. Soc. Pacific), p. 153.

Baliunas, S. L., Sokoloff, D., and Soon, W. H. 1996. *Astrophys. J.*, **457**, L99.

Balthasar, H. 1999. *Sol. Phys.*, **187**, 389.

Balthasar, H. 2003. *Sol. Phys.*, **218**, 85.

Balthasar, H., Lustig, G., Stark, D., and Wöhl, H. 1986. *Astron. Astrophys.*, **160**, 277.

Balthasar, H., Schleicher, H., Bendlin, C., and Volkmer, R. 1996. *Astron. Astrophys.*, **315**, 603.

Balthasar, H., and Wöhl, H. 1983. *Sol. Phys.*, **88**, 71.

Barnes, J. A., Sargent, H. H. III, and Tryon, P. V. 1980. In *The Ancient Sun*, ed. R. O. Pepin, J. A. Eddy and R. B. Merrill (New York: Pergamon), p. 159.

Barnes, J. R., Collier Cameron, A., Donati, J.-F., *et al.* 2005. *Mon. Not. Roy. Astron. Soc.*, **357**, L1.

Barnes, J. R., Collier Cameron, A., James, D. J., and Donati, J.-F. 2000. *Mon. Not. Roy. Astron. Soc.*, **314**, 162.

Barnes, J. R., Collier Cameron, A., Unruh, Y. C., Donati, J.-F., and Hussain, G. A. J. 1998. *Mon. Not. Roy. Astron. Soc.*, **299**, 904.

Barnes, S. A., Sofia, S., Prosser, C. F., and Stauffer, J. R. 1999. *Astrophys. J.*, **516**, 263.

Basu, S. and Antia, H. M. 2003. *Astrophys. J.*, **585**, 553.

Beck, J. G., Gizon, L., and Duvall, T. L. Jr. 2002. *Astrophys. J.*, **575**, L47.

Beckers, J. M. 1968. *Sol. Phys.*, **3**, 258.

Beckers, J. M. 1969a. In *Plasma Instabilities in Astrophysics*, ed. D. G. Wentzel and D. A. Tidman (New York: Gordon and Breach), p. 139.

Beckers, J. M. 1969b. *A Table of Zeeman Multiplets*. Sacramento Peak Observatory Rep. AFCRL-69-0115.

Beckers, J. M. 1971. *Appl. Opt.*, **10**, 973.

Beckers, J. M., and Schneeberger, T. J. 1977. *Astrophys. J.*, **215**, 356.

Beckers, J. M., and Schröter, E. H. 1968. *Sol. Phys.*, **4**, 303.

Beckers, J. M., and Schröter, E. H. 1969. *Sol. Phys.*, **10**, 384.

Beckers, J. M., and Schultz, R. B. 1972. *Sol. Phys.*, **27**, 61.

Beckers, J. M., and Tallant, P. E. 1969. *Sol. Phys.*, **7**, 351.

Beer, J. 2000. *Space Sci. Rev.*, **94**, 53.

Beer, J. 2001. *Spatium*, **8**, 3.

Beer, J., Baumgartner, B., Dittrich-Hannen, J., *et al.* 1994. In *The Sun as a Variable Star*, ed. J. M. Pap, C. Fröhlich, H. S. Hudson and S. K. Solanki (Cambridge: Cambridge University Press), p. 291.

Beer, J., Tobias, S. M., and Weiss, N. O. 1998. *Sol. Phys.*, **181**, 237.

Bello González, N., Okunev, O. V., Domínguez Cerdeña, I., Kneer, S., and Puschmann, K. G. 2005. *Astron. Astrophys.*, **434**, 317.

Bellot Rubio, L. R. 2003. In ASP Conf. Ser. 307, *3rd International Workshop on Solar Polarization*, ed. J. Trujillo Bueno and J. Sánchez Almeida (San Francisco: Astron. Soc. Pacific), p. 301.

Bellot Rubio, L. R., Balthasar, H., and Collados, M. 2004. *Astron. Astrophys.*, **427**, 319.

Bellot Rubio, L. R., Balthasar, H., Collados, M., and Schlichenmaier, R. 2003. *Astron. Astrophys.*, **403**, L47.

Bellot Rubio, L. R., Collados, M., Ruiz Cobo, B., and Rodríguez Hidalgo, I. 2000. *Astrophys. J.*, **534**, 989.

Bellot Rubio, L. R., Langhans, K., and Schlichenmaier, R. 2005. *Astron. Astrophys.*, **443**, L7.

Bellot Rubio, L. R., Schlichenmaier, R., and Tritschler, A. 2006. *Astron. Astrophys.*, **453**, 1117.

Bellot Rubio, L. R., Tsuneta, S., Lites, B. W., *et al.* 2007. *Astrophys. J.*, **668**, L91.

Berdyugina, S. V. 2002. *Astron. Nachr.*, **323**, 192.

Berdyugina, S. V. 2005. *Living Revs Sol. Phys.*, **2**, 8. (www.livingreviews.org/lrsp-2005-8)

Berdyugina, S. V., Berdyugin, A. V., Ilyin, I., and Tuominen, I. 1998. *Astron. Astrophys.*, **340**, 437.

Berdyugina, S. V., Berdyugin, A. V., Ilyin, I., and Tuominen, I. 1999. *Astron. Astrophys.*, **350**, 626.

Berdyugina, S. V., and Solanki, S. K. 2002. *Astron. Astrophys.*, **385**, 701.

Berdyugina, S. V., Solanki, S. K., and Frutiger, C. 2003. *Astron. Astrophys.*, **412**, 513.

Berger, M. A., and Field, G. B. 1984. *J. Fluid Mech.*, **147**, 143.

Berger, T. E., and Berdyugina, S. V. 2003. *Astrophys. J.*, **589**, L117.

Berger, T. E., Löfdahl, M. G., Shine, R. S., and Title, A. M. 1998. *Astrophys. J.*, **495**, 439.

Berger, T. E., Rouppe van der Voort, L., Löfdahl, M. G., *et al.* 2004. *Astron. Astrophys.*, **428**, 613.

Berger, T. E., Rouppe van der Voort, L., and Löfdahl, M. G. 2007. *Astrophys. J.*, **661**, 1272.

Berger, T. E., Schrijver, C. J., Shine, R. S., *et al.* 1995. *Astrophys. J.*, **454**, 531.

Berger, T. E., and Title, A. M. 1996. *Astrophys. J.*, **463**, 797.

Bernstein, I. B., Frieman, E. A., Kruskal, M. D., and Kulsrud, R. M. 1958. *Proc. Roy. Soc. Lond. A*, **244**, 17.

Bhatnagar, A., Livingston, W. C., and Harvey, J. W. 1972. *Sol. Phys.*, **27**, 80.

Bhatnagar, A., and Tanaka, K. 1972. *Sol. Phys.*, **24**, 87.

Biazzo, K., Frasca, A., Henry, G. W., Catalano, S., and Marilli, E. 2007. *Astrophys. J.*, **656**, 474.

Biermann, L. 1941. *Vierteljahresschr. Astron. Ges.*, **76**, 194.

Biermann, L. 1951. *Zeits. Astrophys.*, **29**, 274.

Bigelow, F. H. 1891. *Pub. Astron. Soc. Pacific*, **3**, No. 14, 34.

Bjerknes, V. 1926. *Astrophys. J.*, **64**, 93.

Blackwell, D. E., Dewhirst, D. W., and Dollfus, A. 1957. *Observatory*, **77**, 20.

Blackwell, D. E., Dewhirst, D. W., and Dollfus, A. 1959. *Mon. Not. Roy. Astron. Soc.*, **119**, 98.

Blanchflower, S. M., Rucklidge, A. M., and Weiss, N. O. 1998. *Mon. Not. Roy. Astron. Soc.*, **301**, 593.

Bogdan, T. J. 1992. In *Sunspots: Theory and Observations*, ed. J. H. Thomas and N. O. Weiss (Dordrecht: Kluwer), p. 345.

Bogdan, T. J. 2000. *Sol. Phys.*, **192**, 373.

Bogdan, T. J., Brown, T. M., Lites, B. W., and Thomas, J. H. 1993. *Astrophys. J.*, **406**, 723.

Bogdan, T. J., and Cally, P. S. 1995. *Astrophys. J.*, **453**, 919.

Bogdan, T. J., Carlsson, M., Hansteen, V. H., *et al.* 2003. *Astrophys. J.*, **599**, 626.

Bogdan, T. J., Gilman, P. A., Lerche, I., and Howard, R. 1988. *Astrophys. J.*, **327**, 451.

Bogdan, T. J., and Judge, P. G. 2006. *Phil. Trans. Roy. Soc. Lond. A*, **364**, 313.

Boldyrev, S., and Cattaneo, F. 2004. *Phys. Rev. Lett.*, **92**, 144501.

Bond, G., Kromer, B., Beer, J., *et al.* 2001. *Science*, **294**, 2130.

Bondar', N. I. 1995. *Astron. Astrophys. Suppl.*, **111**, 259.

Bonet, J. A., Márquez, I., Muller, R., Sobotka, M., and Tritschler, A. 2004. *Astron. Astrophys.*, **423**, 737.

Bonet, J. A., Márquez, I., Muller, R., Sobotka, M., and Tritschler, A. 2005. *Astron. Astrophys.*, **430**, 1089.

Bonet, J. A., Ponz, J. D., and Vázquez, M. 1982. *Sol. Phys.*, **77**, 69.

Bopp, B. W., and Evans, D. S. 1973. *Mon. Not. Astron. Soc.*, **164**, 343.

Bopp, B. W., and Rucinski, S. M. 1981. In IAU Symp. 93, *Fundamental Problems in the Theory of Stellar Evolution*, ed. D. Sugimoto, D. G. Lamb and D. N. Schramm (Dordrecht: Reidel), p. 177.

Bopp, B. W., and Stencel, R. E. 1981. *Astrophys. J.*, **247**, L131.

Börner, P., and Kneer, F. 1992. *Astron. Astrophys.*, **259**, 307.

Borrero, J. M., Lagg, A., Solanki, S. K., and Collados, M. 2005. *Astron. Astrophys.*, **436**, 333.

Borrero, J. M., Solanki, S. K., Bellot Rubio, L. R., Lagg, A., and Mathew, S. K. 2004. *Astron. Astrophys.*, **422**, 1093.

Borrero, J. M., Solanki, S. K., Lagg, A., Socas-Navarro, H., and Lites, B. 2006. *Astron. Astrophys.*, **450**, 383.

Boruta, N. 1977. *Astrophys. J.*, **215**, 364.

Botha, G. J. J., Rucklidge, A. M., and Hurlburt, N. E. 2006. *Mon. Not. Roy. Astron. Soc.*, **369**, 1611.

Botha, G. J. J., Rucklidge, A. M., and Hurlburt, N. E. 2007. *Astrophys. J.*, **662**, L27.

Brandenburg, A. 2005. *Astrophys. J.*, **625**, 539.

Brandenburg, A., Krause, F., Meinel, R., Moss, D., and Tuominen, I. 1989. *Astron. Astrophys.*, **213**, 411.

Brandenburg, A., Saar, S. A., and Turpin, C. R. 1998. *Astrophys. J.*, **498**, L51.

Brandenburg, A., and Schmitt, D. 1998. *Astron. Astrophys.*, **338**, L55.

Brandenburg, A., and Subramaniam, K. 2005. *Phys. Rep.*, **417**, 1.

Brandt, P. N., Schmidt, W., and Steinegger, M. 1990. *Sol. Phys.*, **129**, 191.

Brandt, P. N., Stix, M., and Weinhardt, H. 1994. *Sol. Phys.*, **152**, 119.

Brants, J. J., and Zwaan, C. 1982. *Sol. Phys.*, **80**, 251.

Braun, D. C., Duvall, T. L. Jr., and LaBonte, B. J. 1987. *Astrophys. J.*, **319**, L27.

Braun, D. C., Duvall, T. L. Jr., and LaBonte, B. J. 1988. *Astrophys. J.*, **335**, 1015.

Braun, D. C., Duvall, T. L. Jr., LaBonte, B. J., *et al.* 1992a. *Astrophys. J.*, **391**, L113.

Braun, D. C., and Fan, Y. 1998. *Astrophys. J.*, **508**, L105.

Braun, D. C., and Lindsey, C. 2000. *Sol. Phys.*, **192**, 285.

Braun, D. C., Lindsey, C., Fan, Y., and Fagan, M. 1998. *Astrophys. J.*, **502**, 968.

Braun, D. C., Lindsey, C., Fan, Y., and Jefferies, S. M. 1992b. *Astrophys. J.*, **392**, 739.

Bray, R. J., and Loughhead, R. E. 1964. *Sunspots* (London: Chapman and Hall).

Bray, R. J., and Loughhead, R. E. 1974. *The Solar Chromosphere* (London: Chapman and Hall).

Bray, R. J., Loughhead, R. E., and Durrant, C. J. 1984. *The Solar Granulation*, 2nd edn (Cambridge: Cambridge University Press).

Brekke, P., and Maltby, P. 1963. *Ann. Astrophys.*, **26**, 383.

Brekke, P., Kjeldseth-Moe, O., Bartoe, J.-D. F., and Brueckner, G. E. 1987. In *Proc. 8th ESA Symp. on European Rocket and Balloon Programmes and Related Research* (ESA SP 270), p. 341.

Brickhouse, N. S., and LaBonte, B. J. 1988. *Sol. Phys.*, **115**, 43.

Brosius, J. W. 2005. *Astrophys. J.*, **622**, 1216.

Brosius, J. W., and Landi, E. 2005. *Astrophys. J.*, **632**, 1196.

Brouwer, M. P., and Zwaan, C. 1990. *Sol. Phys.*, **129**, 221.

Brown, D. S., Nightingale, R. W., Alexander, D., *et al.* 2003. *Sol. Phys.*, **216**, 79.

Brown, S. F., Donati, J.-F., Rees, D. E., and Semel, M. 1991. *Astron. Astrophys.*, **250**, 463.

Brown, T. M., Bogdan, T. J., Lites, B. W., and Thomas, J. H. 1992. *Astrophys. J.*, **394**, L65.

Browning, M. K., Miesch, M. S., Brun, A. S., and Toomre, J. 2006. *Astrophys. J.*, **648**, L157.

Broxon, J. W. 1942. *Phys. Rev.*, **62**, 508.

Brueckner, G. E. 1981. In *Solar Active Regions*, ed. F. Q. Orrall (Boulder: Colorado Associated University Press), p. 113.

Brueckner, G. E., and Bartoe, J.-D. F. 1974. *Sol. Phys.*, **38**, 133.

Brummell, N. H., Tobias, S. M., Thomas, J. H., and Weiss, N. O. 2008. *Astrophys. J.*, in press.

Brun, A. S., Miesch, M. S., and Toomre, J. 2004. *Astrophys. J.*, **613**, 1253.

Brunner, W. 1930. *Astr. Mitt. Zürich*, No. 124, 67.

Bruzek, A. 1969. *Sol. Phys.*, **8**, 29.

Brynildsen, N., Kjeldseth-Moe, O., Maltby, P., and Wilhelm, K. 1999b. *Astrophys. J.*, **517**, L159.

Brynildsen, N., Leifsen, T., Kjeldseth-Moe, O., Maltby, P., and Wilhelm, K. 1999a. *Astrophys. J.*, **511**, L121.

Brynildsen, N., Maltby, P., Brekke, P., *et al.* 1998. *Astrophys. J.*, **502**, L85.

Brynildsen, N., Maltby, P., Fredvik, T., Kjeldseth-Moe, O. 2002. *Sol. Phys.*, **207**, 259.

Brynildsen, N., Maltby, P., Fredvik, T., Kjeldseth-Moe, O., and Wilhelm, K. 2001. *Sol. Phys.*, **198**, 89.

Brynildsen, N., Maltby, P., Leifsen, T., Kjeldseth-Moe, O., and Wilhelm, K. 2000. *Sol. Phys.*, **191**, 129.

Budding, E. 1977. *Astrophys. Space Sci.*, **48**, 207.

Bumba, V. 1963. *Bull Astron. Inst. Czech.*, **14**, 91.

Bushby, P. J. 2003. *Mon. Not. Roy. Astron. Soc.*, **342**, L15.

Bushby, P. J. 2005. *Astron. Nachr.*, **326**, 218.

Bushby, P. J. 2006. *Mon. Not. Roy. Astron. Soc.*, **371**, 772.

Bushby, P. J. 2007. In IAU Symp. 239, *Convection in Astrophysics*, ed. F. Kupka, I. Roxburgh and K. Chan (Cambridge: Cambridge University Press), p. 514.

Bushby, P. J., and Houghton, S. M. 2005. *Mon. Not. Roy. Astron. Soc.*, **362**, 313.

Bushby, P. J., Houghton, S. M., Proctor, M. R. E., and Weiss, N. O. 2008. *Mon. Not. Roy. Astron. Soc.*, **387**, 698.

Bushby, P. J., and Mason, J. 2004. *Astron. Geophys.*, **45**, 4.07.

Busse, F. H., 1987. In *The Role of Fine-Scale Magnetic Fields on the Structure of the Solar Atmosphere*, ed. E. H. Schröter, M. Vázquez and A. A. Wyller (Cambridge: Cambridge University Press), p. 187.

Byrne, P. B. 1992. In *Sunspots: Theory and Observations*, ed. J. H. Thomas and N. O. Weiss (Dordrecht: Kluwer), p. 63.

Byrne, P. B. 1996. In IAU Symp. 176, *Stellar Surface Structure*, ed. K. G. Strassmeier and J. L. Linsky (Dordrecht: Kluwer), p. 299.

Cabrera Solana, D., Bellot Rubio, L. R., Beck, C., and del Toro Iniesta, J. C. 2006. *Astrophys. J.*, **649**, L41.

Cabrera Solana, D., Bellot Rubio, L. R., Beck, C., and del Toro Iniesta, J. C. 2007. *Astron. Astrophys.*, **475**, 1067.

Čalagović, J., Arnold, F., Beer, J., *et al.* 2008. In preparation.

Caligari, P., Moreno-Insertis, F., and Schüssler, M. 1995. *Astrophys. J.*, **441**, 886.

Cally, P. S. 1983. *Sol. Phys.*, **88**, 77.

Cally, P. S. 2000. *Sol. Phys.*, **192**, 395.

Cally, P. S. 2005. *Mon. Not. Roy. Astron. Soc.*, **358**, 353.

Cally, P. S., and Adam, J. A. 1983. *Sol. Phys.*, **85**, 97.

Cally, P. S., and Bogdan, T. J. 1993. *Astrophys. J.*, **402**, 721.

Cally, P. S., and Bogdan, T. J. 1997. *Astrophys. J.*, **486**, L67.

Cally, P. S., Bogdan, T. J., and Zweibel, E. G. 1994. *Astrophys. J.*, **437**, 505.

Cally, P. S., Crouch, A. D., and Braun, D. C. 2004. *Mon. Not. Roy. Astron. Soc.*, **346**, 381.

Camp, C. D., and Tung, K. K. 2007. *Geophys. Res. Lett.*, **34**, L14703.

Capocci, E. 1827. *Astron. Nachr.*, **5**, 313.

Carlsson, M., Stein, R. F., Nordlund, Å. and Scharmer, G. B. 2004. *Astrophys. J.*, **610**, L137.

Carrington, R. C. 1863. *Observations of the Spots on the Sun* (London: Williams & Norgate).

Carroll, T. A., Kopf, M., Ilyin, I., and Strassmeier, K. G. 2007. *Astron. Nachr.*, **328**, 1043.

Casanovas, J. 1997. In ASP Conf. Ser. 118, *Advances in the Physics of Sunspots*, ed. B. Schmieder, J. C. del Toro Iniesta and M. Vázquez (San Francisco: Astron. Soc. Pacific), p. 3.

Catalano, S., Biazzo, K., Frasca, A., and Marilli, E. 2002a. *Astron. Astrophys.*, **394**, 1009.

Catalano, S., Biazzo, K., Frasca, A., *et al.* 2002b. *Astron. Nachr.*, **323**, 260.

Catalano, S., and Rodonò, M. 1967. *Mem. Soc. Astron. Ital.*, **38**, 465.

Catalano, S., and Rodonò, M. 1974. *Pub. Astron. Soc. Pacific*, **86**, 398.

Cattaneo, F. 1999. *Astrophys. J.*, **515**, L39.

Cattaneo, F., Emonet, T., and Weiss, N. O. 2003. *Astrophys. J.*, **588**, 1183.

Cattaneo, F., and Hughes, D. W. 1996. *Phys. Rev. E*, **54**, 4532.

Cattaneo, F., and Hughes, D. W. 2006. *J. Fluid Mech.*, **553**, 401.

Cattaneo, F., and Hughes, D. W. 2008. *J. Fluid Mech.*, **594**, 495.

Cattaneo, F., Lenz, D., and Weiss, N. O. 2001. *Astrophys. J.*, **563**, L91.

Centeno, R., Socas-Navarro, H., Lites, B. W., *et al.* 2007. *Astrophys. J.*, **666**, L137.

Chae, J. 2001. *Astrophys. J.*, **560**, L95.

Chandrasekhar, S. 1952. *Phil. Mag.(7)*, **43**, 501.

Chandrasekhar, S. 1961. *Hydrodynamic and Hydromagnetic Stability* (Oxford: Clarendon Press).

Chapman, G. A., Cookson, A. M., and Dobias, J. J. 1994. *Astrophys. J.*, **432**, 403.

Chapman, G. A., Cookson, A. M., and Dobias, J. J. 1997. *Astrophys. J.*, **482**, 541.

Charbonneau, P. 2005. *Living Revs Sol. Phys.*, **2**, 2. (www.livingreviews.org/lrsp-2005-2)

Charbonneau, P., Blais-Laurier, G., and St-Jean, C. 2004. *Astrophys. J.*, **616**, L183.

Charbonneau, P., and MacGregor, K. B. 1997. *Astrophys. J.*, **486**, 502.

Chatterjee, P., Nandy, D., and Choudhuri, A. R. 2004. *Astron. Astrophys.*, **427**, 1019.

Chen, H.-R., Chou, D.-Y., and the TON Team 1996. *Astrophys. J.*, **465**, 985.

Chen, H.-R., Chou, D.-Y., and the TON Team 1997. *Astrophys. J.*, **490**, 452.

Cheung, M. C. M., Schüssler, M., and Moreno-Insertis, F. 2007. *Astron. Astrophys.*, **467**, 703.

Chevalier, S. 1916. *Ann. Obs. Astron. Zô Sè*, **9**, B1.

Chitre, S. M. 1963. *Mon. Not. Roy. Astron. Soc.*, **126**, 431.

Chitre, S. M. 1992. In *Sunspots: Theory and Observations*, ed. J. H. Thomas and N. O. Weiss (Dordrecht: Kluwer), p. 333.

Chitre, S. M., and Davila, J. M. 1991. *Astrophys. J.*, **371**, 785.

Chitre, S. M., and Shaviv, G. 1967. *Sol. Phys.*, **2**, 150.

Choudhuri, A. R. 1989. *Sol. Phys.*, **123**, 217.

Choudhuri, A. R. 1992. In *Sunspots: Theory and Observations*, ed. J. H. Thomas and N. O. Weiss (Dordrecht: Kluwer), p. 243.

Choudhuri, A. R. 1998. *The Physics of Fluids and Plasmas* (Cambridge: Cambridge University Press).

Choudhuri, A. R. 2003. In *Dynamic Sun*, ed. B. N. Dwivedi (Cambridge: Cambridge University Press), p. 103.

Choudhuri, A. R., and Gilman, P. A. 1987. *Astrophys. J.*, **316**, 788.

Choudhuri, A. R., Schüssler, M., and Dikpati, M. 1995. *Astron. Astrophys.*, **303**, L29.

Christensen-Dalsgaard, J., and Thompson, M. J. 2007. In *The Solar Tachocline*, ed. D. W. Hughes, R. Rosner and N. O. Weiss (Cambridge: Cambridge University Press), p. 53.

Christopoulou, E. B., Georgakilas, A. A., and Koutchmy, S. 2000. *Astron. Astrophys.*, **354**, 305.

Christopoulou, E. B., Georgakilas, A. A., and Koutchmy, S. 2001. *Astron. Astrophys.*, **375**, 617.

Chugainov, P. F. 1966. *IAU Inf. Bull. Var. Stars*, No. 122, 1.

Chugainov, P. F., Lovkaya, M. N., and Zajtseva, G. V. 1991. *IAU Inf. Bull. Var. Stars*, No. 3680, 1.

Clark, A. Jr. 1965. *Phys. Fluids*, **8**, 644.

Clark, A. Jr. 1966. *Phys. Fluids*, **9**, 485.

Clark, A. Jr. 1979. *Sol. Phys.*, **62**, 305.

Clark, A. Jr., and Johnson, H. K. 1967. *Sol. Phys.*, **2**, 433.

Clune, T. L., and Knobloch, E. 1994. *Physica D*, **74**, 151.

Collados, M., del Toro Iniesta, J. C., and Vázquez, M. 1987. *Sol. Phys.*, **112**, 281.

Collados, M., Martínez Pillet, V., Ruiz Cobo, B., del Toro Iniesta, J. C., and Vázquez, M. 1994. *Astron. Astrophys.*, **291**, 622.

Collier Cameron, A. 2002. *Astron. Nachr.*, **323**, 336.

Collier Cameron, A., and Donati, J.-F. 2002. *Mon. Not. Roy. Astron. Soc.*, **329**, L23.

Collier Cameron, A., and Robinson, R. D. 1989a. *Mon. Not. Roy. Astron. Soc.*, **236**, 57.

Collier Cameron, A., and Robinson, R. D. 1989b. *Mon. Not. Roy. Astron. Soc.*, **238**, 657.

Courvoisier, A., Hughes, D. W., and Tobias, S. M. 2006. *Phys. Rev. Lett.*, **96**, 034503.

Covas, E., Moss, D., and Tavakol, R. 2004. *Astron. Astrophys.*, **416**, 775.

Covas, E., Moss, D., and Tavakol, R. 2005. *Astron. Astrophys.*, **429**, 657.

Covas, E., Tavakol, R., Moss, D., and Tworkowski, A. 2000. *Astron. Astrophys.*, **360**, L21.

Covas, E., Tavakol, R., Tworkowski, A., and Brandenburg, A. 1998. *Astron. Astrophys.*, **329**, 350.

Cowling, T. G. 1934. *Mon. Not. Roy. Astron. Soc.*, **94**, 39.

Cowling, T. G. 1946. *Mon. Not. Roy. Astron. Soc.*, **106**, 218.

Cowling, T. G. 1953. In *The Sun*, ed. G. P. Kuiper (Chicago: University of Chicago Press), p. 532.

Cowling, T. G. 1957. *Magnetohydrodynamics* (New York: Interscience).

Cowling, T. G. 1975. *Nature*, **255**, 189.

Cowling, T. G. 1976a. *Magnetohydrodynamics* (Bristol: Adam Hilger).

Cowling, T. G. 1976b. *Mon. Not. Roy. Astron. Soc.*, **177**, 409.

Cowling, T. G. 1981. *Ann. Rev. Astron. Astrophys.*, **19**, 115.

Cowling, T. G. 1985. *Ann. Rev. Astron. Astrophys.*, **23**, 1.

Cram, L. E., and Thomas, J. H. (eds) 1981. *The Physics of Sunspots* (Sunspot, NM: Sacramento Peak Observatory).

Crouch, A. D., and Cally, P. S. 2003. *Sol. Phys.*, **214**, 201.

Crouch, A. D., Cally, P. S., Charbonneau, P., Braun, D. C., and Desjardins, M. 2005. *Mon. Not. Roy. Astron. Soc.*, **363**, 1188.

Damon, P. E., and Sonett, C. P. 1991. In *The Sun in Time*, ed. C. P. Sonett, M. S. Giampapa and M. S. Matthews (Tucson: University of Arizona Press), p. 360.

Danielson, R. E. 1961a. *Astrophys. J.*, **134**, 275.

Danielson, R. E. 1961b. *Astrophys. J.*, **134**, 289.

Danielson, R. E. 1964. *Astrophys. J.*, **139**, 45.

Danielson, R. E. 1965. In IAU Symp. 22, *Stellar and Solar Magnetic Fields*, ed. R. Lüst (Amsterdam: North Holland), p. 314.

Degenhardt, D. 1989. *Astron. Astrophys.*, **222**, 297.

Degenhardt, D. 1991. *Astron. Astrophys.*, **248**, 637.

Degenhardt, D., and Lites, B. W. 1993a. *Astrophys. J.*, **404**, 383.

Degenhardt, D., and Lites, B. W. 1993b. *Astrophys. J.*, **416**, 875.

Degenhardt, D., and Wiehr, E. 1991. *Astron. Astrophys.*, **252**, 821.

Deinzer, W. 1965. *Astrophys. J.*, **141**, 548.

Deinzer, W., Hensler, G., Schüssler, M., and Weisshaar, E. 1984. *Astron. Astrophys.*, **139**, 435.

del Toro Iniesta, J. C. 2003. *Astron. Nachr.*, **324**, 383.

del Toro Iniesta, J. C., Bellot Rubio, L. R., and Collados, M. 2001. *Astrophys. J.*, **549**, L139.

del Toro Iniesta, J. C., Tarbell, T. D., and Ruiz Cobo, B. 1994. *Astrophys. J.*, **436**, 400.

DeLuca, E. E., Fan, Y., and Saar, S. H. 1997. *Astrophys. J.*, **481**, 369.

Démoulin, P., Mandrini, C. H., van Driel-Gesztelyi, L., *et al.* 2002. *Astron. Astrophys.*, **382**, 650.

Denker, C. 1998. *Sol. Phys.*, **180**, 81.

Dere, K. P. 1982. *Sol. Phys.*, **77**, 77.

Deslandres, H.-A. 1910. *Ann. Obs. Astrophys. (Meudon)*, **4**, 119.

Deutsch, A. J. 1958. In IAU Symp. 6, *Electromagnetic Phenomena in Cosmical Physics*, ed. B. Lehnert (Cambridge: Cambridge University Press), p. 209.

Deutsch, A. J. 1970. *Astrophys. J.*, **159**, 985.

DeVore, C. R. 2000. *Astrophys. J.*, **539**, 944.

Diamond, P. H., Hughes, D. W., and Kim, E. 2005. In *Fluid Dynamics and Dynamos in Astrophysics and Geophysics*, ed. A. M. Soward, C. A. Jones, D. W. Hughes and N. O. Weiss (Boca Raton: CRC Press), p. 145.

Dikpati, M., and Charbonneau, P. 1999. *Astrophys. J.*, **518**, 508.

Dikpati, M., and Gilman, P. A. 2006. *Astrophys. J.*, **649**, 498.

Dikpati, M., de Toma, G., Gilman, P. A., Arge, C. N., and White, O. R. 2004. *Astrophys. J.*, **601**, 1136.

Ding, M. D., and Fang, C. 1989. *Astron. Astrophys.*, **225**, 204.

Dobbie, J. C. 1939. *Observatory*, **62**, 289.

Domínguez Cerdeña, I. 2003. *Astron. Astrophys.*, **412**, L65.

Donahue, R. A., Saar, S. A., and Baliunas, S. L. 1996. *Astrophys. J.*, **466**, 384.

Donati, J.-F. 1999. *Mon. Not. Roy. Astron. Soc.*, **302**, 457.

Donati, J.-F., and Brown, S. F. 1997. *Astron. Astrophys.*. **326**, 1135.

Donati, J.-F., and Collier Cameron, A. 1997. *Mon. Not. Roy. Astron. Soc.*, **291**, 1.

Donati, J.-F., Collier Cameron, A., Hussain, G. A. J., and Semel, M. 1999. *Mon. Not. Roy. Astron. Soc.*, **302**, 437.

Donati, J.-F., Collier Cameron, A., Semel, M., *et al.* 2003. *Mon. Not. Roy. Astron. Soc.*, **345**, 1145.

Donati, J.-F., Forveille, T., Collier Cameron, A., *et al.* 2006b. *Science*, **311**, 633.

Donati, J.-F., Howarth, I. D., Jardine, M. M., *et al.* 2006a. *Mon. Not. Roy. Astron. Soc.*, **370**, 629.

Donati, J.-F., Jardine, M. M., Gregory, S. G., *et al.* 2007. *Mon. Not. Roy. Astron. Soc.*, **380**, 1297.

Donati, J.-F., Semel, M., Carter, B. D., Rees, D. E., and Collier Cameron, A. 1997. *Mon. Not. Roy. Astron. Soc.*, **291**, 658.

Donati, J.-F., Wade, G. A., Babel, J., *et al.* 2001. *Mon. Not. Roy. Astron. Soc.*, **326**, 1265.

Dorch, S. B. F., and Nordlund, A. 2001. *Astron. Astrophys.*, **365**, 562.

Dormy, E., and Soward, A. M. (eds) 2007. *Mathematical Aspects of Natural Dynamos* (Boca Raton: CRC Press).

Dorren, J. D. 1987. *Astrophys. J.*, **320**, 756.

Dorren, J. D., and Guinan, E. F. 1994a. *Astrophys. J.*, **428**, 805.

Dorren, J. D., and Guinan, E. F. 1994b. In IAU Coll. 143, *The Sun as a Variable Star: Solar and Stellar Irradiance Variations*, ed. J. M. Pap, C. Fröhlich, H. S. Hudson and S. Solanki (Cambridge: Cambridge University Press), p. 206.

Dorren, J. D., Guinan, E. F., and McCook, G. P. 1984. *Pub. Astron. Soc. Pacific*, **96**, 250.

Drake, S. 1957. *Discoveries and Opinions of Galileo* (New York: Doubleday).

Drobyshevski, E. M., and Yuferev, V. S. 1974. *J. Fluid Mech.*, **65**, 33.

D'Silva, S. 1994. *Astrophys. J.*, **435**, 881.

D'Silva, S., and Choudhuri, A. R. 1993. *Astron. Astrophys.*, **272**, 621.

Dunn, R. B., and Zirker, J. B. 1973. *Sol. Phys.*, **33**, 281.

Duquennoy, A., and Mayor, M. 1991. *Astron. Astrophys.*, **248**, 485.

Durney, B. R., De Young, D. S. and Roxburgh, I. W. 1993. *Sol. Phys.*, **145**, 207.

Durney, B. R., and Latour, J. 1978. *Geophys. Astrophys. Fluid Dyn.*, **9**, 241.

Durrant, C. J., Turner, J. P. R., and Wilson, P. R. 2004. *Sol. Phys.*, **222**, 345.

Duvall, T. L. Jr., D'Silva, S., Jefferies, S. M., Harvey, J. W., and Schou, J. 1996. *Nature*, **379**, 235.

Duvall, T. L. Jr., Jefferies, S. M., Harvey, J. W., and Pomerantz, M. A. 1993. *Nature*, **362**, 430.

Duvall, T. L. Jr., and Kosovichev, A. G. 2001. In IAU Symp. 203, *Recent Insights into the Physics of the Sun and Heliosphere: Highlights from SOHO and Other Space Missions*, ed. P. Brekke, B. Fleck and J. B. Gurman (San Francisco: Astron. Soc. Pacific), p. 159.

Eaton, J. A., and Hall, D. S. 1979. *Astrophys. J.*, **227**, 907.

Eberhard, G., and Schwarzschild, K. 1913. *Astrophys. J.*, **38**, 292.

Eddy, J. A. 1976. *Science*, **192**, 1189.

Eddy, J. A., Gilman, P. A., and Trotter, D. E. 1976. *Sol. Phys.*, **46**, 3.

Eddy, J. A., Gilman, P. A., and Trotter, D. E. 1977. *Science*, **198**, 824.

Eddy, J. A., Stephenson, F. R., and Yau, K. K. C. 1989. *Quart. J. Roy. Astron. Soc.*, **30**, 65.

Efremov, V. I., and Parfinenko, L. D. 1996. *Astron. Rep.*, **40**, 89.

Eschrich, K.-O., and Krause, F. 1977. *Astron. Nachr.*, **298**, 1.

Evans, D. S. 1959. *Mon. Not. Roy. Astron. Soc.*, **119**, 526.

Evans, D. S. 1971. *Mon. Not. Roy. Astron. Soc.*, **154**, 329.

Evershed, J. 1909a. *Mon. Not. Roy. Astron. Soc.*, **69**, 454.

Evershed, J. 1909b. *Observatory*, **32**, 291.

Evershed, J. 1909c. *Kodaikanal Obs. Bull.*, No. 15, p. 63.

Ewell, M. W. Jr. 1992. *Sol. Phys.*, **137**, 215.

Falk, A. E., and Wehlau, W. H. 1974. *Astrophys. J.*, **192**, 409.

Fan, Y. 2001. *Astrophys. J.*, **554**, L111.

Fan, Y. 2004. *Living Revs Sol. Phys.*, **1**, 1. (www.livingreviews.org/lrsp-2004-1)

Ferraro, V. C. A. 1937. *Mon. Not. Roy. Astron. Soc.*, **97**, 458.

Ferraro, V. C. A., and Plumpton, C. 1958. *Astrophys. J.*, **127**, 459.

Ferriz-Mas, A., Schmitt, D., and Schüssler, M. 1994. *Astron. Astrophys.*, **289**, 949.

Fisher, G. H., Fan, Y., Longcope, D. W., Linton, M. G., and Pevtsov, A. A. 2000. *Sol. Phys.*, **192**, 119.

Fleck, B., and Keller, C. U. 2003. In *Dynamic Sun*, ed. B. N. Dwivedi (Cambridge: Cambridge University Press), p. 403.

Fludra, A. 1999. *Astron. Astrophys.*, **344**, L75.

Fludra, A. 2001. *Astron. Astrophys.*, **368**, 639.

Foukal, P. V. 1976. *Astrophys. J.*, **210**, 575.

Foukal, P. 1981a. In *The Physics of Sunspots*, ed. L. E. Cram and J. H. Thomas (Sunspot, NM: Sacramento Peak Observatory), p. 191.

Foukal, P. 1981b. In *The Physics of Sunspots*, ed. L. E. Cram and J. H. Thomas (Sunspot, NM: Sacramento Peak Observatory), p. 391.

Foukal, P. 1993. *Sol. Phys.*, **148**, 219.

Foukal, P. 1998. *Astrophys. J.*, **500**, 958.

Foukal, P. V. 2004. *Solar Astrophysics*, 2nd edn (Weinheim: Wiley-VCH).

Foukal, P., Fowler, L. A., and Livshits, M. 1983. *Astrophys. J.*, **267**, 863.

Foukal, P., Fröhlich, C., Spruit, H., and Wigley, T. M. L. 2006. *Nature*, **443**, 161.

Foukal, P. V., Noyes, R. W., Reeves, E. M., *et al.* 1974. *Astrophys. J.*, **193**, L143.

Fowler, L. A., Foukal, P., and Duvall, T. Jr. 1983. *Sol. Phys.*, **84**, 33.

Fox, P. 1908. *Astrophys. J.*, **28**, 253.

Frasca, A., Biazzo, K., Catalano, S., *et al.* 2005. *Astron. Astrophys.*, **432**, 647.

Frazier, E. N. 1972. *Sol. Phys.*, **26**, 130.

Frick, P., Soon, W. H., Popova, E., and Baliunas, S. L. 2004. *New Astron.*, **9**, 599.

Friis-Christensen, E., Fröhlich, C., Haigh, J. D., Schüssler, M., and von Steiger, R. (eds) 2000. *Solar Variability and Climate* (Dordrecht: Kluwer).

Fröhlich, C. 2006. *Space Sci. Rev.*, **125**, 53.

Fröhlich, C., and Lean, J. 2004. *Astron. Astrophys. Rev.*, **12**, 273.

Fröhlich, H.-E., Tschäpe, R., Rüdiger, G., and Strassmeier, K. G. 2002. *Astron. Astrophys.*, **391**, 659.

Galloway, D. J. 1975. *Sol. Phys.*, 44, 409.

Galloway, D. J. 1978. *Mon. Not. Roy. Astron. Soc.*, **184**, 49P.

Galloway, D. J., and Weiss, N. O. 1981. *Astrophys. J.*, **243**, 945.

Galsgaard, K., Moreno-Insertis, F., Archontis, V., and Hood, A. 2005. *Astrophys. J.*, **618**, L153.

Garcia de la Rosa, J. I. 1987a. *Sol. Phys.*, **112**, 49.

Garcia de la Rosa, J. I. 1987b. In *The Role of Fine-Scale Magnetic Fields on the Structure of the Solar Atmosphere*, ed. E. H. Schröter, M. Vázquez and A. Wyller (Cambridge: Cambridge University Press), p. 140.

Gelfreikh, G. B., Grechnev, V., Kosugi, T., and Shibasaki, K. 1999. *Sol. Phys.*, **185**, 177.

Gilman, P. A. 1979. *Astrophys. J.*, **231**, 284.

Gilman, P. A. 1983. *Astrophys. J. Suppl.*, **53**, 243.

Gilman, P. A., and Cally, P. S. 2007. In *The Solar Tachocline*, ed. D. W. Hughes, R. Rosner and N. O. Weiss (Cambridge: Cambridge University Press), p. 243.

Gilman, P. A., and Howard, R. 1984. *Astrophys. J.*, **283**, 385.

Gilman, P. A., and Howard, R. 1985. *Astrophys. J.*, **295**, 233.

Giovanelli, R. G. 1972. *Sol. Phys.*, **27**, 71.

Giovanelli, R. G., and Jones, H. P. 1982. *Sol. Phys.*, **79**, 267.

Gizon, L., and Birch, A. C. 2002. *Astrophys. J.*, **571**, 966.

Gizon, L., Duvall, T. L. Jr., and Larsen, R. M. 2000. *J. Astrophys. Astron.*, **21**, 339.

Gizon, L., Duvall, T. L. Jr., and Larsen, R. M. 2001. In IAU Symp. 203, *Recent Insights into the Physics of the Sun and Heliosphere*, ed. P. Brekke, B. Fleck and J. B. Gurman (San Francisco: Astron. Soc. Pacific), p. 189.

Glatzmaier, G. A. 1985. *Astrophys. J.*, **291**, 300.

Gleissberg, W. 1939. *Observatory*, **62**, 158.

Gleissberg, W. 1945. *Observatory*, **66**, 123.

Gnevyshev, M. N. 1938. *Pulkovo Obs. Circ.*, **24**, 37.

Godoli, G. 1968. In *Mass Motions in Solar Flares and Related Phenomena*, ed. Y. Öhman (Stockholm: Almqvist and Wiksell), p. 211.

Gokhale, M. H., and Zwaan, C. 1972. *Sol. Phys.*, **26**, 52.

Golub, L. 1980. *Phil. Trans. Roy. Soc. Lond. A*, **297**, 595.

Golub, L., and Pasachoff, J. M. 1997. *The Solar Corona* (Cambridge: Cambridge University Press).

Golub, L., Rosner, R., Vaiana, G., and Weiss, N. O. 1981. *Astrophys. J.*, **243**, 309.

Goncharskii, A. V., Stepanov, V. V., Kokhlova, V. L., and Yagola, A. G. 1977. *Soviet Astron. Lett.*, **3**, 147.

Goossens, M., and Poedts, S. 1992. *Astrophys. J.*, **384**, 348.

Gough, D. O. 1981. In *Variations of the Solar Constant*, ed. S. Sofia (Washington: NASA CP-2191), p. 185.

Granzer, Th., Schüssler, M., Caligari, P., and Strassmeier, K. G. 2000. *Astron. Astrophys.*, **355**, 1087.

Gray, D. F. 1994. *Pub. Astron. Soc. Pacific*, **106**, 1248.

Gray, D. F. 1996. In IAU Symp. 176, *Stellar Surface Structure*, ed. K. G. Strassmeier and J. L. Linsky (Dordrecht: Kluwer), p. 227.

Gray, D. F., Baliunas, S. L., Lockwood, G. W., and Skiff, B. A. 1996. *Astrophys. J.*, **465**, 945.

Green, L. M., López Fuentes, M. C., Mandrini, C. H., *et al.* 2002. *Sol. Phys.*, **208**, 43.

Greenwich, Royal Observatory. 1925. *Mon. Not. Roy. Astron. Soc.*, **85**, 553.

Grossmann-Doerth, U., Schmidt, W., and Schröter, E. H. 1986. *Astron. Astrophys.*, **156**, 347.

Grossmann-Doerth, U., Schüssler, M., and Steiner, O. 1998. *Astron. Astrophys.*, **337**, 928.

Grove, J. M. 1988. *The Little Ice Age* (London: Routledge).

Guckenheimer, J., and Holmes, P. 1986. *Nonlinear Oscillations, Dynamical Systems and Bifurcations of Vector Fields*, 2nd Printing (New York: Springer).

Gurman, J. B. 1993. *Astrophys. J.*, **412**, 865.

Gurman, J. B., and House, L. L. 1981. *Sol. Phys.*, **71**, 5.

Gurman, J. B., and Leibacher, J. W. 1984. *Astrophys. J.*, **283**, 859.

Hackman, T., and Jetsu, L. 2003. In *12th Cambridge Workshop on Cool Stars, Stellar Systems, and the Sun: The Future of Cool-Star Astrophysics*, ed. A. Brown, G. M. Harper and T. R. Ayres (Boulder: University of Colorado Press), p. 935.

Hagenaar, H. J. 2001. *Astrophys. J.*, **555**, 448.

Hagenaar, H. J., Schrijver, C. J., and Title, A. M. 2003. *Astrophys. J.*, **584**, 1107.

Hagenaar, H. J., and Shine, R. A. 2005. *Astrophys. J.*, **635**, 659.

Haigh, J. D. 2003. *Phil. Trans. Roy. Soc. Lond. A*, **361**, 95.

Haigh, J. D. 2007. *Living Revs Sol. Phys.*, **4**, 2. (www.livingreviews.org/lrsp-2007-2)

Hale, G. E. 1908a. *Astrophys. J.*, **28**, 100.

Hale, G. E. 1908b. *Astrophys. J.*, **28**, 315.

Hale, G. E. 1925. *Pub. Astron. Soc. Pacific*, **37**, 268.

Hale, G. E., Ellerman, F., Nicholson, S. B., and Joy, A. H. 1919. *Astrophys. J.*, **49**, 153.

Hale, G. E., and Nicholson, S. B. 1925. *Astrophys. J.*, **62**, 270.

Hale, G. E., and Nicholson, S. B. 1938. *Magnetic Observations of Sunspots 1917–1924. Part I* (Washington: Carnegie Institution, Publ. No. 498).

Hall, D. S. 1972. *Pub. Astron. Soc. Pacific*, **84**, 323.

Hall, D. S. 1990. In *Active Close Binaries*, ed. C. Ibanoglu (Dordrecht: Kluwer), p. 99.

Hall, D. S., and Busby, M. R. 1990. In *Active Close Binaries*, ed. C. Ibanoglu (Dordrecht: Kluwer), p. 377.

Hansen, C. J., and Kawaler, S. D. 1994. *Stellar Interiors: Physical Principles, Structure, and Evolution* (New York: Springer).

Hansen, J. E. 2000. *Space Sci. Rev.*, **94**, 349.

Hansen, J. E., Sato, M., Kharecha, P., *et al.* 2007. *Phil. Trans. Roy. Soc. Lond. A*, **365**, 1925.

Harmon, R. O., and Crews, L. J. 2000. *Astron. J.*, **120**, 3274.

Harrison, R. G. 2000. *Space Sci. Rev.*, **94**, 381.

Harrison, R. G., Bingham, R., Aplin, K., *et al.* 2007. *Astron. Geophys.*, **48**, 2.7.

Hart, A. B. 1954. *Mon. Not. Roy. Astron. Soc.*, **114**, 17.

Hart, A. B. 1956. *Mon. Not. Roy. Astron. Soc.*, **116**, 38.

Hart, J. E., Glatzmaier, G. A., and Toomre, J. 1986. *J. Fluid Mech.*, **173**, 519.

Hart, J. E., Toomre, J., Deane, A. E., *et al.* 1986. *Science*, **234**, 61.

Hartmann, L., Bopp, B. W., Dussault, M., Noah, P. V., and Klimke, A. 1981. *Astrophys. J.*, **249**, 662.

Hartmann, L., and Noyes, R. W. 1987. *Ann. Rev. Astron. Astrophys.*, **25**, 271.

Harvey, J. W. 1973. *Sol. Phys.*, **28**, 9.

Harvey, J. W., Branston, D., Henney, C. J., and Keller, C. U. 2007. *Astrophys. J.*, **659**, L177.

Harvey, K., and Harvey, J. 1973. *Sol. Phys.*, **28**, 61.

Harvey, K. L., and Martin, S. F. 1973. *Sol. Phys.*, **32**, 389.

Harvey-Angle, K. L. 1993. *Magnetic bipoles on the Sun*. Ph.D. dissertation, University of Utrecht.

Hasan, S. S. 1985. *Astrophys. J.*, **143**, 39.

Hasan, S. S. 1991. *Astrophys. J.*, **366**, 328.

Hathaway, D. H. 1996. *Astrophys. J.*, **460**, 1027.

Hathaway, D. H., Nandy, D., Wilson, R. M., and Reichmann, E. J. 2003. *Astrophys. J.*, **589**, 665.

Hatzes, A. P. 1995. *Astrophys. J.*, **451**, 784.

Hatzes, A. P. 1998. *Astron. Astrophys.*, **330**, 541.

Hatzes, A. P., and Vogt, S. S. 1992. *Mon. Not. Roy. Astron. Soc.*, **258**, 387.

Havnes, O. 1970. *Sol. Phys.*, **13**, 323.

Hays, J. D., Imbrie, J., and Shackleton, N. J. 1976. *Science*, **194**, 1121.

Heinemann, T., Nordlund, Å., Scharmer, G. B., and Spruit, H. C. 2007. *Astrophys. J.*, **669**, 1390.

Hempelmann, A., Schmitt, J. H. M. M., Schultz, M., Rüdiger, G., and Stepień, K. 1995. *Astron. Astrophys.*, **294**, 515.

Hendry, M. A., Bryce, H. M., and Valls-Gabaud, D. 2002. *Mon. Not. Roy. Astron. Soc.*, **335**, 539.

Hendry, P. D., and Mochnacki, S. W. 2000. *Astrophys. J.*, **531**, 467.

Henry, G. W., Eaton, J. A., Hamer, J., and Hall, D. S. 1995. *Astrophys. J. Suppl.*, **97**, 513.

Henry, J., and Alexander, S. 1846. *Phil. Mag.*, **28**, 230.

Henry, T. J., Soderblom, D. R., Donahue, R. A., and Baliunas, S. L. 1996. *Astron. J.*, **111**, 439.

Herschel, J. F. W. 1847. *Results of Astronomical Observations at the Cape of Good Hope* (London: Smith, Elder).

Herschel, W. 1795. *Phil. Trans. Roy. Soc. Lond*, **85**, 46.

Herschel, W. 1801. *Phil. Trans. Roy. Soc. Lond*, **91**, 265.

Hindman, B. W., and Brown, T. M. 1998. *Astrophys. J.*, **504**, 1029.

Hindman, B. W., Jain, R., and Zweibel, E. G. 1997. *Astrophys. J.*, **476**, 392.

Hiremath, K. M. 2002. *Astron. Astrophys.*, **386**, 674.

Hirzberger, J., and Wiehr, E. 2005. *Astron. Astrophys.*, **438**, 1059.

Hofmann, J., Deubner, F.-L., Fleck, B., and Schmidt, W. 1994. *Astron. Astrophys.*, **284**, 269.

Hollweg, J. V. 1988. *Astrophys. J.*, **335**, 1005.

Holzhauser, H., Magny, M., and Zumbühl, H. J. 2005. *Holocene*, **15**, 789.

Holzwarth, V. and Schüssler, M. 2002. *Astron. Nachr.*, **323**, 399.

Hormes, A., Beer, J., and Schlüchter, C. 2006. *Geografiska Ann.*, **88A**, 281.

Horn, T., Staude, J., and Landgraf, V. 1997. *Sol. Phys.*, **172**, 69.

Hoskin, M. 2001. In *Encyclopedia of Astronomy and Astrophysics* (London: Nature Publishing Group), p. 3038.

Houghton, J. 2004. *Global Warming* (Cambridge: Cambridge University Press).

Howard, R. F. 1992. *Sol. Phys.*, **137**, 51.

Howard, R., and Gilman, P. A. 1986. *Astrophys. J.*, **307**, 389.

Howard, R., Gilman, P. A., and Gilman, P. I. 1984. *Astrophys. J.*, **283**, 373.

Howard, R., and Harvey, J. W. 1964. *Astrophys. J.*, **139**, 1328.

Howard, R., and Labonte, B. J. 1980. *Astrophys. J.*, **239**, L33.

Howard, R., Tanenbaum, A. S., and Wilcox, J. M. 1968. *Sol. Phys.*, **4**, 286.

Howe, R., Christensen-Dalsgaard, J., Hill, F., *et al.* 2005. *Astrophys. J.*, **634**, 1405.

Hoyle. F. 1949. *Some Recent Researches in Solar Physics* (Cambridge: Cambridge University Press).

Hoyt, D. V., and Schatten, K. H. 1997. *The Role of the Sun in Climate Change* (Oxford: Oxford University Press).

Hoyt, D. V., and Schatten, K. H. 1998. *Sol. Phys.*, **181**, 491.

Hudson, H. S., Silva, S., Woodard, M., and Willson, R. C. 1982. *Sol. Phys.*, **76**, 211.

Huenemoerder, D. P., and Ramsey, L. W. 1987. *Astrophys. J.*, **319**, 392.

Huenemoerder, D. P., Ramsey, L. W., and Buzasi, D. L. 1989. *Astron. J.*, **98**, 2264.

Hughes, D. W. 1992. In *Sunspots: Theory and Observations*, ed. J. H. Thomas and N. O. Weiss (Dordrecht: Kluwer), p. 371.

Hughes, D. W. 2007a. In *The Solar Tachocline*, ed. D. W. Hughes, R. Rosner and N. O. Weiss (Cambridge: Cambridge University Press), p. 275.

Hughes, D. W. 2007b. In *Mathematical Aspects of Natural Dynamos*, ed. E. Dormy and A. M. Soward (Boca Raton: CRC Press), p. 79.

Hughes, D. W., and Proctor, M. R. E. 1988. *Ann. Rev. Fluid Mech.*, **20**, 187.

Hurlburt, N., and Alexander, D. 2002. In *COSPAR Colloquia Ser. 14: Solar-Terrestrial Magnetic Activity and Space Environment*, ed. H. Wang and R. Xu (Oxford: Pergamon), p. 19.

Hurlburt, N. E., Matthews, P. C., and Proctor, M. R. E. 1996. *Astrophys. J.*, **457**, 933.

Hurlburt, N. E., Matthews, P. C., and Rucklidge, A. M. 2000. *Sol. Phys.*, **192**, 109.

Hurlburt, N. E., and Rucklidge, A. M. 2000. *Mon. Not. Roy. Astron. Soc.*, **314**, 793.

Hussain, G. A. J. 2002. *Astron. Nachr.*, **323**, 349.

Hussain, G. A. J., Donati, J.-F., Collier Cameron, A., and Barnes, J. R. 2000. *Mon. Not. Roy. Astron. Soc.*, **318**, 961.

Hussain, G. A. J., Jardine, M., Donati, J.-F., *et al.* 2007. *Mon. Not. Roy. Astron. Soc.*, **377**, 1488.

Ichimoto, K., Shine, R. A., Lites, B., *et al.* 2007b. *Pub. Astron. Soc. Japan*, **59**, S593.

Ichimoto, K., Suematsu, Y., Tsuneta, S., *et al.* 2007a. *Science*, **318**, 1597.

Intergovernmental Panel on Climate Change. 2007., *Climate Change 2007: The Physical Science Basis* (Cambridge: Cambridge University Press).

Isakov, A. B., Schekochihin, A. A., Cowley, S. C., McWilliams, J. C., and Proctor, M. R. E. 2007. *Phys. Rev. Lett.*, **98**, 208501.

Işik, E., Schüssler, M., and Solanki, S. K. 2007. *Astron. Astrophys.*, **464**, 1049.

Ivanova, T. S., and Ruzmaikin, A. A. 1977. *Astron. Zh.*, **54**, 846 (trans. *Soviet Astron.*, **21**, 479).

Jahn, K. 1989. *Astron. Astrophys.*, **222**, 264.

Jahn, K. 1992. In *Sunspots: Theory and Observations*, ed. J. H. Thomas and N. O. Weiss (Dordrecht: Kluwer), p. 139.

Jahn, K., and Schmidt, H. U. 1994. *Astron. Astrophys.*, **290**, 295.

Jain, R., and Haber, D. 2002. *Astron. Astrophys.*, **387**, 1092.

Jakimiec, J. 1965. *Acta Astron.*, **15**, 145.

Jakimiec, J., and Zabza, M. 1966. *Acta Astron.*, **16**, 73.

James, D. J., and Jeffries, R. D. 1997. *Mon. Not. Roy. Astron. Soc.*, **292**, 252.

Järvinen, S. P., Berdyugina, S. V., Korhonen, H., Ilyin, I., and Tuominen, I. 2007. *Astron. Astrophys.*, **472**, 887.

Järvinen, S. P., Berdyugina, S. V., and Strassmeier, K. G. 2005. *Astron. Astrophys.*, **440**, 735.

Jeffers, S. V., Donati, J.-F., and Collier Cameron, A. 2007. *Mon. Not. Roy. Astron. Soc.*, **375**, 567.

Jensen, E. 1955. *Ann. Astrophys.*, **18**, 127.

Jensen, E., Nordø, J., and Ringnes, T. S. 1955. *Astrophys. Norvegica*, **5**, 167.

Jepps, S. A. 1975. *J. Fluid Mech.*, **67**, 625.

Jetsu, L. 1996. *Astron. Astrophys.*, **314**, 153.

Jetsu, L., Pelt, J., and Tuominen, I. 1993. *Astron. Astrophys.*, **278**, 449.

Jiang, J., Chatterjee, P., and Choudhuri, A. R. 2007. *Mon. Not. Roy. Astron. Soc.*, **381**, 1527.

Joncour, I., Bertout, C., and Bouvier, J. 1994. *Astron. Astrophys.*, **291**, L19.

Joncour, I., Bertout, C., and Ménard, F. 1994. *Astron. Astrophys.*, **285**, L25.

Jones, C. A., Weiss, N. O., and Cattaneo, F. 1985. *Physica*, **14D**, 161.

Jones, P. D., and Mann, M. E. 2004. *Rev. Geophys.*, **42**, RG2002.

Joy, A. H., and Wilson, R. E. 1949. *Astrophys. J.*, **109**, 231.

Juckes, M. N., Allen, M. R., Briffa, K. R., *et al.* 2007. *Clim. Past*, **3**, 591.

Julien, K., Knobloch, E., and Tobias, S. M. 1999. *Physica D*, **128**, 105.

Julien, K., Knobloch, E., and Tobias, S. M. 2000. *J. Fluid Mech.*, **410**, 285.

Julien, K., Knobloch, E., and Tobias, S. 2003. In *Advances in Nonlinear Dynamos*, ed. A. Ferriz-Mas and M. Núñez-Jimenez (Bristol: Taylor & Francis), p. 195.

Jurčák, J., Bellot Rubio, L., Ichimoto, K., *et al.* 2007. *Pub. Astron. Soc. Japan*, **59**, S601.

Jurčák, J., Martínez Pillet, V., and Sobotka, M. 2006. *Astron. Astrophys.*, **453**, 1079.

Katsukawa, K., Tsuneta, S., Berger, T. E., *et al.* 2007a. *Science*, **318**, 1594.

Katsukawa, Y., Yokoyama, T., Berger, T. E., *et al.* 2007b. *Pub. Astron. Soc. Japan*, **59**, S577.

Kawai, G., Kurokawa, H., Tsuneta, S., *et al.* 1992. *Pub. Astron. Soc. Japan*, **44**, L193.

Kawakami, H. 1983. *Pub. Astron. Soc. Japan*, **35**, 459.

Keil, S. L., Balasubramaniam, K. S., Smaldone, L. A., and Reger, B. 1999. *Astrophys. J.*, **510**, 422.

Keller, C. U., Schüssler, M., Vögler, A., and Zakharov, V. 2004. *Astrophys. J.*, **607**, L59.

Keller, C. U., and von der Lühe, O. 1992. *Astron. Astrophys.*, **261**, 321.

Keppens, R., Bogdan, T. J., and Goossens, M. 1994. *Astrophys. J.*, **436**, 372.

Keppens, R., and Martínez Pillet, V. 1996. *Astron. Astrophys.*, **316**, 229.

Kersalé, E., Hughes, D. W., and Tobias, S. M. 2007. *Astrophys. J.*, **663**, L113.

Kiepenheuer, K. O. 1953. In *The Sun*, ed. G. P. Kuiper (Chicago: University of Chicago Press), p. 322.

Kinman, T. D. 1952. *Mon. Not. Roy. Astron. Soc.*, **112**, 425.

Kippenhahn, R., and Weigert, A. 1990. *Stellar Structure and Evolution* (Berlin: Springer).

Kitai, R., Watanabe, H., Nakamura, T., *et al.* 2007. *Pub. Astron. Soc. Japan*, **59**, S585.

Kitchatinov, L. L., Pipin, V. V., Makarov, V. I., and Tlatov, A. G. 1999. *Sol. Phys.*, **189**, 227.

Kjeldseth Moe, O., and Maltby, P. 1969. *Sol. Phys.*, **8**, 275.

Kjeldseth Moe, O., and Maltby, P. 1974. *Sol. Phys.*, **36**, 101.

Kleeorin, N. I., and Ruzmaikin, A. A. 1991. *Sol. Phys.*, **131**, 211.

Kneer, F., Mattig, W., and von Uexküll, M. 1981. *Astron. Astrophys.*, **102**, 147.

Knobloch, E., Rosner, R., and Weiss, N. O. 1981. *Mon. Not. Roy. Astron. Soc.*, **197**, 45P.

Knobloch, E., Tobias, S. M., and Weiss, N. O. 1998. *Mon. Not. Roy. Astron. Soc.*, **297**, 1123.

Knobloch, E., and Weiss, N. O. 1984. *Mon. Not. Roy. Astron. Soc.*, **207**, 203.

Knölker, M., and Schüssler, M. 1988. *Astron. Astrophys.*, **202**, 275.

Knölker, M., Schüssler, M., and Weisshaar, E. 1988. *Astron. Astrophys.*, **202**, 275.

Knox, K. T., and Thompson, B. J. 1974. *Astrophys. J.*, **193**, L45.

König, B., Guenther, E. W., Woitas, J., and Hatzes, A. P. 2005. *Astron. Astrophys.*, **435**, 215.

Kopal, Z. 1982. *Astrophys. Space Sci.*, **87**, 149.

Kopp, G., and Rabin, D. 1992. *Sol. Phys.*, **141**, 253.

Korhonen, H., Berdyugina, S. V., Hackman, T., Strassmeier, K. G., and Tuominen, I. 2000. *Astron. Astrophys.*, **360**, 1067.

Kosovichev, A. G. 2002. *Astron. Nachr.*, **323**, 186.

Kosovichev, A. G. 2006. *Adv. Space Res.*, **38**, 876.

Koutchmy, S., and Adjabshirzadeh, A. 1981. *Astron. Astrophys.*, **99**, 111.

Kővari, Zs., Bartus, J., Strassmeier, K. G., *et al.* 2007a. *Astron. Astrophys.*, **463**, 1071.

Kővari, Zs., Bartus, J., Strassmeier, K. G., *et al.* 2007b. *Astron. Astrophys.*, **474**, 165.

Krause, F., and Rädler, K.-H. 1980. *Mean-field Magnetohydrodynamics and Dynamo Theory* (Berlin: Akademie-Verlag).

Krause, F., and Rüdiger, G. 1975. *Sol. Phys.*, **42**, 107.

Krieger, A. S., Vaiana, G. S., and Van Speybroeck, L. P. 1971. In IAU Symp. 43, *Solar Magnetic Fields*, ed. R. Howard (Dordrecht: Reidel), p. 397.

Krivova, N. A., Balmaceda, L., and Solanki, S. K. 2007. *Astron. Astrophys.*, **467**, 335.

Kron, G. E. 1947. *Pub. Astron. Soc. Pacific*, **59**, 261.

Kron, G. E. 1950a. *Astron. J.*, **55**, 69.

Kron, G. E. 1950b. *Astron. Soc. Pacific Leaflets*, **6**, 52.

Krzeminski, W. 1969. In *Low Luminosity Stars*, ed. S. S. Kumar (New York: Gordon and Breach), p. 57.

Krzeminski, W., and Kraft, R. P. 1967. *Astron. J.*, **72**, 307.

Kubo, M., Ichimoto, K., Shimizu, T., *et al.* 2007. *Pub. Astron. Soc. Japan*, **59**, S607.

Kubo, M., Shimizu, T., and Lites, B. W. 2003. *Astrophys. J.*, **595**, 465.

Kubo, M., Shimizu, T., and Tsuneta, S. 2007. *Astrophys. J.*, **659**, 812.

Kuhn, T. S. 1957. *The Copernican Revolution; Planetary Astronomy in the Development of Western Thought* (Cambridge, MA: Harvard University Press).

Küker, M., Arlt, R., and Rüdiger, G. 1999. *Astron. Astrophys.*, **343**, 977.

Küker, M., Rüdiger, G., and Schultz, M. 2001. *Astron. Astrophys.*, **374**, 301.

Kupke, R., LaBonte, B. J., and Mickey, D. L. 2000. *Sol. Phys.*, **191**, 97.

Kürster, M., Hatzes, A. P., Pallavicini, R., and Randich, S. 1992. In ASP Conf. Ser. 26, *Cool Stars, Stellar Systems, and the Sun*, ed. M. S. Giampapa and J. A. Bookbinder (San Francisco: Astron. Soc. Pacific), p. 249.

Kürster, M., Schmitt, J. H. M. M., and Cutispoto, G. 1994. *Astron. Astrophys.*, **289**, 899.

Kuznetsov, Y. A. 1998. *Elements of Applied Bifurcation Theory*, 2nd edn (New York: Springer).

Labitzke, K. 2005. *J. Atmos. Sol.-Terr. Phys.*, **67**, 45.

LaBonte, B. J., and Ryutova, M. 1993. *Astrophys. J.*, **419**, 388.

Landgraf, V. 1997. *Astron. Nachr.*, **318**, 129.

Landman, D. A., and Finn, G. D. 1979. *Sol. Phys.*, **63**, 221.

Langhans, K., Scharmer, G. B., Kiselman, D., Löfdahl, M. G., and Berger, T. E. 2005. *Astron. Astrophys.*, **436**, 1087.

Langhans, K., Scharmer, G. B., Kiselman, D., and Löfdahl, M. G. 2007. *Astron. Astrophys.*, **464**, 763.

Larmor, J. 1919. *Rep. Brit. Assoc. Adv. Sci. 1919*, 159.

Lean, J. L. 2000. *Geophys. Res. Lett.*, **27**, 2425.

Lean, J. L. 2001. In ASP Conf. Ser. 223, *The 11th Cambridge Conference on Cool Stars, Stellar Systems and the Sun*, ed. R. J. García López, R. Rebolo and M. R. Zapatero Osorio (San Francisco: Astron. Soc. Pacific), p. 109.

Lee, J. W. 1992. *Sol. Phys.*, **139**, 267.

Leighton, R. B. 1969. *Astrophys. J.*, **156**, 1.

Leighton, R. B., Noyes, R. W., and Simon, G. W. 1962. *Astrophys. J.*, **135**, 474.

Leka, K. D. 1997. *Astrophys. J.*, **484**, 900.

Leka, K. D., and Skumanich, A. 1998. *Astrophys. J.*, **507**, 454.

Lin, C.-H., Banerjee, D., Doyle, J. G., and O'Shea, E. 2005. *Astron. Astrophys.*, **444**, 585.

Lin, H. and Rimmele, T. 1999. *Astrophys. J.*, **514**, 448.

Lindsey, C., and Braun, D. C. 1990. *Sol. Phys.*, **126**, 101.

Lindsey, C., and Braun, D. C. 1997. *Astrophys. J.*, **485**, 895.

Lites, B. W. 1992. In *Sunspots: Theory and Observations*, ed. J. H. Thomas and N. O. Weiss (Dordrecht: Kluwer), p. 261.

Lites, B. W. 2002. *Astrophys. J.*, **573**, 431.

Lites, B. W., Bida, T. A., Johannesson, A., and Scharmer, G. B. 1991. *Astrophys. J.*, **373**, 683.

Lites, B. W., Elmore, D. F., Seagraves, P., and Skumanich, A. P. 1993. *Astrophys. J.*, **418**, 928.

Lites, B. W., Leka, K. D., Skumanich, A., Martínez Pillet, V., and Shimizu, T. 1996. *Astrophys. J.*, **460**, 1019.

Lites, B. W., Low, B. C., Martínez Pillet, V., *et al.* 1995. *Astrophys. J.*, **446**, 877.

Lites, B. W., Scharmer, G. B., Berger, T. E., and Title, A. M. 2004. *Sol. Phys.*, **221**, 65.

Lites, B. W., and Skumanich, A. 1990. *Astrophys. J.*, **348**, 747.

Lites, B. W., Skumanich, A., Rees, D. E., Murphy, G. A., and Carlsson, M. 1987. *Astrophys. J.*, **318**, 930.

Lites, B. W., Socas-Navarro, H., Kubo, M., *et al.* 2007. *Pub. Astron. Soc. Japan*, **59**, S571.

Lites, B. W., Socas-Navarro, H., Skumanich, A., and Shimizu, T. 2002. *Astrophys. J.*, **575**, 1131.

Lites, B. W., and Thomas, J. H. 1985. *Astrophys. J.*, **294**, 682.

Lites, B. W., Thomas, J. H., Bogdan, T. J., and Cally, P. S. 1998. *Astrophys. J.*, **497**, 464.

Livingston, W. 1991. *Nature*, **350**, 45.

Livingston, W., and Harvey, J. W. 1971. In IAU Symp. 43, *Solar Magnetic Fields*, ed. R. Howard (Dordrecht: Reidel), p. 51.

Livingston, W., Harvey, J. W., Malanushenko, O. V., and Webster. L. 2006. *Sol. Phys.*, **239**. 41.

Livingston, W., and Mahaffey, C. 1981. In *The Physics of Sunspots*, ed. L. E. Cram and J. H. Thomas (Sunspot, NM: Sacramento Peak Observatory), p. 312.

Lockwood, M. 2003. *J. Geophys. Res.*, **108**, 1128.

Lockwood, M. J., and Hapgood, M. A. 2007. *Astron. Geophys.*, **48**, 6.11.

Longcope, D. W., and Welsch, B. T. 2000. *Astrophys. J.*, **545**, 1089.

Lorente, R., and Montesinos, B. 2005. *Astrophys. J.*, **632**, 1104.

Lou, Y.-Q. 1990. *Astrophys. J.*, **350**, 452.

Low, B. C. 1980. *Sol. Phys.*, **67**, 57.

Lustig, G., and Wöhl, H. 1995. *Sol. Phys.*, **157**, 389.

Lyne, A. G. 2000. *Phil. Trans. Roy. Soc. Lond. A*, **358**, 831.

Lyot, B. 1933. *C. R. Acad. Sci.*, **197**, 1593.

Mackay, D. H., Jardine, M., Collier Cameron, A., Donati, J.-F., and Hussain, G. A. J. 2004. *Mon. Not. Roy. Astron. Soc.*, **354**, 737.

Magara, T. 2001. *Astrophys. J.*, **549**, 608.

Magara, T., and Longcope, D. W. 2001. *Astrophys. J.*, **559**, L55.

Magara, T., and Longcope, D. W. 2003. *Astrophys. J.*, **586**, 630.

Makita, M., and Morimoto, M. 1960. *Pub. Astron. Soc. Japan*, **12**, 63.

Malkus, W. V. R., and Proctor, M. R. E. 1975. *J. Fluid Mech.*, **67**, 417.

Maltby, P. 1964. *Astrophys. Norvegica*, **8**, 205.

Maltby, P. 1992. In *Sunspots: Theory and Observations*, ed. J. H. Thomas and N. O. Weiss (Dordrecht: Kluwer), p. 103.

Maltby, P., Avrett, E. H., Carlsson, M., *et al.* 1986. *Astrophys. J.*, **306**, 284.

Maltby, P., Brynildsen, N., Fredvik, T., Kjeldseth-Moe, O., and Wilhelm, K. 1999. *Sol. Phys.*, **190**. 437.

Maltby, P., Brynildsen, N., Kjeldseth-Moe, O., and Wilhelm, K. 2001. *Astron. Astrophys.*, **373**, L1.

Manchester, W. IV, Gombosi, T., DeZeeuw, T., and Fan, Y. 2004. *Astrophys. J.*, **610**, 588.

Mann, M. E., and Jones, P. D. 2003. *Geophys. Res. Lett.*, **30**, 1820.

Marco, E., Aballe Villero, M. A., Vázquez, M., and García de la Rosa, J. I. 1996. *Astron. Astrophys.*, **309**, 284.

Markiel, J. A., and Thomas, J. H. 1999. *Astrophys. J.*, **523**, 827.

Márquez, I., Sánchez Almeida, J., and Bonet, J. A. 2006. *Astrophys. J.*, **638**, 553.

Marsden, S. C., Donati, J.-F., Semel, M., Petit, P., and Carter, B. D. 2006. *Mon. Not. Roy. Astron. Soc.*, **370**, 468.

Marsh, N. D., and Svensmark, H. 2000. *Phys. Rev. Lett.*, **85**, 5004.

Martínez Pillet, V. 1997. In ASP Conf. Ser. 118, *Advances in the Physics of Sunspots*, ed. B. Schmieder, J. C. del Toro Iniesta and M. Vázquez (San Francisco: Astron. Soc. Pacific), p. 212.

Martínez Pillet, V. 2002. *Astron. Nachr.*, **323**, 342.

Martínez Pillet, V., Lites, B. W., and Skumanich, A. 1997. *Astrophys. J.*, **474**, 810.

Martínez Pillet, V., Moreno-Insertis, F., and Vázquez, M. 1993. *Astron. Astrophys.*, **274**, 521.

Martínez Pillet, V., and Vázquez, M. 1990. *Astrophys. Space Sci.*, **170**, 75.

Martínez Pillet, V., and Vázquez, M. 1993. *Astron. Astrophys.*, **270**, 494.

Mason, J., Hughes, D. W., and Tobias, S. M. 2002. *Astrophys. J.*, **580**, L89.

Massaglia, S., Bodo, G., and Rossi, P. 1989. *Astron. Astrophys.*, **209**, 399.

Mathew, S. K., Lagg, A., Solanki, S. K., *et al.* 2003. *Astron. Astrophys.*, **410**, 695.

Mathew, S. K., Martínez Pillet, V., Solanki, S. K., and Krivova, N. A. 2007. *Astron. Astrophys.*, **465**, 291.

Mathys, G., Hubrig, S., Landstreet, J. D., Lanz, T., and Manfroid, J. 1997. *Astron. Astrophys. Suppl.*, **123**, 353.

Matsumoto, R., Tajima, T., Shibata, K., and Kaisig, M. 1993. *Astrophys. J.*, **414**, 357.

Matthews, P. C., Hurlburt, N. E., Proctor, M. R. E., and Brownjohn, D. P. 1992. *J. Fluid Mech.*, **240**, 559.

Maunder, E. W. 1890. *Mon. Not. Roy. Astron. Soc.*, **50**, 251.

Maunder, E. W. 1904. *Mon. Not. Roy. Astron. Soc.*, **64**, 747.

Maunder, E. W. 1922a. *Mon. Not. Roy. Astron. Soc.*, **82**, 534.

Maunder, E. W. 1922b. *J. Brit. Astron. Assoc.*, **32**, 140.

McCracken, K. G., McDonald, F. B., Beer, J., Raisbeck, G., and Yiou, F. 2004. *J. Geophys. Res.*, **109**, A12103.

McIntosh, P. S. 1981. In *The Physics of Sunspots*, ed. L. E. Cram and J. H. Thomas (Sunspot, NM: Sacramento Peak Observatory), p. 7.

Mehltretter, J. P. 1974. *Sol. Phys.*, **38**, 43.

Messina, S., Cutispoto, G., Guinan, E. F., Lanza, A. F., and Rodonò, M. 2006. *Astron. Astrophys.*, **447**, 293.

Messina, S., and Guinan, E. F. 2002. *Astron. Astrophys.*, **393**, 225.

Mestel, L. 1999. *Stellar Magnetism* (Oxford: Clarendon Press).

Mestel, L. 2003. In *Stellar Astrophysical Fluid Dynamics*, ed. M. J. Thompson and J. Christensen-Dalsgaard (Cambridge: Cambridge University Press), p. 75.

Mestel, L., and Landstreet, J. D. 2005. In Lecture Notes in Physics, **664**, *Cosmic Magnetic Fields,* ed. R. Wielebinski and R. Beck (Berlin: Springer), p. 183.

Meyer, F., and Schmidt, H. U. 1968. *Zeits. Ang. Math. Mech.*, **48**, T218.

Meyer, F., Schmidt, H. U., and Weiss, N. O. 1977. *Mon. Not. Roy. Astron. Soc.*, **179**, 741.

Meyer, F., Schmidt, H. U., Weiss, N. O., and Wilson, P. R. 1974. *Mon. Not. Roy. Astron. Soc.*, **169**, 35.

Miesch, M. S. 2005. *Living Revs Sol. Phys.*, **2**, 1. (www.livingreviews.org/lrsp-2005-1)

Miesch, M. S., Elliott, J. R., Toomre, J., *et al.* 2000. *Astrophys. J.*, **533**, 546.

Mihalas, D. 1978. *Stellar Atmospheres*, 2nd edn (San Francisco: Freeman).

Minnaert, M., and Wanders, A. J. M. 1932. *Zeits. Astrophys.*, **5**, 297.

Moberg, A., Sonechkin, D. M., Holmgren, K., Datsenko, N. M., and Kartén, W. 2005. *Nature*, **433**, 613.

Moffatt, H. K. 1978. *Magnetic Field Generation in Electrically Conducting Fluids* (Cambridge: Cambridge University Press).

Moffatt, H. K. 1983. *Rep. Prog. Phys.*, **46**, 621.

Mogilevskii, E. I., Obridko, V. N., and Shel'ting, B. D. 1973. *Radiofizika*, **16**, 1357.

Montesinos, B., and Thomas, J. H. 1989. *Astrophys. J.*, **337**, 977.

Montesinos, B., and Thomas, J. H. 1993. *Astrophys. J.*, **402**, 314.

Montesinos, B., and Thomas, J. H. 1997. *Nature*, **390**, 485.

Montesinos, B., Thomas, J. H., Ventura, P., and Mazziteli, I. 2001. *Mon. Not. Roy. Astron. Soc.*, **326**, 877.

Moon, Y.-J., Chae, J., Choe, G. S., *et al.* 2002. *Astrophys. J.*, **574**, 1066.

Moore, R. L. 1973. *Sol. Phys.*, **30**, 403.

Moore, R. L. 1981. In *The Physics of Sunspots*, ed. L. E. Cram and J. H. Thomas (Sunspot, NM: Sacramento Peak Observatory), p. 259.

Moore, R. L., and Rabin, D. M. 1985. *Ann. Rev. Astron. Astrophys.*, **23**, 239.

Moreno-Insertis, F. 1983. *Astron. Astrophys.*, **122**, 241.

Moreno-Insertis, F. 1986. *Astron. Astrophys.*, **166**, 291.

Moreno-Insertis, F. 1992. In *Sunspots: Theory and Observations*, ed. J. H. Thomas and N. O. Weiss (Dordrecht: Kluwer), p. 385.

Moreno-Insertis, F., Caligari, P., and Schüssler, M. 1995. *Astrophys. J.*, **452**, 894.

Moreno-Insertis, F., and Spruit, H. C. 1989. *Astrophys. J.*, **342**, 1158.

Moss, D., and Brooke, J. 2000. *Mon. Not. Roy. Astron. Soc.*, **315**, 521.

Moss, D., and Sokoloff, D. 2007. *Mon. Not. Roy. Astron. Soc.*, **377**, 1597.

Mossman, J. E. 1989. *Quart. J. Roy. Astron. Soc.*, **30**, 59.

Muglach, K. 2003. *Astron. Astrophys.*, **401**, 685.

Muglach, K., Solanki, S. K., and Livingston, W. C. 1994. In *Solar Surface Magnetism*, ed. R. J. Rutten and C. J. Schrijver (Dordrecht: Kluwer), p. 127.

Mulders, E. S. 1943. *Pub. Astron. Soc. Pacific*, **55**, 21.

Mullan, D. J., and Yun, H. S. 1973. *Sol. Phys.*, **30**, 83.

Muller, R. 1973a. *Sol. Phys.*, **29**, 55.

Muller, R. 1973b. *Sol. Phys.*, **32**, 409.

Muller, R. 1983. *Sol. Phys.*, **85**, 113.

Muller, R. 1992. In *Sunspots: Theory and Observations*, ed. J. H. Thomas and N. O. Weiss (Dordrecht: Kluwer), p. 175.

Muller, R., and Mena, B. 1987. *Sol. Phys.*, **112**, 295.

Musman, S. 1967. *Astrophys. J.*, **149**, 201.

Musman, S., Nye, A. H., and Thomas, J. H. 1976. *Astrophys. J.*, **206**, L175.

Nagashima, K., Sekii, T., Kosovichev, A. G., *et al.* 2007. *Pub. Astron. Soc. Japan*, **59**, S631.

Nakagawa, Y., and Raadu, M. A. 1972. *Sol. Phys.*, **25**, 127.

Nakagawa, Y., Raadu, M. A., Billings, D. E., and McNamara, D. 1971. *Sol. Phys.*, **19**, 72.

Needham, J. 1959. *Science and Civilisation in China. Vol. 3. Mathematics and the Sciences of the Heavens and the Earth* (Cambridge: Cambridge University Press).

Neff, J. E., O'Neal, D., and Saar, S. H. 1995. *Astrophys. J.*, **452**, 879.

Nesme-Ribes, E., Ferreira, E. N., and Mein, P. 1993. *Astron. Astrophys.*, **274**, 563.

Newcomb, W. A. 1961. *Phys. Fluids*, **4**, 391.

Newton, H. W. 1955. *Vistas in Astron.*, **1**, 666.

Nicholson, S. B. 1933. *Pub. Astron. Soc. Pacific*, **45**, 51.

Nicolas, K. R., Kjeldseth-Moe, O., Bartoe, J.-D. F., and Brueckner, G. E. 1982. *Sol. Phys.*, **82**, 253.

Nindos, A., Alissandrakis, C. E., Gelfreikh, G. B., Bogod, V. M., and Gontikakis, C. 2002. *Astron. Astrophys.*, **386**, 658.

Nindos, A., Zhang, J., and Zhang, H. 2003. *Astrophys. J.*, **594**, 1033.

Nordlund, Å. 1983. In IAU Symp. 102, *Solar and Stellar Magnetic Fields: Origins and Coronal Effects*, ed. J. O. Stenflo (Dordrecht: Reidel), p. 79.

Nordlund, Å. 1986. In *Small Scale Magnetic Flux Concentrations in the Solar Photosphere*, ed. W. Deinzer, M. Knölker and H. H. Voigt (Göttingen: Vanderhoeck und Ruprecht), p. 83.

Nordlund, Å., Brandenburg, A., Jennings, R. L., *et al.* 1992. *Astrophys. J.*, **392**, 647.

Nordlund, Å., and Stein, R. F. 1990. In IAU Symp. 138, *Solar Photosphere: Structure, Convection, and Magnetic Fields*, ed. J. O. Stenflo (Dordrecht: Kluwer), p. 191.

Norton, A. A., and Gilman, P. A. 2004. *Astrophys. J.*, **603**, 348.

Norton, A. A., Ulrich, R. K., Bush, R. L., and Tarbell, T. D. 1999. *Astrophys. J.*, **518**, L123.

November, L. J., and Simon, G. W. 1988. *Astrophys. J.*, **333**, 427.

November, L. J., Toomre, J., Gebbie, K. B., and Simon, G. W. 1981. *Astrophys. J.*, **245**, L123.

Noyes, R. W., Hartmann, L. W., Baliunas, S. L., Duncan, D. K., and Vaughan, A. H. 1984. *Astrophys. J.*, **279**, 763.

Noyes, R. W., Raymond, J. C., Doyle, J. G., and Kingston, A. E. 1985. *Astrophys. J.*, **297**, 805.

Noyes, R. W., Vaughan, A. H., and Weiss, N. O. 1984. *Astrophys. J.*, **287**, 769.

Nozawa, S., Shibata, K., Matsumoto, R., *et al.* 1992. *Astrophys. J. Suppl.*, **78**, 267.

Nye, A. H., and Thomas, J. H. 1974. *Sol. Phys.*, **38**, 399.

Nye, A. H., and Thomas, J. H. 1976. *Astrophys. J.*, **204**, 582.

Nye, A. H., Thomas, J. H., and Cram, L. E. 1984. *Astrophys. J.*, **285**, 381.

Obridko, V. N. 1985. *Sunspots and Activity Complexes* (Moscow: Nauka), in Russian.

Oláh, K., Kolláth, Z., and Strassmeier, K. G. 2000. *Astron. Astrophys.*, **346**, 643.

Oláh, K., Panov, K. P., Pettersen, B. R., Valtaoja, E., and Valtaoja, L. 1989. *Astron. Astrophys.*, **218**, 192.

Oláh, K., and Strassmeier, K. G. 2002. *Astron. Nachr.*, **323**, 361.

Oláh, K., Strassmeier, K. G., Granzer, T., Soon, W., and Baliunas, S. L. 2007. *Astron. Nachr.*, **328**, 1072.

O'Neal, D. 2006. *Astrophys. J.*, **645**, 659.

O'Neal, D., and Neff, J. E. 1997. *Astron. J.*, **113**, 1129.

O'Neal, D., Neff, J. E., and Saar, S. H. 1998. *Astrophys. J.*, **507**, 919.

O'Neal, D., Neff, J. E., Saar, S., and Cuntz, M. 2004a. *Astron. J.*, **128**, 1802.

O'Neal, D., Neff, J. E., Saar, S., and Mines, J. K. 2001. *Astron. J.*, **122**, 1954.

O'Neal, D., Saar, S. H., Aufdenberg, J., and Neff, J. E. 2004b. In IAU Symp. 219, *Stars as Suns: Activity, Evolution, and Planets*, ed. A. K. Dupree and A. O. Benz (San Francisco: Astron. Soc. Pacific), p. 957.

O'Neal, D., Saar, S. H., and Neff, J. E. 1996. *Astrophys. J.*, **463**, 766.

Orozco Suárez, D., Bellot Rubio, L. R., del Toro Iniesta, J. C., *et al.* 2007. *Astrophys. J.*, **670**, L61.

Ortiz, A., Solanki, S. K., and Domingo, V. 2002. *Astron. Astrophys.*, **388**, 1036.

O'Shea, E., Muglach, K., and Fleck, B. 2002. *Astron. Astrophys.*, **387**, 642.

Oskanyan, V. S., Evans, D. S., Lacy, C., and McMillan, R. S. 1977. *Astrophys. J.*, **214**, 430.

Ossendrijver, A. J. H. M. 1997. *Astron. Astrophys.*, **323**, 151.

Ossendrijver, M. 2000. *Astron. Astrophys.*, **359**, 364.

Ossendrijver, M. 2003. *Astron. Astrophys. Rev.*, **11**, 287.

Ossendrijver, M., Stix, M., and Brandenburg, A. 2001. *Astron. Astrophys.*, **376**, 713.

Ossendrijver, M., Stix, M., Brandenburg, A., and Rüdiger, G. 2002. *Astron. Astrophys.*, **394**, 735.

Parchevsky, K. V., and Kosovichev, A. 2007. *Astrophys. J.*, **666**, L53.

Parker, E. N. 1955a. *Astrophys. J.*, **121**, 491.

Parker, E. N. 1955b. *Astrophys. J.*, **122**, 293.

Parker, E. N. 1958. *Astrophys. J.*, **128**, 664.

Parker, E. N. 1960. *Astrophys. J.*, **132**, 175.

Parker, E. N. 1963. *Astrophys. J.*, **138**, 552.

Parker, E. N. 1966. *Astrophys. J.*, **145**, 811.

Parker, E. N. 1974. *Sol. Phys.*, **36**, 249.

Parker, E. N. 1975. *Sol. Phys.*, **40**, 291.

Parker, E. N. 1978. *Astrophys. J.*, **221**, 368.

Parker, E. N. 1979a. *Cosmical Magnetic Fields: their Origin and their Activity* (Oxford: Clarendon Press).

Parker, E. N. 1979b. *Astrophys. J.*, **230**, 905.

Parker, E. N. 1984. *Astrophys. J.*, **280**, 423.

Parker, E. N. 1992. *Astrophys. J.*, **390**, 290.

Parker, E. N. 1993. *Astrophys. J.*, **408**, 707.

Parnell, C. 2001. *Sol. Phys.*, **200**, 23.

Pastorff, J. W. 1828. *Astron. Nachr.*, **6**, 291.

Penn, M. J., and Kuhn, J. R. 1995. *Astrophys. J.*, **441**, L51.

Penn, M. J., and LaBonte, B. J. 1993. *Astrophys. J.*, **415**, 383.

Penn, M. J., and Livingston, W. 2006. *Astrophys. J.*, **649**, L45.

Penn, M. J., Walton, S., Chapman, G., Ceja, J., and Plick, W. 2003. *Sol. Phys.*, **213**, 55.

Peter, H. 1996. *Mon. Not. Roy. Astron. Soc.*, **278**, 821.

Petit, P., Donati, J.-F., and Collier Cameron, A. 2004. *Astron. Nachr.*, **325**, 221.

Petit, P., Donati, J.-F., Wade, G. A., *et al.* 2004a. *Mon. Not. Roy. Astron. Soc.*, **348**, 1175.

Petit, P., Donati, J.-F., Oliveira, J. M., *et al.* 2004b. *Mon. Not. Roy. Astron. Soc.*, **351**, 826.

Petrovay, K., and Moreno-Insertis, F. 1997. *Astrophys. J.*, **485**, 398.

Petrovay, K., and van Driel-Gesztelyi, L. 1997. *Sol. Phys.*, **176**, 249.

Pevtsov, A., Maleev, V. M., and Longcope, D. W. 2003. *Astrophys. J.*, **593**, 1217.

Phillips, M. J., and Hartmann, L. 1978. *Astrophys. J.*, **224**, 182.

Pickering, E. C. 1880. *Proc. Am. Acad. Arts Sci.*, **16**, 257.

Pipin, V. V. 1999. *Astron. Astrophys.*, **346**, 295.

Piskunov, N. E. 1991. In IAU Coll. 130, *The Sun and Cool Stars. Activity, Magnetism, Dynamos*, ed. I. Tuominen, D. Moss and G. Rüdiger (Berlin: Springer), p. 309.

Piskunov, N. E., Huenemoerder, D. P., and Saar, S. H. 1994. In ASP Conf. Ser. 64, *8th Cambridge Workshop on Cool Stars, Stellar Systems, and the Sun*, ed. J. P. Caillault (San Francisco: Astron. Soc. Pacific), p. 658.

Pizzo, V. J. 1990. *Astrophys. J.*, **365**, 764.

Plaskett, H. H. 1936. *Mon. Not. Roy. Astron. Soc.*, **96**, 402.

Poe, C. H., and Eaton, J. A. 1985. *Astrophys. J.*, **289**, 644.

Popper, D. M. 1953. *Pub. Astron. Soc. Pacific*, **65**, 278.

Priest, E. R. 1982. *Solar Magnetohydrodynamics* (Dordrecht: Reidel).

Proctor, M. R. E. 1992. In *Sunspots: Theory and Observations*, ed. J. H. Thomas and N. O. Weiss (Dordrecht: Kluwer), p. 221.

Proctor, M. R. E. 2005. In *Fluid Dynamics and Dynamos in Astrophysics and Geophysics*, ed. A. M. Soward, C. A. Jones, D. W. Hughes and N. O. Weiss (Boca Raton: CRC Press), p. 235.

Proctor, M. R. E., and Weiss, N. O. 1982. *Rep. Prog. Phys.*, **45**, 1317.

Prosser, C. F. 1992. *Astron. J.*, **103**, 488.

Pulkkinen, P., and Tuominen, I. 1998. *Astron. Astrophys.*, **332**, 748.

Pulkkinen, T. 2007. *Living Revs Sol. Phys.*, **4**, 1. (www.livingreviews.org/lrsp-2007-1)

Puschmann, K. G., Ruiz Cobo, B., Vázquez, M., Bonet, J. A., and Hanslmeier, A. 2005. *Astron. Astrophys.*, **441**, 1157.

Radick, R. R., Thompson, G. T., Lockwood, G. W., Duncan, D. K., and Baggett, W. E. 1987. *Astrophys. J.*, **321**, 459.

Rädler, K.-H. 1968. *Zeits. Naturforsch.*, **23**, 1851.

Raisbeck, G. M., Yiou, F., Jouzel, J., and Petit, J. R. 1990. *Phil. Trans. Roy. Soc. Lond. A*, **330**, 463.

Ramsey, L. W., and Nations, H. L. 1980. *Astrophys. J.*, **239**, L121.

Rast, M. P., Fox, P. A., Lin, H., *et al.* 1999. *Nature*, **401**, 678.

Rast, M. P., Meisner, R. W., Lites, B. W., Fox, P. A., and White, O. R. 2001. *Astrophys. J.*, **557**, 864.

Raymond, J. C., and Foukal, P. 1982. *Astrophys. J.*, **253**, 323.

Reid, G. C. 1991. *J. Geophys. Res.*, **96**, 2835.

Reid, G. C. 2000. *Space Sci. Rev.*, **94**, 1.

Reimer, P. J., Baillie, M. G. L., Bard, E., *et al.* 2004. *Radiocarbon*, **46**, 1029.

Rempel, M. 2006. *Astrophys. J.*, **647**, 662.

Rempel, M., and Schüssler, M. 2001. *Astrophys. J.*, **552**, L171.

Rezaei, R., Schlichenmaier, R., Beck, C., and Bellot Rubio, L. R. 2006. *Astron. Astrophys.*, **454**, 975.

Ribárik, G., Oláh, K., and Strassmeier, K. G. 2002. In *Sunspots and Starspots*, Poster Proceedings,
    ed. K. G. Strassmeier and A. Washuettl (Potsdam: Astrophysical Institute Potsdam), p. 7.

Ribes, J. C., and Nesme-Ribes, E. 1993. *Astron. Astrophys.*, **276**, 549.

Rice, J. B. 2002. *Astron. Nachr.*, **323**, 220.

Rice, J. B., and Gaizauskas, V. 1973. *Sol. Phys.*, **32**, 421.

Rice, J. B., and Strassmeier, K. G. 1996. *Astron. Astrophys.*, **316**, 164.

Richardson, R. S. 1941. *Astrophys. J.*, **93**, 24.

Rimmele, T. R. 1994. *Astron. Astrophys.*, **290**, 972.

Rimmele, T. R. 1995a. *Astron. Astrophys.*, **298**, 260.

Rimmele, T. R. 1995b. *Astrophys. J.*, **445**, 511.

Rimmele, T. R. 1997. *Astrophys. J.*, **490**, 458.

Rimmele, T. R. 2004. *Astrophys. J.*, **604**, 906.

Rimmele, T., and Marino, J. 2006. *Astrophys. J.*, **646**, 593.

Rincon, F., Lignières, F., and Rieutord, M. 2005. *Astron. Astrophys.*, **430**, L57.

Ringnes, T. S. 1964. *Astrophys. Norvegica*, **8**, 303.

Roberts, B. 1992. In *Sunspots: Theory and Observations*, ed. J. H. Thomas and N. O. Weiss (Dordrecht: Kluwer),
    p. 303.

Roberts, P. H. 1967. *An Introduction to Magnetohydrodynamics* (London: Longmans).

Roberts, P. H. 1994. In *Lectures on Solar and Planetary Dynamos*, ed. M. R. E. Proctor and A. D. Gilbert
    (Cambridge: Cambridge University Press), p. 1.

Roberts, P. H., and Stix, M. 1972. *Astron. Astrophys.*, **18**, 453.

Robinson, R. D. 1980. *Astrophys. J.*, **239**, 961.

Robinson, R. D., Worden, S. P., and Harvey, J. W. 1980. *Astrophys. J.*, **236**, L155.

Rodonò, M., Cutispoto, G., Pazzani, V., *et al.* 1986. *Astron. Astrophys.*, **165**, 135.

Rösch, J. 1959. *Ann. Astrophys.*, **22**, 571.

Rosen, E. 1965. *Kepler's Conversation with Galileo's Sidereal Messenger* (New York: Johnson Reprint Corp.).

Rosenthal, C. S. 1990. *Sol. Phys.*, **130**, 313.

Rosenthal, C. S. 1992. *Sol. Phys.*, **139**, 25.

Rosenthal, C. S., and Julien, K. A. 2000. *Astrophys. J.*, **532**, 1230.

Rosner, R. 2000. *Phil. Trans. Roy. Soc. Lond. A*, **358**, 689.

Rossbach, M., and Schröter, E. H. 1970. *Sol. Phys.*, **12**, 95.

Rouppe van der Voort, L. H. M. 2002. *Astron. Astrophys.*, **389**, 1020.

Rouppe van der Voort, L. H. M. 2003. *Astron. Astrophys.*, **397**, 757.

Rouppe van der Voort, L. H. M., Hansteen, V. H., Carlsson, M., *et al.* 2005. *Astron. Astrophys.*, **435**, 327.

Rouppe van der Voort, L. H. M., Löfdahl. M. G., Kiselman, D., and Scharmer, G. 2004. *Astron. Astrophys.*,
    414, 717.

Rouppe van der Voort, L. H. M., Rutten, R. J., Sütterlin, P., Sloover, P. J., and Krijger, J. M. 2003. *Astron.
    Astrophys.*, **403**, 277.

Roy, J.-R. 1973. *Sol. Phys.*, **28**, 95.

Rucklidge, A. M., Schmidt, H. U., and Weiss, N. O. 1995. *Mon. Not. Roy. Astron. Soc.*, **273**, 491.

Rucklidge, A. M., Weiss, N. O., Brownjohn, D. P., Matthews, P. C., and Proctor, M. R. E. 2000. *J. Fluid Mech.*,
    419, 283.

Rüdiger, G., and Brandenburg, A. 1995. *Astron. Astrophys.*, **296**, 557.

Rüdiger, G., and Hollerbach, R. 2004. *The Magnetic Universe* (Weinheim: Wiley-VCH).

Rüdiger, G., and Kitchatinov, L. L. 2000. *Astron. Nachr.*, **321**, 75.

Rüdiger, G., and Küker, M. 2002. *Astron. Astrophys.*, **385**, 308.

Rüedi, I., and Cally, P. S. 2003. *Astron. Astrophys.*, **410**, 1023.

Rüedi, I., Solanki, S. K., and Livingston, W. 1995. *Astron. Astrophys.*, **302**, 543.

Rüedi, I., Solanki, S. K., Stenflo, J. O., Tarbell, T., and Scherrer, P. H. 1998. *Astron. Astrophys.*, **335**, L97.

Rutten, R. J., de Wijn, A. G., and Sütterlin, P. 2004. *Astron. Astrophys.*, **416**, 333.

Ruzmaikin, A., and Feynman, J. 2002. *J. Geophys. Res.*, **107**, 4209.

Ruzmaikin, A., Feynman, J., Jiang, X., *et al.* 2004. *Geophys. Res. Lett.*, **31**, L12201.

Ryutova, M., Kaisig, M., and Tajima, T. 1991. *Astrophys. J.*, **380**, 268.

Saar, S. H. 1996. In *Magnetohydrodynamic Phenomena in the Solar Atmosphere: Prototypes of Stellar Magnetic Activity*, ed. Y. Uchida, T. Kosugi and H. S. Hudson (Dordrecht: Kluwer), p. 367.

Saar, S. H. 2002. In ASP Conf. Ser. 277, *Stellar Coronae in the Chandra and XMM-NEWTON Era*, ed. F. Favata and J. J. Drake (San Francisco: Astron. Soc. Pacific), p. 311.

Saar, S. H., and Brandenburg, A. 1999. *Astrophys. J.*, **524**, 295.

Sainz Dalda, A., and Martínez Pillet, V. 2005. *Astrophys. J.*, **632**, 1176.

Sakurai, T., Goossens, M., and Hollweg, J. V. 1991. *Sol. Phys.*, **133**, 247.

Sams, B. J. III, Golub, L., and Weiss, N. O. 1992. *Astrophys. J.*, **399**, 313.

Sánchez Almeida, J., and Bonet, J. A. 1998. *Astrophys. J.*, **505**, 1010.

Sánchez Almeida, J., Márquez, I., and Bonet, J. A. 2004. *Astrophys. J.*, **609**, L91.

Savage, B. D. 1969. *Astrophys. J.*, **156**, 707.

Scharmer, G. B., Gudiksen, B. V., Kiselman, D., Löfdahl, M. G., and Rouppe van der Voort, L. H. M. 2002. *Nature*, **420**, 151.

Scharmer, G. B., and Spruit, H. C. 2006. *Astron. Astrophys.*, **460**, 605.

Scheuer, M. A., and Thomas, J. H. 1981. *Sol. Phys.*, **71**, 21.

Schlichenmaier, R. 2002. *Astron. Nachr.*, **323**, 303.

Schlichenmaier, R., Bellot Rubio, L. R., and Tritschler, A. 2005. *Astron. Astrophys.*, **326**, 301.

Schlichenmaier, R., Jahn, K., and Schmidt, H. U. 1998a. *Astrophys. J.*, **493**, L121.

Schlichenmaier, R., Jahn, K., and Schmidt, H. U. 1998b. *Astron. Astrophys.*, **337**, 897.

Schlichenmaier, R., and Schmidt, W. 1999. *Astron. Astrophys.*, **349**, L37.

Schlichenmaier, R., and Schmidt, W. 2000. *Astron. Astrophys.*, **358**, 1122.

Schlüter, A., and Temesváry, S. 1958. In IAU Symp. 6, *Electromagnetic Phenomena in Cosmical Physics*, ed. B. Lehnert (Cambridge: Cambridge University Press), p. 263.

Schmidt, H. U. 1991. *Geophys. Astrophys. Fluid Dyn.*, **62**, 249.

Schmidt, H. U., and Wegmann, R. 1983. In *Dynamical Problems in Mathematical Physics*, ed. B. Brosowski and E. Martensen (Frankfurt: Lang), p. 137.

Schmidt, W., Hofmann, A., Balthasar, H., Tarbell, T. D., and Frank, Z. A. 1992. *Astron. Astrophys.*, **264**, L27.

Schmidt, W., and Schlichenmaier, R. 2000. *Astron. Astrophys.*, **364**, 829.

Schmieder, B., del Toro Iniesta, J. C., and Vázquez, M. (eds) 1997. *Advances in the Physics of Sunspots*. Astron. Soc. Pacific Conf. Series, **118**.

Schmieder, B., Raadu, M. A., Démoulin, P., and Dere, K. P. 1989. *Astron. Astrophys.*, **213**, 402.

Schmitt, D. 1987. *Astron. Astrophys.*, **174**, 281.

Schmitt, D., Schüssler, M., and Ferriz-Mas, A. 1996. *Astron. Astrophys.*, **311**, L1.

Schmitt, J. H. M. M., and Liefke, C. 2004. *Astron. Astrophys.*, 417, 651.

Schrijver, C. J. 2002. *Astron. Nachr.*, **323**, 157.

Schrijver, C. J., and Title, A. M. 2001. *Astrophys. J.*, **551**, 1099.

Schrijver, C. J., Title, A. M., Berger, T. E., *et al.* 1999. *Sol. Phys.*, **187**, 261.

Schrijver, C. J., Title, A. M., Harvey, K. L., *et al.* 1998. *Nature*, **394**, 152.

Schrijver, C. J., and Zwaan, C. 2000. *Solar and Stellar Magnetic Activity* (Cambridge: Cambridge University Press).

Schultz, R. B. 1974. *A spectroscopic study of dynamic phenomena in sunspots*. Ph.D. dissertation, University of Colorado.

Schultz, R. B., and White, O. R. 1974. *Sol. Phys.*, **35**, 309.

Schunker, H., and Cally, P. S. 2006. *Mon. Not. Roy. Astron. Soc.*, **372**, 551.

Schüssler, M. 1980. *Nature*, **288**, 150.

Schüssler, M. 1981. *Astron. Astrophys.*, **94**, L17.

Schüssler, M. 1990. In IAU Symp. 138, *Solar Photosphere: Structure, Convection and Magnetic Fields*, ed. J. O. Stenflo (Dordrecht: Kluwer), p. 161.

Schüssler, M. 1991. *Geophys. Astrophys. Fluid Dyn.*, **62**, 271.

Schüssler, M. 2001. In ASP Conf. Ser. 236, *Avanced Solar Polarimetry: Theory, Observation and Instrumentation*, ed. M. Sigwarth (San Francisco: Astron. Soc. Pacific), p. 343.

Schüssler, M. 2005. *Astron. Nachr.*, **326**, 194.

Schüssler, M., Caligari, P., Ferriz-Mas, A., Solanki, S. K., and Stix, M. 1996. *Astron. Astrophys.*, **314**, 503.

Schüssler, M., and Knölker, M. 2001. In ASP Conf. Ser. 248, *Magnetic Fields across the Hertzsprung-Russell Diagram*, ed. G. Mathys, S. K. Solanki and D. T. Wickramasinghe (San Francisco: Astron. Soc. Pacific), p. 115.

Schüssler, M., and Solanki, S. K. 1992. *Astron. Astrophys.*, **264**, L13.

Schüssler, M., and Vögler, A. 2006. *Astrophys. J.*, **641**, L73.

Schuster, A. 1891. *Rep. Brit. Assoc. Adv. Sci. 1891*, p. 634.

Schwabe, H. 1843. *Astron. Nachr.*, **20**, No. 295.

Schwarzschild, K., and Villiger, W. 1906. *Astrophys. J.*, **23**, 284.

Schwarzschild, M. 1959. *Astrophys. J.*, **130**, 345.

Schwenn, R. 2006. *Living Revs Sol. Phys.*, **3**, 2.
(www.livingreviews.org/lrsp-2006-2)

Secchi, P. A. 1870. *Le Soleil* (Paris: Gauthier-Villars); also *Die Sonne* (Braunschweig: Westermann Verlag).

Semel, M. 1989. *Astron. Astrophys.*, **225**, 456.

Settele, A., Sigwarth, M., and Muglach, K. 2002. *Astron. Astrophys.*, **392**, 1095.

Severino, G., Gomez, M. T., and Caccin, B. 1994. In *Solar Surface Magnetism*, ed. R. J. Rutten and C. J. Schrijver (Dordrecht: Kluwer), p. 169.

Sheeley, N. R. Jr. 1969. *Sol. Phys.*, **9**, 347.

Sheeley, N. R. Jr. 1972. *Sol. Phys.*, **25**, 98.

Sheeley, N. R. Jr. 2005. *Living Revs Sol. Phys.*, **2**, 5.
(www.livingreviews.org/lrsp-2005-5)

Sheeley, N. R. Jr., and Bhatnagar, A. 1971. *Sol. Phys.*, **19**, 338.

Shi, J. H., and Zhao, G. 2003. *Acta Astron. Sin.*, **44** Suppl., 179.

Shibasaki, K. 2001. *Astrophys. J.*, **550**, 1113.

Shibata, K., Nozawa, S., Matsumoto, R., Sterling, A. C., and Tajima, T. 1990. *Astrophys. J.*, **351**, L25.

Shibata, K., Tajima, T., Steinolfson, R. S., and Matsumoto, R. 1989. *Astrophys. J.*, **345**, 584.

Shimizu, T., Shine, R. A., Title, A. M., Tarbell, T. D., and Frank, Z. 2002. *Astrophys. J.*, **574**, 1074.

Shine, R. A., Simon, G. W., and Hurlburt, N. E. 2000. *Sol. Phys.*, **193**, 313.

Shine, R. A., and Title, A. M. 2001. In *Encyclopedia of Astronomy and Astrophysics* (London: Nature Publishing Group), p. 3209.

Shine, R. A., Title, A. M., Tarbell, T. D., *et al.* 1994. *Astrophys. J.*, **430**, 413.

Shine, R. A., Title, A. H., Tarbell, T. D., and Topka, K. P. 1987. *Science*, **238**, 1264.

Sider, D., and Brunschön, C. W. 2007. *Theophrastus of Eresus on Weather Signs* (Leiden: Brill).

Simon, G. W., and Leighton, R. B. 1964. *Astrophys. J.*, **140**, 1120.

Simon, G. W., Title, A. M., Topka, K. P., *et al.* 1988. *Astrophys. J.*, **327**, 964.

Simon, G. W., Title, A. M., and Weiss, N. O. 1995. *Astrophys. J.*, **442**, 886.

Simon, G. W., Title, A. M., and Weiss, N. O. 2001. *Astrophys. J.*, **561**, 427.

Simon, G. W., and Weiss, N. O. 1970. *Sol. Phys.*, **13**, 85.

Simon, G. W., Weiss, N. O., and Nye, A. H. 1983. *Sol. Phys.*, **87**, 65.

Simon, G. W., and Wilson, P. R. 1985. *Astrophys. J.*, **295**, 241.

Skumanich, A. 1972. *Astrophys. J.*, **171**, 565.

Skumanich, A. 1992. In *Sunspots: Theory and Observations*, ed. J. H. Thomas and N. O. Weiss (Dordrecht: Kluwer), p. 121.

Skumanich, A. 1999. *Astrophys. J.*, **512**, 975.

Skumanich, A., Smythe, C., and Frazier, E. N. 1975. *Astrophys. J.*, **200**, 747.

Small, L. M., and Roberts, B. 1984. In *Hydromagnetics of the Sun*, ESA SP-220, p. 257.

Snodgrass, H. B. 1984. *Sol. Phys.*, **94**, 13.

Sobotka, M. 1988. *Bull. Astron. Inst. Czechosl.*, **39**, 236.

Sobotka, M. 1997. In ASP Conf. Ser. 184, *3rd Advances in Solar Physics Euroconference: Magnetic Fields and Oscillations*, ed. B. Schmieder, A. Hofmann and J. Staude (San Francisco: Astron. Soc. Pacific), p. 155.

Sobotka, M. 2002. In *Solar Variability: from Core to Outer Frontiers*, ed. A. Wilson (Noordwijk: ESA SP-506), p. 381.

Sobotka, M., Bonet, J. A., and Vázquez, M. 1993. *Astrophys. J.*, **415**, 832.

Sobotka, M., Bonet, J. A., Vázquez, M., and Hanslmeier, A. 1995. *Astrophys. J.*, **447**, L133.

Sobotka, M., Brandt, P. N., and Simon, G. W. 1997a. *Astron. Astrophys.*, **328**, 682.

Sobotka, M., Brandt, P. N., and Simon, G. W. 1997b. *Astron. Astrophys.*, **328**, 689.

Sobotka, M., Brandt, P. N., and Simon, G. W. 1999. *Astron. Astrophys.*, **348**, 621.

Sobotka, M., and Hanslmeier, A. 2005. *Astron. Astrophys.*, **442**, 323.

Sobotka, M., and Sütterlin, P. 2001. *Astron. Astrophys.*, **380**, 714.

Sobotka, M., Vázquez, M., Bonet, J. A., Hanslmeier, A., and Hirzberger, J. 1999. *Astrophys. J.*, **511**, 436.

Socas-Navarro, H. 2001. In ASP Conf. Ser. 236, *Advanced Solar Polarimetry: Theory, Observation, and Instrumentation*, ed. M. Sigwarth (San Francisco: Astron. Soc. Pacific), p. 487.

Socas-Navarro, H., Martínez Pillet, V., and Lites, B. W. 2004. *Astrophys. J.*, **611**, 1139.

Socas-Navarro, H., Martínez Pillet, V., Sobotka, M., and Vázquez, M. 2004. *Astrophys. J.*, **614**, 448.

Soderblom, D. R. 1985. *Astron. J.*, **90**, 2103.

Soderblom, D. R., and Clements, S. D. 1987. *Astron. J.*, **93**, 920.

Soderblom, D. R., Jones, B. F., Balachandran, S., *et al.* 1993c. *Astron. J.*, **106**, 1059.

Soderblom, D. R., Jones, B. F., and Fischer, D. 2001. *Astrophys. J.*, **563**, 334.

Soderblom, D. R., and Mayor, M. 1993. *Astrophys. J.*, **402**, L5.

Soderblom, D. R., Stauffer, J. R., Hudon, J. D., and Jones, B. F. 1993a. *Astrophys. J. Suppl.*, **85**, 315.

Soderblom, D. R., Stauffer, J. R., MacGregor, K. G., and Jones, B. F. 1993b. *Astrophys. J.*, **409**, 624.

Solanki, S. K. 2002. *Astron. Nachr.*, **323**, 165.

Solanki, S. K. 2003. *Astron. Astrophys. Rev.*, **11**, 153.

Solanki, S. K., Finsterle, W., Rüedi, I., and Livingston, W. 1999. *Astron. Astrophys.*, **347**, L27.

Solanki, S. K., Inhester, B., and Schüssler, M. 2006. *Rep. Prog. Phys.*, **69**, 563.

Solanki, S. K., and Montavon, C. A. P. 1993. *Astron. Astrophys.*, **275**, 283.

Solanki, S. K., and Montavon, C. A. P. 1994. In *Solar Surface Magnetism*, ed. R. J. Rutten and C. J. Schrijver (Dordrecht: Kluwer), p. 239.

Solanki, S. K., Montavon, C. A. P., and Livingston, W. 1994. *Astron. Astrophys.*, **283**, 221.

Solanki, S. K., Rüedi, I., and Livingston, W. 1992. *Astron. Astrophys.*, **263**, 339.

Solanki, S. K., and Unruh, Y. C. 2004. *Mon. Not. Roy. Astron. Soc.*, **348**, 307.

Solanki, S. K., Usoskin, I. G., Kromer, M., Schüssler, M., and Beer, J. 2004. *Nature*, **431**, 1084.

Solanki, S. K., Walther, U., and Livingston, W. 1993. *Astron. Astrophys.*, **277**, 639.

Soltau, D., Schröter, E. H., and Wöhl, H. 1976. *Astron. Astrophys.*, **50**, 367.

Spiegel, E. A., and Weiss, N. O. 1980. *Nature*, **287**, 616.

Spörer, G. 1889. *Verh. Kais. Leopold. Akad. Naturforscher*, **53**, 283.

Spruit, H. C. 1976. *Sol. Phys.*, **50**, 269.

Spruit, H. C. 1977. *Sol. Phys.*, **55**, 3.

Spruit, H. C. 1979. *Sol. Phys.*, **61**, 363.

Spruit, H. C. 1981a. *Astron. Astrophys.*, **98**, 155.

Spruit, H. C. 1981b. In *The Physics of Sunspots*, ed. L. E. Cram and J. H. Thomas (Sunspot, NM: Sacramento Peak Observatory), p. 98.

Spruit, H. C. 1981c. In *The Physics of Sunspots*, ed. L. E. Cram and J. H. Thomas (Sunspot, NM: Sacramento Peak Observatory), p. 359.

Spruit, H. C. 1982a. *Astron. Astrophys.*, **108**, 348.

Spruit, H. C. 1982b. *Astron. Astrophys.*, **108**, 356.

Spruit, H. C. 1991. In Lecture Notes in Physics **388**, *Challenges to Theories of the Structure of Moderate Mass Stars*, ed. J. Toomre and D. O. Gough (Berlin: Springer), p. 121.

Spruit, H. C. 1992. In *Sunspots: Theory and Observations*, ed. J. H. Thomas and N. O. Weiss (Dordrecht: Kluwer), p. 163.

Spruit, H. C., and Bogdan, T. J. 1992. *Astrophys. J.*, **391**, L109.

Spruit, H. C., Nordlund, Å., and Title, A. M. 1990. *Ann. Rev. Astron. Astrophys.*, **28**, 263.

Spruit, H. C., and Roberts, B. 1983. *Nature*, **304**, 401.

Spruit, H. C., and Scharmer, G. B. 2006. *Astron. Astrophys.*, **447**, 343.

Spruit, H. C., Title, A. M., and van Ballegooijen, A. A. 1987. *Sol. Phys.*, **110**, 115.

Spruit, H. C., and Weiss, A. 1986. *Astron. Astrophys.*, **166**, 167.

Spruit, H. C., and Zweibel, E. G. 1979. *Sol. Phys.*, **62**, 15.

St. John, C. E. 1913. *Astrophys. J.*, **37**, 322.

Stanchfield, D. C. H. II, Thomas, J. H., and Lites, B. W. 1997. *Astrophys. J.*, **477**, 485.

Staude, J. 1981. *Astron. Astrophys.*, **100**, 284.

Staude, J. 1994. In *Solar Surface Magnetism*, ed. R. J. Rutten and C. J. Schrijver (Dordrecht: Kluwer), p. 189.

Staude, J. 1999. In ASP Conf. Ser. 184, *3rd Advances in Solar Physics Euroconference: Magnetic Fields and Oscillations*, ed. B. Schmieder, A. Hofmann and J. Staude (San Francisco: Astron. Soc. Pacific), p. 113.

Staude, J. 2002. *Astron. Nachr.*, **323**, 317.

Stauffer, J. R., and Hartmann, L. W. 1987. *Astrophys. J.*, **318**, 337.

Stauffer, J. R., Hartmann, L. W., Burnham, J. N., and Jones, B. F. 1985. *Astrophys. J.*, **289**, 247.

Stauffer, J. R., Hartmann, L. W., and Jones, B. F. 1989. *Astrophys. J.*, **346**, 160.

Stauffer, J. R., Hartmann, L. W., Prosser, C. F., *et al.* 1997. *Astrophys. J.*, **479**, 776.

Steenbeck, M., and Krause, F. 1969. *Astron. Nachr.*, **291**, 49.

Steenbeck, M., Krause, F., and Rädler, K.-H. 1965. *Zeits. Naturforsch.*, **21a**, 369.

Stein, R. F., and Nordlund, Å. 1998. *Astrophys. J.*, **499**, 914.

Stein, R. F., and Nordlund, Å. 2006. *Astrophys. J.*, **642**, 1246.

Steiner, O. 2005. *Astron. Astrophys.*, **430**, 691.

Steiner, O., Grossmann-Doerth, U., Knölker, M., and Schüssler, M. 1998. *Astrophys. J.*, **495**, 468.

Stellmacher, G., and Wiehr, E. 1988. *Astron. Astrophys.*, **191**, 149.

Stenflo, J. O. 1973. *Sol. Phys.*, **32**, 41.

Stenflo, J. O. 1994. *Solar Magnetic Fields: Polarized Radiation Diagnostics* (Dordrecht: Kluwer).

Stephenson, F. R. 1990. *Phil. Trans. Roy. Soc. Lond. A*, **330**, 499.

Stephenson, F. R., and Willis, D. M. 1999. *Astron. Geophys.*, **40**, 6.21.

Stix, M. 1974. *Astron. Astrophys.*, **37**, 121.

Stix, M. 2002. *The Sun*, 2nd edn (Berlin: Springer).

Strassmeier, K. G. 1988. *Astrophys. Space Sci.*, **140**, 223.

Strassmeier, K. G. 1994. *Astron. Astrophys.*, **281**, 395.

Strassmeier, K. G. 2002. *Astron. Nachr.*, **323**, 309.

Strassmeier, K. G., Bartus, J., Cutispoto, G., and Rodonò, M. 1997. *Astron. Astrophys. Suppl.*, **125**, 11.

Strassmeier, K. G., and Hall, D. S. 1988. *Astrophys. J. Suppl.*, **67**, 453.

Strassmeier, K. G., Hall, D. S., Boyd, L. J., and Genet, R. M. 1989. *Astrophys. J. Suppl.*, **69**, 141.

Strassmeier, K. G., Pichler, T., Weber, M., and Granzer, T. 2003. *Astron. Astrophys.*, **411**, 595.

Strassmeier, K. G., and Rice, J. B. 1998a. *Astron. Astrophys.*, **330**, 685.

Strassmeier, K. G., and Rice, J. B. 1998b. *Astron. Astrophys.*, **339**, 497.

Strassmeier, K. G., and Rice, J. B. 2006. *Astron. Astrophys.*, **460**, 751.

Strassmeier, K. G., Washuettl, A., and Schwope, A. 2002. *Astron. Nachr.*, **323**, 155.

Strassmeier, K. G., Welty, A. D., and Rice, J. B. 1994. *Astron. Astrophys.*, **285**, L17.

Strous, L. H., Scharmer, G., Tarbell, T. D., Title, A. M., and Zwaan, C. 1996. *Astron. Astrophys.*, **306**, 947.

Stuiver, M., and Braziunas, T. F. 1988. In *Secular Solar and Geomagnetic Variations in the Last 10,000 Years*, ed. F. R. Stephenson and A. W. Wolfendale (New York: Springer), p. 245.

Stuiver, M., and Braziunas, T. F. 1993. *Holocene*, **3**, 289.

Sütterlin, P. 1998. *Astron. Astrophys.*, **333**, 305.

Sütterlin, P. 2001. *Astron. Astrophys.*, **374**, L21.

Sütterlin, P., Bellot Rubio, L. R., and Schlichenmaier, R. 2004. *Astron. Astrophys.*, **424**, 1049.

Sütterlin, P., Schröter, E. H., and Muglach, K. 1996. *Sol. Phys.*, **164**, 311.

Svensmark, H. 1998. *Phys. Rev. Lett.*, **81**, 5027.

Svensmark, H. 2007. *Astron. Geophys.*, **48**, 1.18.

Svensmark, H., and Friis-Christensen, E. 1997. *J. Atmos. Sol.-Terr. Phys.*, **59**, 1225.

Sweet, P. A. 1955. *Vistas in Astron.*, **1**, 675.

Tao, L., Proctor, M. R. E., and Weiss, N. O. 1998. *Mon. Not. Roy. Astron. Soc.*, **300**, 907.

Tassoul, J.-L., and Tassoul, M. 2004. *A Concise History of Solar and Stellar Physics* (Princeton: Princeton University Press).

Tayler, R. J. 1997. *The Sun as a Star* (Cambridge: Cambridge University Press).

Terndrup, D. M., Pinsonneault, M., Jeffries, R. D., *et al.* 2002. *Astrophys. J.*, **576**, 950.

Thelen, J.-C. 2000a. *Mon. Not. Roy. Astron. Soc.*, **315**, 155.

Thelen, J.-C. 2000b. *Mon. Not. Roy. Astron. Soc.*, **315**, 165.

Thiessen, G. 1950. *Observatory*, **70**, 234.

Thomas, J. H. 1978. *Astrophys. J.*, **225**, 275.

Thomas, J. H. 1981. In *The Physics of Sunspots*, ed. L. E. Cram and J. H. Thomas (Sunspot, NM: Sacramento Peak Observatory), p. 345.

Thomas, J. H. 1983. *Ann. Rev. Fluid Mech.*, **15**, 321.

Thomas, J. H. 1984a. *Astron. Astrophys.*, **135**, 188.

Thomas, J. H. 1984b. In *Small-scale Dynamical Processes in Quiet Stellar Atmospheres*, ed. S. L. Keil (Sunspot, NM: National Solar Observatory), p. 276.

Thomas, J. H. 1985. *Australian J. Phys.*, **38**, 811.

Thomas, J. H. 1988. *Astrophys. J.*, **333**, 407.

Thomas, J. H. 1990. In *Physics of Magnetic Flux Ropes*, Geophys. Monograph No. 58, ed. C. T. Russell, E. R. Priest and L. C. Lee (Washington: American Geophysical Union), p. 133.

Thomas, J. H. 1994. In *Solar Surface Magnetism*, ed. R. J. Rutten and C. J. Schrijver (Dordrecht: Kluwer), p. 219.

Thomas, J. H. 2005. *Astron. Astrophys.*, **440**, L29.

Thomas, J. H., Cram, L. E., and Nye, A. H. 1982. *Nature*, **297**, 485.

Thomas, J. H., Cram, L. E., and Nye, A. H. 1984. *Astrophys. J.*, **285**, 368.

Thomas, J. H., Lites, B. W., Gurman, J. B., and Ladd, E. F. 1987. *Astrophys. J.*, **312**, 457.

Thomas, J. H., and Montesinos, B. 1990. *Astrophys. J.*, **359**, 550.

Thomas, J. H., and Montesinos, B. 1991. *Astrophys. J.*, **375**, 404.

Thomas, J. H., and Montesinos, B. 1993. *Astrophys. J.*, **407**, 398.

Thomas, J. H., and Nye, A. H. 1975. *Phys. Fluids*, **18**, 490.

Thomas, J. H., and Scheuer, M. A. 1982. *Sol. Phys.*, **79**, 19.

Thomas, J. H., and Stanchfield, D. C. H. II. 2000. *Astrophys. J.*, **537**, 1086.

Thomas, J. H., and Weiss, N. O. (eds) 1992a. *Sunspots: Theory and Observations*, NATO ASI Series C375 (Dordrecht: Kluwer).

Thomas, J. H., and Weiss, N. O. 1992b. In *Sunspots: Theory and Observations*, ed. J. H. Thomas and N. O. Weiss (Dordrecht: Kluwer), p. 3.

Thomas, J. H., and Weiss, N. O. 2004. *Ann. Rev. Astron. Astrophys.*, **42**, 517.

Thomas, J. H., Weiss, N. O., Tobias, S. M., and Brummell, N. H. 2002. *Nature*, **420**, 390.

Thompson, M. J. 2006a. *An Introduction to Astrophysical Fluid Dynamics* (London: Imperial College Press).

Thompson, M. J., Christensen-Dalsgaard, J., Miesch, M. S., and Toomre, J. 2003. *Ann. Rev. Astron. Astrophys.*, **41**, 599.

Thompson, S. D. 2005. *Mon. Not. Roy. Astron. Soc.*, **360**, 1290.

Thompson, S. D. 2006b. *Modelling magnetoconvection in sunspots*. Ph.D. dissertation, University of Cambridge.

Thompson, W. B. 1951. *Phil. Mag. (7th Ser.)*, **42**, 1417.

Tildesley, M. J. 2003. *Mon. Not. Roy. Astron. Soc.*, **338**, 497.

Tildesley, M. J., and Weiss, N. O. 2004. *Mon. Not. Roy. Astron. Soc.*, **350**, 657.

Title, A. M. 2000. *Phil. Trans. Roy. Soc. Lond. A*, **358**, 657.

Title, A. M., Frank, Z. A., Shine, R. A., *et al.* 1992. In *Sunspots: Theory and Observations*, ed. J. H. Thomas and N. O. Weiss (Dordrecht: Kluwer), p. 195.

Title, A. M., Frank, Z. A., Shine, R. A., *et al.* 1993. *Astrophys. J.*, **403**, 780.

Title, A. M., and Schrijver, C. J. 1998. In ASP Conf. Ser. 154, *Cool Stars, Stellar Systems and the Sun*, ed. R. A. Donahue and J. A. Bookbinder (San Francisco: Astron. Soc. Pacific), p. 345.

Tobias, S. M. 1996. *Astron. Astrophys.*, **307**, L21.

Tobias, S. M. 1997a. *Geophys. Astrophys. Fluid Dyn.*, **86**, 287.

Tobias, S. M. 1997b. *Astron. Astrophys.*, **322**, 1007.

Tobias, S. M. 1998. *Mon. Not. Roy. Astron. Soc.*, **296**, 653.

Tobias, S. M. 2002a. *Phil. Trans. Roy. Soc. Lond. A*, **360**, 2741.

Tobias, S. M. 2002b. *Astron. Nachr.*, **323**, 417.

Tobias, S. M. 2005. In *Fluid Dynamics and Dynamos in Astrophysics and Geophysics*, ed. A. M. Soward, C. A. Jones, D. W. Hughes and N. O. Weiss (Boca Raton: CRC Press), p. 193.

Tobias, S. M., Brummell, N. H., Clune, T. L., and Toomre, J. 1998. *Astrophys. J.*, **502**, L177.

Tobias, S. M., Brummell, N. H., Clune, T. L., and Toomre, J. 2001. *Astrophys. J.*, **549**, 1183.

Tobias, S. M., and Weiss, N. O. 2000. *J. Climate*, **13**, 3745.

Tobias, S. M., and Weiss, N. O. 2004. *Astron. Geophys.*, **45**, 4.28.

Tobias, S. M., and Weiss, N. O. 2007a. In *The Solar Tachocline*, ed. D. W. Hughes, R. Rosner and N. O. Weiss (Cambridge: Cambridge University Press), p. 319.

Tobias, S. M., and Weiss, N. O. 2007b. In *Mathematical Aspects of Natural Dynamos*, ed. E. Dormy and A. M. Soward (Boca Raton: CRC Press), p. 281.

Tobias, S. M., Weiss, N. O., and Kirk, V. 1995. *Mon. Not. Roy. Astron. Soc.*, **273**, 1150.

Toner, C. G., and Gray, D. F. 1988. *Astrophys. J.*, **334**, 1008.

Toner, C. G., and LaBonte, B. J. 1993. *Astrophys. J.*, **415**, 847.

Topka, K. P., Tarbell, T. D., and Title, A. M. 1986. *Astrophys. J.*, **306**, 304.

Topka, K. P., Tarbell, T. D., and Title, A. M. 1997. *Astrophys. J.*, **484**, 479.

Torres, C. A. O., and Ferraz Mello, S. 1973. *Astron. Astrophys.*, **27**, 231.

Tritschler, A., and Schmidt, W. 2002a. *Astron. Astrophys.*, **382**, 1093.

Tritschler, A., and Schmidt, W. 2002b. *Astron. Astrophys.*, **388**, 1048.

Tritschler, A., Schmidt, W., and Rimmele, T. 2002. In *Proc. 10th European Solar Physics Meeting, Solar Variability: From Core to Outer Frontiers*, ESA SP-506, p. 477.

Tsiropoula, G., Alissandrakis, C. E., Dialetis, D., and Mein, P. 1996. *Sol. Phys.*, **167**, 79.

Tsiropoula, G., Alissandrakis, C. E., and Mein, P. 2000. *Astron. Astrophys.*, **355**, 375.

Tuominen, J. 1962. *Zeits. Astrophys.*, **55**, 110.

Uchida, Y., and Sakurai, T. 1975. *Pub. Astron. Soc. Japan*, **27**, 259.

Ulrich, R. K., Boyden, J. E., Webster, L., Padilla, S. P., and Snodgrass, H. B. 1988. *Sol. Phys.*, **117**, 291.

Unno, W., and Ando, H. 1979. *Geophys. Astrophys. Fluid Dyn.*, **12**, 107.

Unruh, Y. C., Collier Cameron, A., and Cutispoto, G. 1995. *Mon. Not. Roy. Astron. Soc.*, **277**, 1145.

Usoskin, I. G., Schüssler, M., Solanki, S. K., and Mursula, K. 2005. *J. Geophys. Res.*, **110**, A10102.

Usoskin, I. G., Solanki, S. K., and Korte, M. 2006. *Geophys. Res. Lett.*, **33**, L08103.

Vainshtein, S. I., and Cattaneo, F. 1992. *Astrophys. J.*, **393**, 165.

Valenti, J. A., Marcy, G. W., and Basri, G. 1995. *Astrophys. J.*, **439**, 939.

van Ballegooijen, A. A. 1982. *Astron. Astrophys.*, **113**, 99.

van Driel-Gesztelyi, L., Malherbe, J.-M., and Démoulin, P. 2000. *Astron. Astrophys.*, **364**, 845.

van Driel-Gesztelyi, L., van der Zalm, E. B. J., and Zwaan, C. 1992. In ASP Conf. Ser. 127, *The Solar Cycle*, ed. K. L. Harvey (San Francisco: Astron. Soc. Pacific), p. 89.

van Loon, H., and Labitzke, K. 2000. *Space Sci. Rev.*, **94**, 259.

van Loon, H., Meehl, G. A., and Arblaster, J. M. 2004. *J. Atmos. Sol.-Terr. Phys.*, **66**, 1767.

Vargas Domínguez, S., Bonet, J., Martínez Pillet, V., *et al.* 2007. *Astrophys. J.*, **660**, L165.

Vaughan, A. H., and Preston, G. W. 1980. *Pub. Astron. Soc. Pacific*, **92**, 385.

Vázquez, M. 1973. *Sol. Phys.*, **31**, 377.

Venkatakrishnan, P. 1985. *J. Astron. Astrophys.*, **6**, 21.

Vernazza, J. E., Avrett, E. H., and Loeser, R. 1981. *Astrophys. J. Suppl.*, **45**, 635.

Vilhu, O., Gustafsson, B., and Edvardsson, B. 1987. *Astrophys. J.*, **320**, 850.

Vimeux, F., Cuffey, K. M., and Jouzel, J. 2002. *Earth Planet. Sci. Lett.*, **203**, 829.

Vögler, A., and Schüssler, M. 2007. *Astron. Astrophys.*, **465**, L43.

Vögler, A., Shelyag, S., Schüssler, M., *et al.* 2005. *Astron. Astrophys.*, **429**, 335.

Vogt, S. S. 1975. *Astrophys. J.*, **199**, 418.

Vogt, S. S. 1979. *Pub. Astron. Soc. Pacific*, **91**, 616.

Vogt, S. S. 1981a. *Astrophys. J.*, **247**, 975.

Vogt, S. S. 1981b. *Astrophys. J.*, **250**, 327.

Vogt, S. S. 1981c. In *The Physics of Sunspots*, ed. L. E. Cram and J. H. Thomas (Sunspot, NM: Sacramento Peak Observatory), p. 455.

Vogt, S. S., and Hatzes, A. P. 1996. In IAU Symp. 176, *Stellar Surface Structure*, ed. K. G. Strassmeier and J. L. Linsky (Dordrecht: Kluwer), p. 245.

Vogt, S. S., Hatzes, A. P., Misch, A. A., and Kürster, M. 1999. *Astrophys. J. Suppl.*, **121**, 547.

Vogt, S. S., and Penrod, G. D. 1983. *Pub. Astron. Soc. Pacific*, **95**, 565.

Vogt, S. S., Penrod, G. D., and Hatzes, A. P. 1987. *Astrophys. J.*, **321**, 496.

von der Lühe, O. 1983. *Astron. Astrophys.*, **119**, 85.

von Klüber, H. 1955. *Vistas in Astron.*, **1**, 751.

Vonmoos, M., Beer, J., and Muscheler, R. 2006. *J. Geophys. Res.*, **111**, A10105.

Vorontsov, S. V., Christensen-Dalsgaard, J., Schou, J., Strakhov, V. N., and Thompson, M. J. 2002. *Science*, **296**, 101.

Vrabec, D. 1971. In IAU Symp. 43, *Solar Magnetic Fields*, ed. R. Howard (Dordrecht: Reidel), p. 329.

Vrabec, D. 1974. In IAU Symp. 56, *Chromospheric Fine Structure*, ed. R. G. Athay (Dordrecht: Reidel), p. 201.

Wade, G. A., Aurière, M., Bagnulo, S., *et al.* 2006. *Astron. Astrophys.*, **451**, 293.

Wagner, G., Beer, J., Masarik, J., *et al.* 2001a. *Geophys. Res. Lett.*, **28**, 303.

Wagner, G., Livingstone, D. M., Masarik, J., Muscheler, R., and Beer, J. 2001b. *J. Geophys. Res.*, **106**, 3381.

Waldmeier, M. 1937. *Zeits. Astrophys.*, **14**, 91.

Waldmeier, M. 1939. *Astron. Mitt. Zürich*, No. 138, p. 439.

Waldmeier, M. 1947. *Pub. Zurich Obs.*, **9**, 1.

Waldmeier, M. 1955. *Ergebnisse und Probleme der Sonnenforschung*, 2nd edn (Leipzig: Akademische Verlagsgesellschaft).

Waldmeier, M. 1961. *The Sunspot-Activity in the Years 1610-1960* (Zurich: Schulthess).

Walén, C. 1949. *On the Vibratory Rotation of the Sun* (Stockholm: Henrik Lindstahls Bokhandel).

Wallenhorst, S. G., and Howard, R. 1982. *Sol. Phys.*, **76**, 203.

Wallenhorst, S. G., and Topka, K. P. 1982. *Sol. Phys.*, **81**, 33.

Wang, H., and Zirin, H. 1992. *Sol. Phys.*, **140**, 41.

Wang, J., Wang, H., Tang, F., *et al.* 1995. *Sol. Phys.*, **160**, 277.

Wang, Y.-M., Lean, J. L., and Sheeley, N. R. Jr. 2005. *Astrophys. J.*, **625**, 522.

Wang, Y.-M., and Sheeley, N. R. Jr. 1991. *Astrophys. J.*, **375**, 761.

Weart, S. R. 1970. *Astrophys. J.*, **162**, 987.

Weart, S. 1972. *Astrophys. J.*, **177**, 271.

Webb, A. R., and Roberts, B. 1978. *Sol. Phys.*, **59**, 249.

Weiss, J. E., and Weiss, N. O. 1979. *Quart. J. Roy. Astron. Soc.*, **20**, 115.

Weiss, N. O. 1966. *Proc. Roy. Soc. Lond. A*, **293**, 310.

Weiss, N. O. 1991. *Geophys. Astrophys. Fluid Dyn.*, **62**, 229.

Weiss, N. O. 1994. In *Lectures on Solar and Planetary Dynamos*, ed. M. R. E. Proctor and A. D. Gilbert (Cambridge: Cambridge University Press), p. 59.

Weiss, N. O. 2002. *Astron. Nachr.*, **323**, 371.

Weiss, N. O. 2003. In *Stellar Astrophysical Fluid Dynamics*, ed. M. J. Thompson and J. Christensen-Dalsgaard (Cambridge: Cambridge University Press), p. 329.

Weiss, N. O., Brownjohn, D. P., Hurlburt, N. E., and Proctor, M. R. E. 1990. *Mon. Not. Roy. Astron. Soc.*, **245**, 434.

Weiss, N. O., Brownjohn, D. P., Matthews, P. C., and Proctor, M. R. E. 1996. *Mon. Not. Roy. Astron. Soc.*, **283**, 1153.

Weiss, N. O., Cattaneo, F., and Jones, C. A. 1984. *Geophys. Astrophys. Fluid Dyn.*, **30**, 305.

Weiss, N. O., Proctor, M. R. E., and Brownjohn, D. P. 2002. *Mon. Not. Roy. Astron. Soc.*, **337**, 293.

Weiss, N. O., Thomas, J. H., Brummell, N. H., and Tobias, S. M. 2004. *Astrophys. J.*, **600**, 1073.

Wentzel, D. G. 1992. *Astrophys. J.*, **388**, 211.

Westendorp Plaza, C., del Toro Iniesta, J. C., Ruiz Cobo, B., *et al.* 1997. *Nature*, **389**, 47.

Westendorp Plaza, C., del Toro Iniesta, J. C., Ruiz Cobo, B., *et al.* 2001. *Astrophys. J.*, **547**, 1130.

White, W. R., Lean, J., Cayan, D. R., and Dettinger, M. D. 1997. *J. Geophys. Res.*, **102**, 3255.

Wiehr, E. 1996. *Astron. Astrophys.*, **309**, L4.

Wiehr, E., and Degenhardt, D. 1992. *Astron. Astrophys.*, **259**, 313.

Wiehr, E., Stellmacher, G., Knölker, M., and Grosser, H. 1986. *Astron. Astrophys.*, **155**, 402.

Wild, W. J. 1989. *Pub. Astron. Soc. Pacific*, **101**, 844.

Wild, W. J. 1991. *Astrophys. J.*, **368**, 622.

Willson, R. C., and Hudson, H. S. 1991. *Nature*, **351**, 42.

Wilson, O. C. 1978. *Astrophys. J.*, **226**, 379.

Wilson, P. R. 1973. *Sol. Phys.*, **32**, 435.

Wilson, P. R. 1994. *Solar and Stellar Activity Cycles* (Cambridge: Cambridge University Press).

Wilson, P. R., and Cannon, C. J. 1968. *Sol. Phys.*, **4**, 3.

Wilson, P. R., and McIntosh, P. S. 1969. *Sol. Phys.*, **10**, 370.

Wilson, P. R., and Simon, G. W. 1983. *Astrophys. J.*, **273**, 805.

Winebarger, A. R., DeLuca, E. E., and Golub, L. 2001. *Astrophys. J.*, **553**, L81.

Winebarger, A. R., Warren, H., van Ballegooijen, A., DeLuca, E. E., and Golub, L. 2002. *Astrophys. J.*, **567**, L89.

Wittmann, A. 1969. *Sol. Phys.*, **7**, 366.

Wittmann, A. D., and Xu, Z. T. 1987. *Astron. Astrophys. Suppl.*, **70**, 83.

Wöhl, E., and Brajša, R. 2001. *Sol. Phys.*, **198**, 57.

Wood, F. B. 1947. *Astron. J.*, **52**, 133.

Woods, D. T., and Cram, L. E. 1981. *Sol. Phys.*, **69**, 233.

Wright, J. T., Marcy, G. W., Butler, R. P., and Vogt, S. S. 2004. *Astrophys. J. Suppl.*, **152**, 261.

Yallop, B. D., Hohenkerk, C., Murdin, L., and Clark, D. H. 1982. *Quart. J. Roy. Astron. Soc.*, **23**, 213.

Yang, G., Xu, Y., Wang, H., and Denker, C. 2003. *Astrophys. J.*, **597**, 1190.

Yashiro, S., Shibata, K., and Shimojo, M. 1998. *Astrophys. J.*, **493**, 970.

Yau, K. K. C. 1988. In *Secular Solar and Geomagnetic Variations in the Last 10,000 Years*, ed. F. R. Stephenson and A. W. Wolfendale (Dordrecht: Kluwer), p. 161.

Yau, K. K. C., and Stephenson, F. R. 1988. *Quart. J. Roy. Astron. Soc.*, **29**, 175.

Yokoyama, T., and Shibata, K. 1995. *Nature*, **375**, 42.

Yokoyama, T., and Shibata, K. 1996. *Pub. Astron. Soc. Japan*, **48**, 353.

Yoshimura, H. 1978a. *Astrophys. J.*, **220**, 692.

Yoshimura, H. 1978b. *Astrophys. J.*, **226**, 706.

Yoshimura, H. 1981. *Astrophys. J.*, **247**, 1102.

Yoshimura, H. 1983. *Sol. Phys.*, **87**, 251.

Young, C. A. 1881. *The Sun* (Akron, Ohio: Werner).

Yun, H. S. 1970. *Astrophys. J.*, **162**, 975.

Yun, H. S., Beebe, H. A., and Baggett, W. E. 1984. *Sol. Phys.*, **92**, 145.

Yurchyshyn, V. B., Wang, H., and Goode, P. R. 2001. *Astrophys. J.*, **550**, 470.

Zeilik, M., De Blasi, C., and Rhodes, M. 1988. *Astrophys. J.*, **332**, 293.

Zeilik, M., Gordon, S., Jaderlund, E., *et al.* 1994. *Astrophys. J.*, **421**, 303.

Zeldovich, Ya. B. 1956. *JETP*, **31**, 154 (trans. *Sov. Phys. JETP*, **4**, 460, 1957).

Zeldovich, Ya. B., Ruzmaikin, A. A., and Sokoloff, D. D. 1983. *Magnetic Fields in Astrophysics* (New York: Gordon and Breach).

Zhang, J., Solanki, S. K., and Wang, J. 2003. *Astron. Astrophys.*, **399**, 755.

Zhao, J., and Kosovichev, A. G. 2004. *Astrophys. J.*, **603**, 776.

Zhao, J., Kosovichev, A. G., and Duvall, T. L. Jr. 2001. *Astrophys. J.*, **557**, 384.

Zharkov, S., Nicholas, C. J., and Thompson, M. J. 2007. *Astron. Nachr.*, **328**, 240.

Zhugzhda, Y. D., Locans, V., and Staude, J. 1983. *Sol. Phys.*, **82**, 369.

Zhugzhda, Y. D., Locans, V., and Staude, J. 1987. *Astron. Nachr.*, **308**, 257.

Zhugzhda, Y. D., Staude, J., and Locans, V. 1984. *Sol. Phys.*, **91**, 219.

Zirin, H. 1972. *Sol. Phys.*, **22**, 34.

Zirin, H. 1974. In IAU Symp. 56, *Chromospheric Fine Structure*, ed. R. G. Athay (Dordrecht: Reidel), p. 161.

Zirin, H., and Stein, A. 1972. *Astrophys. J.*, **178**, L85.

Zirin, H., and Wang, H. 1991. *Adv. Space Res.*, **11**, 225.

Zwaan, C. 1965. *Rech. Astron. Obs. Utrecht*, **17**, No. 4.

Zwaan, C. 1968. *Ann. Rev. Astron. Astrophys.*, **6**, 135.

Zwaan, C. 1978. *Sol. Phys.*, **60**, 213.

Zwaan, C. 1985. *Sol. Phys.*, **100**, 397.

Zwaan, C. 1987. *Ann. Rev. Astron. Astrophys.*, **25**, 83.

Zwaan, C. 1992. In *Sunspots: Theory and Observations*, ed. J. H. Thomas and N. O. Weiss (Dordrecht: Kluwer), p. 75.

Zwaan, C., Brants, J. J., and Cram, L. E. 1985. *Sol. Phys.*, **95**, 3.

Index

accretion disc, 9, 11, 13, 176, 177
acoustic halos, 122
active longitudes, starspots, 172, 173, 179
active region, 123
   decay, 143–144
   ephemeral, 144–146, 210
   evolution, 128–136
     magnetic configuration, 141–142
     magnetic flux emergence, 129–133
activity cycles, stellar, 191–193, 206
   periods, 192–193, 206
adaptive optics, 234–235
Alfvén, Hannes, 37, 41, 239
Alfvén speed, 241
Alfvén's theorem, 30, 239
α-effect, 31, 33, 195–202
   saturation, 199
Ångström, Anders, 26
Ap stars, 34, 156, 162
arch filament system, 129
Aristotle, 16, 18

Babcock, Harold, 27, 236
Babcock, Horace, 27, 31, 34, 37, 55, 156, 198, 200, 236
Biermann, Ludwig, 30, 31, 37, 41
Bigelow, Frank, 29
Boulliau, Ismael, 33
Boussinesq approximation, 71, 89, 90, 105, 149, 210, 244–245
buoyancy braking, 74, 77, 92
butterfly diagram, 23, 202–205
BY Dra stars, 162, 172–173, 193

Ca II emission, 191, 193
   solar, 4, 29, 86, 106, 111–112, 117, 150, 151
   stellar, 10–11, 34, 150–153, 191, 193
Capocci, Ernesto, 23, 37
Carrington, Richard, 20, 37, 189
Cassini, Jean-Dominique, 19, 21
Castelli, Benedetto, 17
chromospheric emission, *see* Ca II emission

chromospheric filament, 216
collar flow, 65
continuity equation, 239
convective collapse, 133–135
convective intensification, 135–136
convective overshoot, 198
Coriolis force, 31, 52, 132, 141, 196, 198, 199, 206, 207, 243
coronal mass ejection, 31, 217, 219
cosmic rays, 186, 223, 227
   cloud formation, 227
Cowling, Thomas, 30, 31, 37, 60, 72, 138, 139, 194–196, 201, 238
Cowling's theorem, 195

Danielson, Robert, 25, 31, 37, 41, 68, 89, 92
de la Rue, Warren, 23
δ-sunspot, 76, 103
Deslandres, Henri, 26
differential rotation
   solar, 7, 20, 33, 142, 189
   stellar, 175, 177–179, 193
diffusivity, turbulent, 196
dMe stars, 156
Doppler imaging, 11, 36, 162–167
dynamo
   chaotic, 204–205
   flux-transport, 200–201
   interface, 201–203
   numerical simulation, 210
   small-scale, 149, 209–210
   solar, 33, 197–205
   stellar, 205–208
   theory, 31, 33, 183, 194–197, 230
   waves, 199, 202

Earth
   climate change, 221–228
   global warming, 227–228
   quasi-biennial oscillation, 224
   temperature variation, 224–226

271

1-MONTH